T0269480

Regulation and Investments in Energy Markets

Solutions for the Mediterranean

Regulation and Investments in Energy Markets

Solutions for the Mediterranean

Edited by

Alessandro Rubino

Maria Teresa Costa Campi

Veronica Lenzi

Ilhan Ozturk

AMSTERDAM • BOSTON • HEIDELBERG • LONDON
NEW YORK • OXFORD • PARIS • SAN DIEGO
SAN FRANCISCO • SINGAPORE • SYDNEY • TOKYO
Academic Press is an Imprint of Elsevier

Academic Press is an imprint of Elsevier
125, London Wall, EC2Y 5AS, UK
525 B Street, Suite 1800, San Diego, CA 92101-4495, USA
225 Wyman Street, Waltham, MA 02451, USA
The Boulevard, Langford Lane, Kidlington, Oxford OX5 1GB, UK

British Library Cataloguing-in-Publication Data
A catalogue record for this book is available from the British Library

Library of Congress Cataloging-in-Publication Data
A catalog record for this book is available from the Library of Congress

ISBN: 978-0-12-804436-0

For information on all Academic Press publications
visit our website at http://store.elsevier.com/

Contents

3 Renewable Energy in the Southern and Eastern Mediterranean: Current Trends and Future Developments

Simone Tagliapietra

4 Scaling Up Renewable Energy Deployment in North Africa

Georgeta Vidican

Contributors

Emma Åberg, Independent Consultant, Stockholm, Sweden

Ernesto Bonafé, Energy Charter Secretariat, Brussels, Belgium

Bernhard Brand, Research Group 1 "Future Energy and Mobility Structures" Wuppertal Institute for Climate, Environment and Energy, Wuppertal, Germany

Carlo Cambini, Politecnico di Torino, DIGEP Department of Management, Torino; IEFE, Bocconi University, Milan, Italy

Pantelis Capros, Department of Electrical and Computer Engineering, National Technical University of Athens, Athens, Greece

Cambini Carlo, Politecnico di Torino, DIGEP Department of Management, Torino, Italy

Tareq Emtairah, International Institute for Industrial Environmental Economics (IIIEE), Lund University, Lund, Sweden

Gonzalo Escribano, Energy Programme, the Elcano Royal Institute; Applied Economics, Spanish Open University (UNED), Spain

Panagiotis Fragkos, Department of Electrical and Computer Engineering, National Technical University of Athens, Athens, Greece

Pedro Mejía Gómez, OMIE, Madrid, Spain

Manfred Hafner, Fondazione Eni Enrico Mattei, Milan, Italy

Riku Huttunen, Energy Department, Ministry of Employment and the Economy, Finland

Houda Ben Jannet Allal, General Directorate, OME

Nikos Kouvaritakis, Department of Electrical and Computer Engineering, National Technical University of Athens, Athens, Greece

Abrardi Laura, Politecnico di Torino, DIGEP Department of Management, Torino, Italy

Rondi Laura, Politecnico di Torino, DIGEP Department of Management, Torino, Italy

Nurzat Myrsalieva, Regional Center for Renewable Energy and Energy Efficiency (RCREEE), Cairo, Egypt

Alberto Ponti, Pan-European Utilities Team, Societé Générale – Cross Assets Research, London

Branislav Prelevic, Energy Community Regulatory Board; the Regulatory Energy Agency of Montenegro; Energy Economics, University of Donja Gorica, Montenegro

Alessandro Rubino, DISAG Department of Business and Law Studies, University of Bari Aldo Moro, Bari, Italy

Ernesto Somma, DISAG Department of Business and Law Studies, University of Bari Aldo Moro, Bari, Italy

Simone Tagliapietra, Fondazione Eni Enrico Mattei, Milan, Italy

Matteo Urbani, Electricity Division, OME

Francesca Pia Vantaggiato, School of Politics, Philosophy, Language and Communication Studies (PPL), University of East Anglia (UEA), Norwich, UK

Georgeta Vidican, Department of Sustainable Economic and Social Development, Deutsches Institut für Entwicklungspolitik/German Development Institute, Bonn, Germany

Biographies

Alessandro Rubino

Alessandro Rubino is Lecturer in Economics at Bari University. Before completing his PhD he has worked at Ofgem as Regulatory Economist. In this capacity he worked on the design and implementation of the incentive scheme for the national transmission system operators for gas and electricity in United Kingdom. From June 2009 to July 2012 he worked as Research Assistant at the Florence School of Regulation (European University Institute) within the Robert Schuman Centre, a research unit specialized in applied economics. From July 2012 to January 2014 he has been head of Capacity building and knowledge dissemination at the Enel Foundation and is currently Senior Policy Advisor at RES4MED and Member of the Scientific Committee of the MEDREG Forum.

Maria Teresa Costa Campi

Maria Teresa Costa Campi is a Doctor of Economic Science and Professor of Economics at the University of Barcelona (UB). Currently she is Director of the Chair of Energy Sustainability at the University of Barcelona, a post held since 2012, and has been General Coordinator of the FUNSEAM project since 2013. She has been President of the Spanish National Energy Commission (2005–2011). President of the Association of Energy Regulators (2005–2012), President of the Iberian Electricity Market (MIBEL) (2006–2007), Vice President of MEDREG (2010–2011), Member of the European Regulators Group for Electricity and Gas (ERGEG) (2005–2011), Member of the Board of Regulators of the Agency for Cooperation of Energy Regulators (ACER) (2010–2011).

She regularly publishes research studies and scientific papers in specialist journals and books and participated in several collective works, she has published over 140 research studies and scientific papers. She chairs the Scientific Committee of the MEDREG Forum.

Veronica Lenzi

Veronica Lenzi holds a PhD in Political Systems and Institutional Change from the IMT Institute for Advanced Studies of Lucca, Italy. Her research interests concern geopolitics of energy, structure, and administrative reforms of regulatory authorities and international economy of energy. Prior to joining the Association of Mediterranean Energy Regulators (MEDREG) as Research and Training Manager, Dr Lenzi has spent visiting periods at the Observatoire Méditerranéen de l'Energie (OME) and at the Ludwig Maximilian University of Munich.

Ilhan Ozturk

Ilhan Ozturk is currently working as Associate Professor at Cag University, Mersin, Turkey. His research interests include energy economics and international economics. He published extensively in international journals and participated in many international conferences also as keynote speaker. He is the editor of *International Journal of Energy Economics and Policy* and *International Journal of Economics and Financial Issues* and he has been member of the editorial board of several international journals.

Foreword

European Union (EU) energy policy is centered on three key objectives: security of supply, sustainability, and competitiveness. To step-up our efforts to reach these long-established goals, earlier this year the European Union launched a strategy to create a resilient Energy Union with a forward-looking climate policy. In today's ever more competitive world, and with the growing need for more resources, it is clear that our Energy Union can only be successful if the EU works closely with its partners around the world.

It should come as no surprise, then, that the Energy Union strategy identified our partnership with the Mediterranean region as a priority. After all, the southern and eastern Mediterranean countries are long-standing partners of the EU. Our regions often have to face similar energy challenges and it is logical that wherever possible we should work together to find solutions. Energy needs in the southern and eastern Mediterranean countries are rapidly growing, as the population rises. Ensuring that the energy supply to households and businesses is secure, and that its production and use are competitive and sustainable, are therefore key challenges we have in common.

To rise to these challenges in the Mediterranean, substantial new investments are needed, be it in infrastructure to transport gas from newly discovered sites or in new technologies to help tap into the enormous potential of renewables in the area.

Yet, for the necessary investments to happen, a stable and harmonized regulatory framework is a must. In an attempt to offer this stability to investors and to facilitate the integration of national markets in the region, the Association of Mediterranean Energy Regulators (MEDREG) was created in 2007. Last November, MEDREG organized its first forum in Barcelona, focusing on the challenges of creating a stable regulatory climate that would attract the much-needed investments in energy infrastructure.

Furthermore, MEDREG entered into a Memorandum of Understanding with the European Commission and the Association of Mediterranean Transmission System Operators for Electricity, with the aim of establishing a platform on regional electricity markets.

We have now embarked on a broader process, designed to inject new impetus into EU–Mediterranean cooperation. Three new Unions for the Mediterranean platforms have been unveiled recently – one for gas, one for a regional electricity market, and one for renewables and energy efficiency. These platforms will also encourage dialog, facilitate partnerships, and strengthen cooperation between the countries of the Union of the Mediterranean.

The main focus of the cooperation platforms will be on market integration, developing new sustainable energy technologies, and removing regulatory, infrastructure, and investment barriers in the region. They will explore the possibilities of developing gas production in North Africa and Eastern Mediterranean countries for domestic markets and export to the EU. They will also identify ways to gradually remove technical, regulatory, and infrastructure barriers for the free trade in electricity across international borders. Furthermore, discussions will focus on how to promote the right regulatory frameworks and markets to encourage investments in renewables and energy efficiency in the region.

MEDREG has a key role to play in the success of all three new platforms, and in particular in the electricity platform. A cornerstone of the EU Energy Union will be a fully functioning and efficient internal market transporting energy from where it is produced to where it is needed. This internal market must be linked up with modern infrastructure and it must ensure a level playing field for all energy companies, including the producers of renewable energy. While price signals will be the main drivers of market development, the role of independent energy regulators will also be vital.

The European Commission sees the integration of regional markets in the Mediterranean region as key to cost-efficient investments and we are therefore encouraging the development of a regional electricity market in the area. Market integration will boost security of supply on both sides of the Mediterranean and provide the right signals to investors. Investors in the energy sector need long-term planning and vision on the choice of fuel mix in a country or a group of countries, and in addition they need to operate in a stable and transparent legal and regulatory framework.

There is strong support from investors in the energy market in the Mediterranean area – the success of last year's MEDREG forum and the results of the public consultation on the Investment Report on infrastructure interconnections, launched by MEDREG in 2014, prove this.

This April, one of the largest AC synchronous grids in the world, ENTSO-E's Continental Europe grid, was extended to Turkey, serving an additional 75 million consumers. This move boosts trade in electricity, allows more sharing of power reserves, more security, and mutual help in emergency situations. This is a major step forward toward the energy market integration of non-EU countries with Europe.

The European Commission welcomes this move to build a common approach between Mediterranean partners. Working together we can create the kind of environment that can trigger the right investments and the common vision we need to cope with the energy challenges we face – from coping with increased energy consumption and supporting new technologies to removing barriers and fostering market integration. The kind of environment that will also present us with new opportunities for revitalized cooperation across the Mediterranean region.

Christopher Jones

Deputy Director-General DG Energy,
European Commission

Introduction

1 THE ENERGY COOPERATION IN THE MEDITERRANEAN – OVERVIEW OF THE MAIN CHALLENGES

The political agenda of the Mediterranean region has been reshaped by recent events concerning climate change, food crises, swinging oil reserves and prices, the potential impact of the shale gas revolution, unstable financial markets, and social unrest. All these issues led to question the current pattern of development and ultimately to shift it toward sustainable development at different geographical scales: global, regional, national, and local. The growing complexity of the actors and layers involved calls for a new and different role for the existing (and traditional) governance structure as well as definition of the missing ones. While multilevel governance is the default choice under these circumstances, the characterization of the institutional actors involved in this process is not trivial and can be described as a dynamic process.

In this context, a careful understanding of the implications of these wider dynamics for the Mediterranean region is of utmost importance for the countries concerned. It is often recalled that the Mediterranean Basin is a zone of trade and cultural exchange, but also of deep-rooted tension between the countries belonging to the three continents bordering it. Therefore, the region shows the main characteristics of the "world economy," while preserving a peculiar national and local dimension. The Mediterranean has, however, huge potential to become a global laboratory for innovation and cooperation in the field of sustainable development, since it is engaged in an open multilateralism that represents a resource toward building a common future.

This composite background has triggered the development of a "Mediterranean approach" for researchers working on this area, watching over the state of the environment in the Mediterranean and its future trends, yet lacking a systematic review in the many facets of the sustainable development debate (Rubino, 2014).

Energy cooperation is one of the main pillars of this composite area of research that has profound ramifications on the economic, social, and environmental domain. The Mediterranean energy sector is currently facing multiple challenges because of the combination of institutional, technical, and social factors.

Natural gas is of critical relevance for the overall Mediterranean energy trade. The region displays a relevant consumption level of over 300 billion cubic meters (bcm)/year. However, the gas traded is around 80 bcm/year, barely a quarter of the regional demand. Infrastructure investments have, therefore, a high potential to boost regional gas trade.

It is however electricity which plays the main role in the energy development of the region. From the technical point of view, there are exist two main electricity interconnection corridors running through the region: the Maghreb block, which includes Algeria, Morocco, and Tunisia; and the eight-country block (Egypt, Iraq, Jordan, Lebanon, Libya, Palestine, Syria, and Turkey – EIJLLPST), which is part of an effort to upgrade these countries' electricity systems to a regional standard. Although the Maghreb and EIJLLPST interconnections have existed for some time, electricity trade among these countries has remained at modest levels especially when considering availability of resources and geographical proximity. This is due to barriers such as limited generation reserve margins, the absence of a harmonized regulatory framework, and institutional weaknesses, both at the national and regional level (MEDREG, 2015).

While increasing penetration of nonprogramable generation is a prevalent feature of modern electricity systems in the last decade, the Middle East and North Africa (MENA) region presents a generation mix dominated by fossil fuel sources. The prevailing technology is a legacy of an energy system that has been unable to keep up with the institutional and regulatory innovation that, on the contrary, has been a common feature in Europe and America. Whereas the final balance of the liberalization process in place in the last two decades is still an open question, it is largely agreed that the organization of the existing electricity systems in most MENA countries appears unsustainable. Those systems are characterized by scarce dynamic efficiency, meaning that the revenues accruing from the retail activities cannot guarantee a level of investment in line with the ever-increasing demand, as well as from an intertemporal point of view. In addition, the quality and security of electricity provision is declining, due to an aging generation fleet. Electricity supply is not able to provide the desired level of revenues for two main reasons: (1) the lack of market dynamic as electricity prices do not adjust to reflect a scarcity signal and (2) the widespread diffusion of subsidies in the supply chain.

The combination of these factors has increased the burden on the public budget, at a time when public spending needs to be committed to other social needs, worsening the overall outlook of macroeconomic indicators. As a result, it is natural to wonder whether the electricity sector in the MENA region has the possibility to maintain its current *status quo* or whether it needs to inaugurate a transition process to ensure the sustainability of the sector, as happened in many western economies in the 1990s.

There is evidence that some sort of transition has begun. Most contributions in this book evaluate the dynamic at play at this stage and the direction this process is taking. However, it seems unclear who are the actors involved in the transition process and one may question if those currently involved are the right ones.

The electricity sector in the MENA countries, and in the Mediterranean region in general, is dominated by strong incumbent utilities, often publicly owned, which are mostly vertically integrated, thus supporting the security of supply of their internal markets. In addition, these utilities often represent a relevant part of the wider mechanism that defines and governs domestic energy policy. This means that the stakeholders called to define the direction that the transition process needs to take are often the same that might lose the most if the existing situation changes. In other words, there might be in place a system of perverse incentives in the selection of the viable process of reforms and of the alternative governance models.

To minimize this potential problem, a number of new and independent actors are now called to play a significant role. In particular, National Regulatory Agencies (NRAs) will play a pivotal role in the reform process. Other actors, such as International Financing Institutions (IFIs), independent transmission system operators (TSOs) (where they exist), and the wider community of energy stakeholder will also become more important in this new phase.

In the book, we evaluate how traditional internal stakeholders (relevant ministries, energy utilities, the political system) combine with new independent players (NRAs, MEDREG, TSOs, Med-TSO, IFIs) and other external actors (such as the European Union, the Union for the Mediterranean (UfM), Energy Charter, energy community) adding an extra complexity to these processes. It is important to evaluate which role each actor will be playing and the wider dynamics that this complex stakeholder matrix is able to determine.

A related area of analysis that mainly concerns the connection between the institutional dimensions explored earlier, the technological, and the social one is the speed at which the reforms can be implemented, accepted, and internalized in the system. Since electricity and gas are now a fundamental component of daily life, they present a significant social dimension that cannot be disregarded. Therefore, measures that are perfectly plausible from an economic and institutional point of view might lack effectiveness if not accompanied by appropriate social compensation measures.

A typical example of this type of interaction is the implementation of energy subsidies reforms. While it is now widely accepted that energy subsidies (and in particular the electricity ones) are highly distortive for the market, unfair in terms of wealth redistribution, and inefficient to ensure social protection, the implementation of an effective subsidies reform remains a challenging task for the governments of the region. In other cases, the social dimension is more dynamic than the institutional one, demanding reforms that fail to be implemented rapidly by the relevant institutions. For instance, the diffusion of microgrids and demand scale renewable energy source (RES) generation is widely accepted by the population, but still faces institutional (and sometimes technical) delays that makes its effective penetration suboptimal. Therefore, our analysis looks at the main challenges that the region is facing and we have been able to identify a number of areas of concern.

The first one is the definition of the appropriate governance set-up that could allow the system to be sustainable in the long run. Investments in systems open to competition are no longer coordinated by the same mechanisms as in the past. The planning activities that enabled a monopolistic vertically integrated producer to adjust base, peak load, and transmission capacities has been replaced by a series of decentralized decisions, partly based on prices. If correctly applied, this new decision-making process, which involves many agents and combines market signals with regulation, should result in an investment level that is consistent with the public interest (social optimum). On the contrary, where this centralized system is still in place, the public budget is unable to guarantee investments of this magnitude while existing price signals are not strong and stable enough to attract private investment.

In addition, it is difficult to determine the generation mix that should be preferable in the long run. Ideally, an efficient power system should provide the optimal level of investment for both generation and transmission, with the goal to minimize the cost of electricity for current and future consumers. Therefore, by way of market dynamics or through a centralized decision-making process, the right amount of transport and generation capacity should emerge. However, available evidence fails to show results anywhere near optimality in most countries belonging to the Mediterranean Basin. Indeed, a good number of governance, institutional, and market problems manifest themselves in the planning, financing, and fulfillment of investment plans. We therefore consider it natural to have a deeper look at the regulation and investment environment in order to grasp the inner determinants of these major challenges and propose tentative solutions.

2 ORIGINS OF THE MEDREG FORUM

The complexity of the dynamics at play suggested that those issues, briefly introduced earlier and extensively analyzed in the following chapters, should be considered with and independent stance, looking at the broader picture, involving independent researchers and experts engaged in energy cooperation in the region. With this aim in mind, during the 15th MEDREG General Assembly held in Alexandria (Egypt) in June 2013, the former President of MEDREG, Mr. Michel Thiollière launched the proposal to establish an annual Forum on energy regulation in the Mediterranean, to be directly organized and managed by the Association. In the light of a renewed discussion on the future of the Mediterranean energy exchanges, both the regulators, members of MEDREG, and the European Commission (EC) endorsed this initiative.

The Association decided that the Forum would involve MEDREG's external partners in a discussion on selected topics of particular relevance for MEDREG. This event had to be conceived and designed in order to increase the visibility of MEDREG's agenda and provide a concrete feedback from regional energy stakeholders. To fulfill these objectives, the MEDREG Secretariat and Presidency Board decided that the setting up of the event was to follow an ambitious strategy and rely on high-level contributors, speakers, and partners in order to make the Forum a landmark in the Mediterranean energy sector. With the aim to timely account for the main changes taking place in the energy debate of the Mediterranean region, MEDREG established that the Forum would be organized every 2 years.

The Association, aware that the coordination of the scientific content of the Forum would require both a strong knowledge of the main debates going on in energy regulation and a long-lasting experience in the Mediterranean area, appointed Ms Maria Teresa Costa Campi, former President of the Spanish energy regulator and Professor of Energy Sustainability at the University of Barcelona, as chair of the scientific committee. The work of Prof. Costa Campi as Chair of the Scientific Committee of the first MEDREG Forum was supported by Hafez El Salmawy (Managing Director, EgyptEra, Egypt), Claude Mandil (former IEA General Director, Member of the Board of Directors, Total France), Ilhan Ozturk (Professor of Energy Economics at Cag University, Turkey), Pippo Ranci Ortigosa (Professor of Economics at the Catholic University of Milan, Italy), and José Sierra (former Commissioner of the Spanish energy regulator).

The Scientific Committee (SC) worked to draft and define the annual Forum's structure and program; select the panel speakers, chairpersons, and discussants; collect and evaluate abstracts and papers to be produced and presented by the speakers at the Forum; and ensure the overall coherence of the event according to high-level standards, in line with MEDREG's objectives and activities as defined in the Action Plan of the Association. In addition, the SC decided, in accordance with MEDREG, and in consideration of the main elements defined earlier, to devote the first edition of the Forum to a discussion on energy infrastructure investments. The topic of energy infrastructure investment has risen prominently among the working priorities of the Association and represents an open issue in the energy cooperation in the region; few key elements have contributed to define the perimeter of the debate during the Forum. We shall examine these factors in the wider context of the relationship between investment and regulation.

3 THE KEY PARTNERSHIP BETWEEN INVESTMENTS AND REGULATION

Fossil energy (hydrocarbons and coal) represents 80% of energy supplied in the Mediterranean, both in the southern (95%) and in the northern shore (70%).[1] In the last 5 years, a citizen of a northern Mediterranean country consumed almost 150% more than a citizen of a southern shore country (2.7 tons of oil equivalent compared with 1.2 tons of oil equivalent). By 2030, these figures will see an additional, sharp increase. Primary energy demand could grow 1.3–1.6 times. However, southern and eastern countries are expected to see a rise five times higher than northern countries, representing almost 50%

1. Source: IPEMED (Institut de Prospective Economique du Monde Méditerranéen).

of the total demand of the Mediterranean Basin. National energy markets today show degrees of maturity that widely differ from one country to another. In the Southern shore, utilities are state-owned and operate either based on vertically integrated service providers or using a single buyer model. The high degree of subsidies distorts price, puts the state budgets under heavy financial pressure, and hampers investments in energy infrastructure. Only one-tenth of the total intra-Mediterranean exchanges concern trade among the southern shore countries, including exchanges with Europe (Morocco–Spain). These reduced quantities derive from the limited capacity of the existing electrical interconnections.

In the EU-Mediterranean countries, investment planning is harmonized by EU legislation. At the national level, every year the TSO submits a 10-year development plan to the regulator, indicating the main transmission infrastructure that needs to be built or upgraded. This plan presents the main projects to be developed or updated for the following 10 years, and lists the investments to be implemented in the coming 3 years. All these aspects are accompanied by a detailed timeline. If the TSO does not implement an investment foreseen for the subsequent 3 years, the regulator may take measures to ensure that the investment in question is made, if it is still considered relevant. Regarding the annual investment plan, regulatory powers differ according to the TSO unbundling model implemented at the national level.[2]

On the contrary, looking south and southeast, regulatory frameworks for investment planning widely differ. For example, in Egypt existing interconnection projects were established before the creation of the regulator. However, when planning future projects, the transmission company that is affiliated to the electricity holding company under the supervision of the ministry of electricity and renewable energy is solely responsible for planning the interconnection projects. In Algeria, the regulator gives the highest priority to securing electricity supply at the national level through new generation projects. Investments in interconnection projects are given a lower priority. Establishing interconnections represents both a regulatory and an investment challenge. Uncertainty on different aspects can influence the decision to invest. The key determinants for interconnection investments can be grouped into three broad categories: financial feasibility, a clear legal and regulatory framework including cross-border cooperation, and the ability to address environmental and social concerns. These challenges have resulted in developing different financing schemes to improve the cross-border interconnection of grids.

2. The EU Directives provide for different unbundling models: the Ownership Unbundling Model and the Independent Transmission Operator Model (where there is no clear disposition in the texts regarding NRA powers in annual investment plans) and the Independent System Operator Model (where the NRA approves the annual investment plan). Each national government decides which model to apply at national level.

Impacted aspects include the following:

Technical aspects. The physical features of the interconnected systems, such as synchronization, magnitudes and directions of the anticipated power flows, the physical distance covered by the interconnection, as well as technical and operating differences between interconnected systems.
Economic and financial aspects. Costs for the purchase and/or production of fuels used in electricity generation, capital costs for building generation facilities, and income from power sales.
Externalities. Indirect financial, social, and environmental benefits, such as employment of labor, impacts of improved power supplies in fostering development of local industry, better power quality, income from power exports, the experience and incentive due to additional cooperative activities between countries, and improvement in reducing pollutant emissions due to the potential optimization of resources.
The complexity of agreements. Cross-border investments can involve a variety of national, subnational, and even international parties to assent to plans for designing, building, and operating interconnections. They are expected to provide frameworks for power purchase and pricing, siting of power lines and related infrastructures, power line operation and security, environmental performance, and liability for power line failure. In particular, Mediterranean countries face five major challenges related to the development of their electricity and gas sectors:

- *Unclear institutional architecture at national level.* Regulators, TSOs, operators, and other actors should cooperate with clear distinction of roles at the national level. Sometimes considerable conflicts of interest occur, heavily affecting the credibility of the country to foreign investors.
- *Lack of sound cost-benefit analysis (CBA).* In some Mediterranean countries there are no effective methodologies for evaluating the estimated costs and benefits of new infrastructure projects. Thus, it is very difficult to have a clear view on the economic profitability of single interconnection projects, which result in less than effective investment plans. The lack of cross-border cost allocation (CBCA) methodologies is also significant. Building on CBA, CBCA has the potential to support the realization of interconnections that have unclear cross-border impacts. In areas such as the Mediterranean one, where financing conditions suffer from lack of transparency, regulatory decisions based on CBCA helps clarifying the benefits and costs for each country involved, thus facilitating appropriate cost allocation among hosting and affected third countries.
- *Lack of innovative financing mechanisms for the successful implementation of new infrastructures.* As the estimated financing needs of the Mediterranean region will be probably higher than the potential contribution of public funding, the key challenge will be to identify

what conditions are necessary to attract investments from IFIs and the private sector for new interconnection projects. Nevertheless, support from IFIs would encourage the establishment of a favorable investment framework and demonstrate, for instance, the economic viability of specific technologies for developing innovative business models.

- *Lack of transparency*. Mediterranean energy markets are mainly managed by state-owned monopolies that influence prices and trading conditions. For this reason, foreign investments tend to be discouraged by scarce information on market prices and available transmission capacity. This problem is coupled with a lack of legal obligations for the monopolist, which makes it increasingly difficult for a third party to access the market.
- *Significant subsidization*. In some non-EU Mediterranean countries, governments tend to heavily subsidize domestic prices, without any market mechanism in place. This hinders the development of cost-reflective energy prices, which are key in fostering private investment in the energy sector.

Overall, these challenges lead to an unclear prioritization of barriers to investments in the different Mediterranean countries. In a recent public consultation[3] held by MEDREG, respondents were asked to create a priority list on the impact of barriers choosing from a provided set. According to the response received, three main barriers to investment appear to dominate the scene. They are: (1) political instability and lack of a clear institutional framework; (2) lack of internal reforms; and (3) insufficient market demand.

This ranking seems to confirm that the absence of a stable and reliable investment framework is the main cause of concern for investors. The mention of insufficient market demand suggests that the lack of a unified solution for the development of regional and subregional markets may hamper the balanced increase of electricity and gas usage by existing consumers, as well as the access of new ones. The first Mediterranean Forum on Energy Regulation offered the opportunity to address real problems concerning energy investments with a rigorous approach combining the analysis of urgent issues that require very short-term solutions, with a commitment to a long-term energy model. In a Euro-Mediterranean energy context marked by security of supply issues, deep structural changes, and climate challenges, regional cooperation plays a major role in the 2030 perspective and beyond. As energy sources and routes diversify, with both the liquefied natural gas and RES becoming more and more affordable options in the future, the role of MEDREG and Mediterranean energy

3. The public consultation on the report "Interconnection Infrastructures in the Mediterranean: A Challenging Environment for Investments" sought the advice of stakeholders in order to provide an input to the conclusions of the document, which aim at assessing a comprehensive set of actions to revive energy infrastructure investments in the Mediterranean region. More info at www.medreg-regulators.org

regulators is fundamental for the progressive evolution and integration of electricity and gas markets at a regional level. This also includes providing a place and time to gather stakeholders working on the different aspects of the energy markets that compose the core of energy regulation.

4 OVERVIEW OF THE TOPICS DISCUSSED AT THE FORUM AND THE STRUCTURE OF THE BOOK

When deciding how to tackle energy investments in the Mediterranean region, it became clear that the topic has multiple facets to be addressed in order to understand why a viable regional approach has yet to emerge and what types of challenges each country is facing. The Forum was therefore organized into three tables, devoted to the institutional side, the mechanics of the markets, and the financial gap, respectively that have been reflected in the structure of this book.

The first part of this book is entitled "A Roadmap for a Mediterranean Energy Community" and discusses how the role of RES and the changing landscape of security of supply, due to new gas discoveries, can impact the institutional approach of Mediterranean countries and the EU as well as the chances to create a Mediterranean Energy Community. The case of the energy community of the Balkan Region and its efforts to incentivize investments is reported as a case study.

Prelevic's contribution, in Chapter 1, provides an overview of the possible regulatory instruments to attract investments in new electricity infrastructure projects as well as recommendations for possible regulatory support options and investment incentives. This chapter tends to underline a regional approach as the key element of such a policy, and thus contains possible solutions for regulatory regime harmonization. According to the author, three main conclusion can be drawn: investments in the transmission network are necessary to create conditions for strengthening the single competitive European market for electricity; national regulators, when creating their tariff methodologies and market rules, need to follow the best European practice; and investment mechanisms on the interconnection capacity and lines of international relevance should be harmonized on a regional basis.

In Chapter 2, Vantaggiato investigates how the EC redefined its Euro-Mediterranean energy relations over time, focusing on how priorities emerged in the EU external energy policy. Through the analysis of the launch of the European Neighbourhood Policy (ENP) and the failed endorsement of the Mediterranean Solar Plan (MSP) she observes that the EC moved Euro-Mediterranean regional energy cooperation from the EU's energy policy to its "external relations" framework, also causing the EC to predicate its policy interest in Euro-Mediterranean energy cooperation on drawbacks in its relations with Russia pushing the EC to fall back on existing policy templates (such as the UfM).

In Chapter 3, Tagliapietra analyses the existing renewable energy potential of the MENA region, which is however accompanied by underdevelopment in terms of solar and wind energy deployment. The author argues that this paradox is mainly due to the fact that the deployment of renewable energy in the region faces four key barriers: commercial, infrastructural, regulatory, and financial. In order to effectively tackle these barriers, a "double-track" approach seems to be essential. These barriers are so resilient that they should be faced both singularly and globally, in a single action. Particularly, in the medium-term all SEMCs should advance an energy subsidy reform process, phasing out universal fossil fuel consumption subsidies in favor of targeted subsidies aimed at effectively addressing the problem of energy poverty.

In Chapter 4, Vidican provides a review of the energy mix diversification taking place in North African countries. While she acknowledges that initial steps toward a modernization and diversification of the national energy mix have been initiated in Egypt, Morocco, Tunisia, and Algeria, she considers that only the interaction between the various dimensions affecting power sector transition at a national level can ensure the creation of a system where renewable energy becomes pivotal. The transition toward a sustainable and decentralized electricity system requires a systemic approach rather than one-off measures. However, the author argues that there are significant obstacles that are likely to impede, or at least to slow down this process: vested interest in the *status quo*, insufficient market dimension, persistency of fossil fuel subsidization, and flawed and fragmented regulatory set-up can together play an adverse effect toward an effective "green" transition. Vidican maintains that a greater effort is needed to understand the political economy behind this process.

In Chapter 5, Brand critically analyzes the national RES targets (also roadmaps) that have recently been published by most MENA countries. He argues that RES targets main rationale was to attract investors and stakeholders attention. Their nonbinding nature coupled with the absence of a coherent set of reforms might fail to provide the appropriate incentives for a transition toward more sustainable energy systems. On the contrary, a lock-in effect could prevail, imposing further barriers for renewable generation diffusion.

After reviewing the regional energy landscape in Chapter 6, Hafner and Tagliapietra identify six main milestones that can foster a new Euro-Mediterranean cooperation: enhancing hydrocarbon cooperation in the region, challenging the persistence of energy subsidies, promoting energy efficiency, unlocking the renewable energy potential, promoting a new interconnected regional market, and financing the sustainable energy transition represent the key priorities to promote sustainable energy development in the SEMCs. Therefore, a coordinated energy roadmap, rather than spot intervention, will be necessary to tackle the main energy challenges that the region is currently facing.

Chapter 7, by Escribano, closes the first part of the book. Escribano approaches the discussion on the development of a Mediterranean Energy Community in three steps. It starts by describing the different pathways that could lead to it, highlighting the need for differentiation both at the geographical (or energy corridor) and energy source levels. Second, he contextualizes this institutional construct within the harsh and volatile southern Mediterranean geopolitical realities. Finally, he presents some elements for a functioning Mediterranean Energy Community that should be based on the definition of a single European energy model. He concludes by calling for a renewal in the EU's energy narrative toward its Mediterranean neighborhood to increase its attractiveness to the region.

The second part is entitled "Challenge of Market-Based Regulation" and evaluates the adequacy of the designs of energy markets in the region, also building on the mistakes made in the past, focusing on the technical and economic advantages that cross-border cooperation can provide. The geographical characteristics of the region call for the exploitation of renewables and the combined increase in efficiency levels. These objectives are coherent and compatible with those of the EU Energy Roadmap, which could therefore be expanded to include a section detailing how the considerable technological knowledge of the EU can be shared with the southern shore in order to ensure that the development of markets is balanced and interesting to investors.

In Chapter 8, Cambini and Rubino address the challenge of market-based regulation in the MENA region through a survey that assesses how energy rule promotion takes place in the Mediterranean region. The authors take into consideration three rule-diffusion patterns: top-down (referring to the EU and other international actors as rule promoters), bottom-up (impact of domestic actors in promoting rule changes and rules adoption), and network approaches (the role of networks in rule diffusion). The results show that the respondents consider direct bilateral actions from the EU as the most effective in rule promotion in the region. It also emerged that MEDREG's role and visibility has increased since 2011, and is the largest when compared with other energy networks in the Mediterranean. Finally, the results also confirm the Euro-Mediterranean region is still characterized by a fragmented scenario where domestic actors play a leading role in promoting rule adoption and institutional change. In this framework, vertically integrated utilities and national champions can exercise a significant veto power and are able to halt or slow down the process required for the creation of an integrated regional market.

In Chapter 9, Capros, Fragkos, and Kouvaritakis evaluate in quantitative terms the implications of alternative strategies regarding the configuration of the energy demand–supply system in the SEM–EU region. The analysis is based on a large-scale energy system modeling both the southeast Mediterranean and the EU regions and assesses two alternative cooperation strategies. The first strategy would propose centralized actions involving large-scale exploitation of renewables in utilities scale, export-oriented installations. The second strategy would give priority to decentralized investment, exploiting dispersed renewable sources, combined with proactive removal of current energy pricing distortions and incentives for energy savings by energy consumers. The final objective of the analysis is to evaluate in quantitative terms the implications of alternative strategies regarding the configuration of the energy demand–supply system in the SEM–EU region. The results show that the current policies in SEM countries are insufficient to achieving the required level of sustainability. However win–win cooperation can be achieved between the SEM region and the EU by extending the EU Emission Trading Scheme (ETS) to the SEM region with free allocation of allowances and, developing a large-scale renewable infrastructure in the SEM region to export electricity to the EU through direct current (DC)-based interconnections.

In Chapter 10, Gómez presents his experience in the European Market Coupling Project, promoted by the power exchanges and TSOs in order to implement a pan-European market coupling algorithm (price coupling of regions, PCR) that integrates the day-ahead markets of 17 countries in the EU. Mejia explains that the main benefit of the market coupling approach is the improvement in market liquidity combined with the beneficial side effect of less volatile electricity prices and best use of the existing cross-border capacity. As market coupling is paving the way toward establishing a solid day-ahead electricity price for Europe, the author posits that a European market coupling project will affect all surrounding countries, including the MENA region, by providing a more robust price formation. Although it is not yet realistic to think about an extension of market coupling in the short term, neighboring countries could benefit by having a reference price upon which they could decide, efficiently, whether to increase imports or start developing their own new-generation resources.

In Chapter 11, Åberg, Myrsalieva, and Emtairah discuss the reforms undertaken by several countries in order to open up their power market to private sector involvement particularly in relation to renewable energy (RE) sourced electricity generation. In particular, the chapter provides a revision of the unbundling regime in MENA countries. The authors observe that the MENA region is characterized by a lack of clear signals from governments. In addition to increased legal certainty, most MENA countries need to improve and diversify the possibility for the private sector to participate in RE power generation activities. Most countries in the region have imposed themselves ambitious RE targets. However, given the current power market structures and schemes, these targets are likely to be undershot unless governments manage to increase the amount of tenders for power purchase agreements (PPAs) and make sure to streamline these processes in order to make them more timely and certain.

In Chapter 12, Huttunen provides another example of market integration. He refers to the electricity and gas market of the Baltic Sea as an interesting comparison to the case of the Mediterranean Basin. The economic and political environments are naturally different, but many features are also common in both regions. For example, both these regional markets are divided by the sea and characterized by the presence of important gas and oil producers. In addition, they fulfill a strategic role for the EU with several member states involved in significant energy trade and exchanges. In light of this, conclusions drawn from the Baltic experience may inspire Mediterranean actors. It is important to develop interconnections crossing the sea, to transport electricity between the southern and the northern shore of the basin. In addition, taking into consideration that investments of this size entail strong political agreements between countries, only a clear and focused regulation generates the necessary stability and predictability. Finally, in some cases subsidies might be necessary to launch large infrastructure investments.

The third part of the book looks at the role and opinions of financial institutions under the topic "Investment for Grids and Generation Projects" exploring the main hurdles that prevent investments and financing mechanisms that can be used to meet the needs in energy generation and infrastructure in the Mediterranean. The growing energy demand of the region requires a comprehensive strategy from governments to improve the activity of financial markets. In order to incentivize the participation of the private sector, governments should resort to an articulated toolbox comprising international and bilateral agreements to foster cross-border exchanges; better and harmonized regulation; phasing-out of the current production and consumption subsidy schemes; and improvement of the overall quality of institutions. Therefore, following the organization of the table at the Forum, we have organized the book into three parts, each focusing on the sections developed during the MEDREG Forum and briefly discussed earlier.

In Chapter 13, Somma and Rubino discuss how global infrastructure investments are evolving after the global crises. They particularly look at private participation in infrastructure for energy investments in MENA countries. Evidence shows that the MENA region is lagging behind both in relative and absolute terms. Only two countries, Morocco and Jordan, are able to attract private participation in infrastructure investment, mostly in RES generation. The authors observe that while a stable political and institutional environment undoubtedly favors a better investment climate, the MENA region should provide investment incentives aligned to the existing financial and institutional uncertainty that most Mediterranean countries experience these days. Public and private partnership (PPP) in a region dominated by political and social uncertainty, requires shorter lead times and returns (opportunity cost, risk adjusted) that are similar, or above, the returns of other investments. The stabilization of the remuneration provided with the most common regulatory tool for RES technologies (feed-in-tarrifs (FiT), establishment of RES quotas and targets) could provide the necessary guarantee. In this context, RES investment then, offers a valid alternative to other more capital-intensive alternatives, also typically dominated by greater irreversibility. However, the necessary institutional and legal framework needs to be established in order to provide an appropriate regulatory design able to offset, at least partially, the higher country risk that investors are likely to face.

In Chapter 14, Abrardi, Cambini, and Rondi posits the importance of NRAs in all Mediterranean countries, with the twofold aim of increasing infrastructure investments and converging toward a homogeneous and harmonized regulatory framework with Europe. The authors developed a model that examines a sample of MENA countries and study the rate of growth of their generation capacity over a time span of 22 years. The model finds that the inception of a freestanding regulatory agency external to the direct ministry's direction seems to have a positive effect on investments. Even when NRAs do not enjoy a high degree of substantial independence from the executive power, the establishment of a regulatory agency distinct from the ministry may lend credibility to good regulatory practices. The presence of a NRA is therefore positively correlated with indicators measuring the enforcement of the rule of law and of the control of corruption. This confirms that the establishment of a regulatory agency is often associated to the institutional endowment of the country.

In Chapter 15, Allal and Urbani explore the issue of financing electricity infrastructures in the SEM, highlighting the existing regulatory and market challenges, as well as socioeconomic opportunities that could provide positive signals to promote investments that are urgently required to deal with the expected increase in energy needs. Allal and Urbani show how financing Mediterranean energy infrastructure is a complex and articulated undertaking, as financial aspects are interrelated with economic, political, institutional, technical, and regulatory features, in a context where geopolitical risks is high. It is therefore imperative that nondiscriminatory conditions are applied, entailing transparency and total reciprocity between countries.

In Chapter 16, Bonafé provides an analysis of the structural reforms that are taking place in MENA countries. The author affirms that there is a generalized tendency toward open and market-based approaches. National interventions are now combined within the activities of three Euro-Mediterranean regional platforms (acting on electricity, gas, and renewables, respectively) that shall be able to provide a regional spin to the otherwise local initiatives. Bonafé argues that the launch of the UfM platforms can be framed in the wider revised external energy policy that takes into consideration the different role and aspirations that regional countries might play. Bonafé considers the role that the Energy Charter might have in accommodating these evolving needs within a framework of legally binding agreements, able to ensure legal certainty for the investment sought in the region.

In Chapter 17, Ponti tries to answer a simple question: what do investors want in order to put their money in new (or existing) energy infrastructures? He provides an answer to this question presenting a historical overview of the utilities sector in Europe. Ponti argues investments in energy utilities are not safe, as they are exposed to both commodity risk and government interventions. The author details the minimum regulatory requirements that should be in place to provide consumers and tax payers with concrete benefits. MENA countries should grant higher returns on investment to offset the country risk. The author maintains that allowing a high level of transparency and consistency in the regulatory process is ultimately, what maximizes the benefits to consumers and minimizes the cost to tax payers in any given country.

5 CONCLUDING REMARKS

Energy cooperation and the wider socioeconomic ramifications that energy has in the Mediterranean region have never been so relevant to policy agendas. While emerging trends appear to be more consolidated and robust, after the standstill experienced in 2011, the direction that modern energy systems are taking in the region and the pace of the observed movement still needs to be clearly defined.

Mediterranean countries are facing significant social, economic, and environmental challenges and they all have heavy repercussions on the energy balance and trade in the region. Those dynamics, in the past mostly national and domestic, are now increasingly regional in their diffusion and present global characteristics. It is therefore possible to identify some common challenges that are widespread in the Mediterranean Basin, and present some interesting communalities among the different countries. The analysis carried out in the following 17 chapters identifies security of supply (SoS), economic and environmental sustainability, and empowerment of demand side/consumers as the main common challenges for the energy sector in the Mediterranean region.

Most pressing challenges are SoS, affordability, and competitiveness. The current and expected increase in demand, highlighted as one of the main drivers for the emerging energy scenarios in the coming decades, requires a correspondent significant increase in available generation capacity. However, in order to tackle the sustainability challenge the capacity to be added into the system needs to be both economically and environmentally sustainable. On the contrary, the existing paradigm does not possess the necessary dynamic efficiency that allows the electricity supply industry to recover fixed and variable costs necessary to finance its long-term viability. In addition, it does not comply with the emerging emission standards. In this regard, RES appear to be able to play a pivotal part in enhancing SoS in the region in a cost-effective way, while conforming to the increasingly stringent environmental measures. Renewable sources therefore, emerge as the second key driver, beside SoS matters, which will shape the present and future energy outlook in the region. A third demanding task for the energy system of the countries belonging to the Mediterranean Basin is to cast a new and more active role for energy consumers. This includes empowerment and protection of consumers, and the possibility to overcome the widespread subsidization policies that support energy consumption. Consumers' empowerment and protection should also aim at facilitating the promotion of the consumer–producer and consumer–investor role, typical of more mature and market-oriented energy systems.

Tackling these complex challenges is a demanding task and requires immediate action. It is now apparent that the existing "*status quo*" is unsustainable, also because current electricity generation is hardly able to cover the existing demand, in most countries, while at the same time the quality of power supply is deteriorating across the region. Although piecemeal solutions can represent an attractive alternative in the short run, in particular to accommodate the most pressing needs, the aim of the MEDREG Forum and the ambition of this publication is to identify a long-term vision able to address the main challenges identified earlier. It is fair to say that the contribution collected in the book and the discussion held at the Forum confirm that there is no single way to tackle, at the same time, SoS, sustainability, and consumer empowerment. However, it is possible to define a set of milestones that should be able to drive the transition of the existing energy systems bringing a more secure and sustainable energy supply to benefit current and future energy consumers that need to be empowered with more opportunities, while receiving an appropriate level of protection.

The main conclusion that arises from this book is that a clear roadmap toward a new regulatory model is needed, in order to create favorable investment conditions. This approach is to be centered on market-based mechanisms. Currently, the existence of a competitive market (segment) is restricted, and often the only space for commercial initiatives is limited to new RES installations. Whereas the presence of the state in the Mediterranean region is a common feature, most MENA countries fail to distinguish clearly the roles and responsibility of state-owned utilities, public bodies (such as relevant ministers and public administrators), and regulators. In particular, all national regulatory bodies of MENA countries should accompany the transition of their energy systems, with stronger power and competences. Transition toward new market-based regulation is expected to enhance the attractiveness of new infrastructure investment in generation, transmission, and distribution, reinforcing the desired link between regulation and investment. Therefore, casting a greater role for competition in sectors where this is possible (typically generation and retail) calls for a stable and robust regulatory framework, able to define a levelled playing field that allows state-owned and private initiatives to compete in the market, and for the market. An enhanced stronger role for NRAs requires a parallel step back in terms of public presence in other sectors of the energy value chain, in particular in the competitive fringe of the market. It also requires granting equal access to transport and distribution infrastructures. In particular, security of supply challenges, as well as problems of affordability and the lack of competitiveness, calls for effective regulation, based on the establishment of a market approach. This regulatory model is expected to deliver, in particular, price signals to all agents. The market approach involves the elimination of final energy price subsidies and the reinforcement of NRA independence. NRAs should be entrusted with clear responsibilities to establish rules and monitor compliance, ensuring the development of necessary infrastructure facilities and progressive liberalization of the market to make room for new private participants, thus facilitating energy exchanges within the region that are based on a market approach.

As mentioned earlier, the correct implementation of this market approach model involves stronger powers for NRAs. Regulatory authorities, in turn, should provide the necessary regulatory tools to create a wholesale market for the electricity trade, introduce the right of free establishment for generators and retailers, design a network access tariff that fully compensates costs, monitor the competition conditions for all the operators in the market, prevent market power, and set-up unbundling between all the stages of generation, transport, distribution, and retail. As experienced in most liberalized energy systems, this lengthy process requires a gradual introduction of the necessary reforms. An approach aiming at introducing, at the same time, the entire reform package could prove to be unsustainable for the delicate institutional ecosystem of the Mediterranean region. Most countries have introduced some interesting reform packages, but the transition process that started in the power sector of most MENA countries in recent years has not yet been completed. Problems including weak security of supply, final price subsidization, and limited energy exports within the region are still current.

As in many other energy system reforms, the transition in MENA countries cannot focus solely on security of supply problems, but is expected also to take into account climate change mitigation. Globally, countries are transforming their power sectors in order to cover the demand, coupling it with measures that allow the fight against climate change. To that aim, renewable energy technologies have evolved steadily in recent years. At the same time, the Mediterranean region has a vast untapped potential to produce clean energy from solar and wind. A strategic plan is needed for the development of all this potential while the energy mix is still dominated by conventional generation technologies (mainly fuel). For renewable energies and energy efficiency initiatives to achieve their market potential, policy frameworks and financial instruments are required to provide the necessary assurances and incentives to shift investment away from carbon-emitting conventional technologies to small, medium, and industrial investment in clean-energy systems and energy efficiency interventions. The benefits expected from increasing the installed capacity of renewable energy are: (1) improvement of SoS; (2) reduction in emissions; and (3) increase of MENA exporting capacity, as long as the development is parallel to the improvement of interconnection lines. To that end, regulation and regional cooperation will be crucial to reach the additional generation required to keep the system in balance and to achieve the emission reduction targets set in each country through infrastructure investment. Additionally, the price signals that the new market-based regulatory model is able to deliver will provide the opportunity to launch cost-effective energy efficiency programs. In addition to being essential to the reduction of emissions, energy efficiency can also contribute to ease SoS problems.

The complementary characteristics of the energy systems in the Mediterranean Basin, coupled with their high potential for RES generation, increases the importance of cooperation among regulators and with the wider community of institutional, social, and industrial stakeholders. Cooperation is needed to establish a mechanism that encourages investments in infrastructure. This investment will depend on the implementation of financial facilities for private capital coming from international financial institutions, industrial partners, and local investors. On the other hand, cooperation between NRAs and MEDREG will allow the definition of the main features of a harmonized Mediterranean energy market capable of benefiting citizens by reducing energy prices and assuring that "the lights will stay on." What is more, a better integration of renewables targets and standards will be obtained for the entire region. Moreover, the role of other institutions, namely the EU, also in relation to the Euro-Mediterranean energy platforms created in the context of the UfM, will be crucial to fully profit from all the positive aspects that the energy sector transition could bring to Mediterranean countries.

All levels of cooperation will need to be developed and reinforced. As the contributions to this book shows, reinforcing the energy exchange for mutual needs and SoS collaboration should begin bilaterally between neighboring countries. Afterward, subregional agreements should be deployed (the so-called "corridor" approach) for countries with similar levels of market development. Finally, the creation of a regional market needs to be defined with the aim of being cost-effective. This book shows several examples of best practice drawn from the European energy market, as well as a number of mistakes, which should be avoided. The lengthy process that led to the opening of the internal EU energy market, a process that is not over yet, represents a useful benchmark, although it can hardly be considered a model for its peculiar political dimension, to get things done properly from the outset in the MENA region.

Stronger cooperation requires that the regulatory and legal frameworks from different countries belonging to the Mediterranean Basin are harmonized. Besides that, the first step will be the creation of national markets in all countries and the construction of cross-border interconnections. The rules under which these markets operate must be similar so that tangible steps can be taken toward the creation of subregional markets, and finally, a regional integrated market. This convergence process, aiming at the creation of a Mediterranean Energy Market, will bring benefits in the form of easy cross-border energy flows, effective opening of national markets, and the financing of the transition to sustainable energy. The role that MEDREG is called to play is clear, while complex: encouraging all the NRAs to adopt a regional vision. The dimension of the national markets and the fragmented economic and political nature of the Mediterranean Basin is a powerful obstacle to attracting investment. However, the creation of a harmonized regulatory framework, the promotion of coherent RES target, and the existence of better cross-border interconnections might all play a positive role for the benefit of each individual energy system, to the advantage of greater regional welfare. MEDREG should remain a permanent voluntary platform enabling Mediterranean regulators to discuss the different matters that concern the implementation of a regional vision. In addition to that, a certain degree of delegation of power, from NRAs to MEDREG, should be considered in the future, in order to allow the definition of a binding commitment in the process of market integration. The definition of a common regional landscape requires that some of the powers currently held at national level are entrusted to supranational organizations, that might provide a better understanding of global dynamics extending well beyond domestic borders. Climate change mitigation guidelines could represent a good example of a policy dimension that requires the combination of national and regional powers, and that could represent a first step toward the definition of a more stable regional set up. In this field, a step forward is needed, with a definition of a more ambitious agenda for MEDREG and its role in the region.

The contributions collected confirm that theoretical evaluation stands at a significant distance from its practical implementation. We must accept that the ideal, of a Utopian energy cooperation, will elude us, but instead we can engage regulators and institutional and industrial stakeholders to elaborate principles of gradual and pragmatic convergence based on more efficient decision-making processes. Such is the hope of this undertaking.

REFERENCES

MEDREG, 2015. Interconnection Infrastructures in the Mediterranean: A Challenging Environment for Investments. MedReg, Milan.

Rubino, A., 2014. A Mediterranean electricity cooperation strategy. Vision and rationale. In: Cambini, C., Rubino, A. (Eds.), Regional Energy Initiatives. MedReg and the Energy Community. Routledge, London, pp. 31–44.

Part I

A Roadmap for a Mediterranean Energy Community

Chapter 1

The Regulatory Framework of the Energy Community in South East Europe: Considerations on the Transferability of the Concept

Branislav Prelevic

Energy Community Regulatory Board; the Regulatory Energy Agency of Montenegro; Energy Economics, University of Donja Gorica, Montenegro

1 INTRODUCTION

Analysis of the success of electricity sector reform in the EU and its transferability into the South East Europe (SEE) Energy Community (EC) is twofold. The first aspect is general – it analyzes the possibility of transferability of the concept. The concept was designed in such a way that certain sectoral reforms and policies were first to be implemented in the European Union (EU) and then replicated through a system of institutions such as the SEE EC by EU accession countries. The second aspect looks at individual solutions, such as sectoral strategies, regulatory mechanisms, and individual legislative provisions implemented in practice. The two perspectives are necessary as they allow us to identify the root of the problem. As a result, failure of the concept cannot be blamed on poor individual solutions and, vice versa.

This chapter aims to analyze the following: first, the mechanisms behind investment incentives in the transmission or, to be more precise, in the interconnection infrastructure of electric power systems; second, the possibility of this set of mechanisms establishing a model or any other formal framework and, last, to provide a discussion on the transferability of such a model or concept. The discussion has a strictly regional perspective since it addresses the interconnector infrastructure (i.e., the mechanisms of investment incentives in the interconnectors are considered as measures that have to be undertaken by two or more countries constituting a given region).

Starting with Pollitt (2007), there have been a number of papers (Petrov, 2007; Monastiriotis, 2008; Uvalic, 2009; Zuokui, 2010; Haney and Pollitt, 2010) that

Regulation and Investments in Energy Markets. http://dx.doi.org/10.1016/B978-0-12-804436-0.00001-1

have analyzed the issue of transferability of the concept of EU energy sector reform, including its application in the SEE EC region. However, our dilemma concerns a set of more specific issues. Namely, one of the objectives of this chapter is to check whether it is possible to identify incentives in the transmission system implemented in the EU as well as those subsequently transposed to the EC that have a regional character from among the mechanisms for investments; whether their system-level application can result in a model; what the necessary level of harmonization for these mechanisms is, and whether the model so established may be transferred elsewhere so as to become a kind of matrix to encourage investment in other developing regions. Therefore, Pollitt's thesis was only a trigger or an inspiration for this kind of approach. It is not the intention in this chapter to discuss Pollitt's thesis itself, which is essentially broader and multivalent.

Studies, materials, and evaluations mainly published from 2008 onward were used to verify this thesis. Earlier assessments or materials were avoided because the EC effectively started its operation in 2006 and the more serious work on incentive mechanisms started only in 2009. Following a review of all applied or recommended incentive mechanisms, the focus was then placed only on those mechanisms that could be applied within a regional framework. Therefore, incentive mechanisms have not been analyzed according to their effects or strengths such as regulatory *ex ante* tests, incentive mechanisms that provide incentives to reduce costs, various exemptions from regulation related to rates of return, tendering for new investments to encourage investment from independent investors, or regulatory asset base (RAB) determination. Our thesis is that these types of incentives may be applied anywhere, at any time, in any country – but it is not natural to expect countries in any region, no matter how compact that region may be, to implement these incentive schemes in the same way, with the same intensity, and to the same extent. Therefore, they have a national character – not a regional one. Naturally, the existence of a harmonized set of incentive schemes would be ideal for investment and we should strive to achieve it, but we believe that cannot be achieved in practice because of the social, political, economic, and, above all, technical differences among energy systems.

2 BACKGROUND OF THE EC CONCEPT

The EC is a unique international institution based on a legally binding framework for contracting parties in accordance with the highest European legal standards and proven EU experience in energy market modeling. By signing the Treaty in 2005, the EU expanded its mission of "a common voice" in the energy sector into SEE countries.

Looking at the energy sector from the perspective of the EU, the SEE region is interesting both in itself and as part of the broader geopolitics of energy. First, this region has significant energy potential, especially when it comes to renewable energy sources, opening new markets, and so on. Second, since this region is one of the routes toward bigger and more significant energy sources and markets, its geographical position secures it an important geopolitical role. Third, the EU does not like to see unstable zones (in either political or energy terms) on its borders. Fourth, most countries in this region aspire to become members of the EU. Given all this, there is no doubt that the EU both needs and has an interest in making this region compatible with the EU.

From the perspective of the small transitional countries of the SEE region, their small and mutually nonconnected markets, developing institutions, and energy policies, there is a need to establish a strong *regional* institution to articulate their numerous problems, strengthen their market position, assist in the establishment of institutions, and finally bring them closer to achieving their political goal of becoming members of the EU. This seems to be a solution to which there is no alternative.

3 TRANSFERRING EU POLICIES AND MECHANISMS TO SEE

The concept of regionalization has been implemented through a number of mechanisms – such as the Central European Free Trade Association (CEFTA) and the EC – which defined a far-reaching development path for SEE countries: Albania, Bosnia and Herzegovina, Kosovo, Former Yugoslav Republic of Macedonia (FYROM), Moldova, Montenegro, Serbia, and Ukraine as contracting parties with Armenia, Norway, and Turkey as observers. Seventeen EU member states have the status of participants. According to the political context in which it appears and develops, the EC was established[1] as a regional actor that has its own political role and whose economic sustainability will be ensured or strengthened through its synergetic effect.

The SEE EC had its origin in the 1999 Stability Pact for the SEE and the Regional Cooperation Council (RCC). The SEE Stability Pact was the mechanism by which the international community tried to help SEE countries overcome problems resulting from the wars in the 1990s. The Stability Pact was a mechanism that encouraged regional cooperation as a nucleus of cohabitation. Just as was the case at the very beginning, these regional initiatives also have a political background today, since the existence of the EC without the current political ambitions of its member states is questionable. Political events linked to the relationship between the EU and Russia could also affect the future format of the EC through its possible expansion to include Armenia, Georgia or Turkey.

1. By the Energy Community Treaty, which was concluded after a period of 10 years. As soon as the European Commission concluded that the Energy Community had proven to be an efficient framework for regional cooperation in the field of energy, the Ministerial Council decided (2011) to extend the duration of the Treaty for a further period of 10 years (Energy Community, 2014).

After this overview of the reform process in SEE, let us return to Pollitt's thesis (2007, p. 3):

> *Thus the SEE is and will be a test of both the transferability of the EU reform model within the EU (from the leading reformers) and also its transferability to a set of developing countries more generally.*

We must evaluate whether other countries or regions would follow this model, bearing in mind the specific nature of the EU as a basic model and the political factors that significantly determine the behavior of countries that are supposed to accept and implement the EU model as a matrix.

Signing of the Memorandum (Athens Memorandum, 2002) for some countries meant paving the way to reform guided by EU best practice, while for others it was just one more step in the agenda of accession to the EU. Whether the governments of SEE countries (bearing in mind the political instability in developing countries, the volatility of the reforms initiated, and the economic and financial crisis, *inter alia*) would have decided at a certain point in time to cancel the announced reforms had there not been firm commitments undertaken toward the EU or not is difficult to prove. We can only guess, but it seems to be quite obvious that political commitments related to integration into the EU and the carrot constantly offered by that integration have been anchor points of the reform process. Without underestimating the awareness and maturity of ruling structures, we are not sure whether in some difficult situations typical of transitional processes, governments would have decided otherwise (as various South American countries did in the late 1980s) had it not been for this "higher goal" named by the EU.

On the other hand, the power of the EC is more formal. The EC Secretariat supervises implementation of the EC Treaty, and thus it can open infringement proceedings vis-à-vis a contracting party. The nature of dispute settlement at the EC level is closer to political and diplomatic action than to any form of international jurisdiction (Kuhlmann, 2014, p. 3).

4 THE REGIONAL CONCEPT AS A PRECONDITION

The political background of the EC should certainly not obscure or diminish its economic importance. Economic recovery in the region was only possible thanks to the provision of huge investments that were supposed to be the momentum or driving force behind it. Stable development of the region would also have been impossible without reconstruction of its energy system. A large part of the energy infrastructure was destroyed during the wars. On top of that, there was a need for it to be modernized and connected with the European transmission network. Studies since January 2007 commissioned by the World Bank (Varadarajan et al., 2007, p. 15) show that €16,748 billion needed to be invested in the region for the renovation and construction of new generation capacities as well as an equivalent amount for network investment.

Affirmation of the regional concept as a basic precondition is motivated by economic logic as well as superiority of the regional over the national market. This is based on acknowledging that regional coordination is the only feasible solution for gaining synergy effects within the small and only partly connected emerging energy markets of the region:

> *In the given geographic set up, the Treaty's goals of attracting investment, enhancing security of supply, improving the environmental situation and creating market liquidity – all for the benefit of customers – can only be reached when taking regionally coordinated action (Energy Community Secretariat, 2011, p. 8).*

The economic motive behind regionalization is clear. Regional cooperation was imposed as the only possible solution that would provide a synergy effect for small and only partly connected energy markets. The economic success of the former Yugoslavia could not be achieved by the states that emerged from it. These countries did not succeed in creating a market that might be attractive to investors. Any market having a few hundred thousand or a few million inhabitants could never be as attractive as a market of 20 million.

Whether the energy investment plans in the region are consistent or inconsistent in the near future is a question that will determine the success of the concept of an EC. For the countries of the region it is desirable but not imperative to have a common legal framework, a common policy, and the same regulatory principles and methodologies. Yet, it is essential for them to have a common market and common investments for crossborder projects. A condition for regional investment is redefinition of the myth about national security of energy supply. This is a critical point that has not been overcome by many EU countries in their policies. The issue of security of energy supply is still viewed from narrow national perspectives. Plans to build new energy capacities are still based on domestic demand or domestic resources no matter how uneconomic such resources are. The EC has a key role to play in the creation and implementation of investment plans.

5 THE ROLE OF INVESTMENTS IN TRANSMISSION NETWORKS

Historically, European transmission systems have been constructed to ensure sufficient electricity to supply national end users. Crossborder interconnectors were set up to allow mutual assistance in case of technical disruptions – not to promote interstate trade in electricity as such (Kapff and Pelkmans, 2010, p. 4). Closed energy systems are costly and subject to the risk of random outages. Therefore, the connectivity of national transmission systems is a prerequisite to efficiency and security of supply. Networking all national transmission systems is a prerequisite to creation of a single European transmission system and market. This allows electric power to be generated where it is most efficient and then transported to where it is most needed.

Generally speaking, there are four main elements underlying the need for new investment:

- The first factor is the need to build new interconnectors, which is driven by the necessity to increase market liquidity. Integration processes, both regional and global, result in an increase in the crossborder trade of all commodities and services. When big national producers regarded their own territory as their own "hunting ground" it was difficult to ensure competition since such producers could rely on loyal transmission companies that did not allow "third-party" access or developed interconnection capacities to allow free access to their market (except to the degree necessary for the provision of minimum security of supply). Therefore, the liquidity of electricity as a prerequisite to market opening is essentially determined by the level of development, security, and efficiency of interconnection capacities.
- The second factor, which in the last two decades has emphasized the expansion of transmission capacities, is the construction of generation capacities based on renewable energy sources. The geographic dispersion of energy sources, distance from the point of consumption, inability to accumulate the energy generated, and generation management, as with conventional sources, make renewable energy sources utterly dependent on the construction and reliability of the transmission network.
- The third factor is modernization of the network driven by technological progress and legal or security concerns.
- The fourth factor is the increase in the volume of energy transmission due to increased domestic consumption and the entry of new major consumers. We believe this is the key reason to expansion of the network or its capacity.

When it comes to Europe, the facts clearly point to a need for a new energy infrastructure, another major challenge faced by the EU (Oettinger, 2012, p. 2). The European Commission has assessed energy grids in the EU are in such a state that they cannot respond to forthcoming challenges laid out in the "Europe 2020" strategy or to the decarbonization targets set in the "Energy Roadmap 2050." In order to achieve these objectives, the EC in its communication *Energy Infrastructure Priorities for 2020 and Beyond* identifies the tremendous need for new investment of approximately €200 billion for energy transmission projects of European interest. This amount does not include investment in national projects or reconstruction, but only in regional and pan-European projects (€140 billion relates to electricity and €60 billion to gas transmission infrastructure).

Having well-developed transmission capacities within its own market and in surrounding countries are equally important to the EU. Most SEE countries will in the near future become members of the EU and thus members of the EU single market. In addition, these neighboring countries are important routes for the flow of energy (i.e., they lead to regions that are particularly important regarding the reliability of the EU's energy supply in the future).

These are clearly capital investments that require special mechanisms (i.e., incentive measures and special sources of financing) as a result of their volume and specificity.

The volume of capital problem stems from the fact that many transmission system operators (TSOs), especially in transitional societies, are currently in the postprivatization phase when states, after either complete or partial privatization, are no longer ready for new investment. This situation came about, we believe, by the transitional illusion of state administration, which used the privatization of national companies, among other things, to eliminate its responsibility to finance nonefficient state companies. By contrast, private investors as a rule are more alert when investing in large infrastructure projects than the state. We think such behavior on the part of private investors is to be expected given that they do not enjoy the support of (in)dependent state administration.

Hence, when there is a need for major investment mostly by private investors under conditions of a financial crisis, a whole set of measures and mechanisms need to be designed to stimulate new investment.

6 THE ROLE OF REGULATORY AGENCIES

To create favorable conditions for new investment is a complex task in itself. A favorable investment mechanism usually means unfavorable circumstances for those who have to pay or repay a given investment. These are the consumers or, as they are termed today, the customers. The task of the EC in all this is particularly complex. The experience and functioning of the EC in the previous 8 years have confirmed this. The fact that there has been no significant investment in generation capacity within this region for the last 30 years proves that there are other obstacles besides wars that discourage investment. Since the beginning of the 1990s, most countries in the region have undergone political, economic, and social tensions; economic blockages; transition processes; and financial crises. It is not possible, or even expected that they might quickly create circumstances that could be suitable for what is typically very sensitive and capital-intensive investment in energy.

Investment still needs to be seen from the point of view of those targeted for investment (i.e., the region, the host country, and the consumer). Depending on the economic situation of the country, available natural resources, the social and political system, development strategies, and social structure, state institutions create investment plans that consider investment size and mechanism (i.e., incentives for successful completion of these plans). Very ambitious investment plans linked to less stable economies bear a higher level of risks for investors. A higher level of risk requires more expensive financing, which is often an insurmountable obstacle for less developed countries or developing countries. Since investment in transmission and distribution networks is recovered through tariffs charged to consumers, a vicious circle can take place as it is assumed that in less developed countries there is a greater need for investment, while the social structure of consumers is such that they will not tolerate increased tariff charges that accompany new investment.

The emergence of national energy systems, be they in the east or west of Europe, is linked to the creation of monopolies. They were initially established with the assistance of the state as legal monopolies (i.e., monopolies installed by the state to protect its infant industry).[2] As the importance of electricity has increased, so has infant industry. It has developed from being a legal to a natural monopoly.[3]

In the early 1990s, Europe was divided into national, monopolized, and inefficient energy systems. The establishment of the European Union went hand in hand with development of the idea that these entirely monopolized national systems, because of their inefficiency, had become an obstacle to further development of the energy system, national economies, and the EU as a whole. The idea of creating a single electricity market also meant that competition must be recognized. Energy policies in the 1980s and 1990s focused on the objectives of privatization, liberalization, and competition (Helm, 2005, p. 7). A free market could not be established in such circumstances without introducing elements of competition and privatization (i.e., liberalization). Instead, it was necessary to ensure the liquidity of electricity, the harmonization of laws, the transparency of processes, and the promotion of personal property rights. System reform has become indispensable along with new legal frameworks, transparent processes, new investment, and most important of all, new independent regulatory authorities.

Sectoral reform has brought an end to monopolies on the generation and supply of electricity (i.e., they are unbundled from vertically integrated monopolies, while the transmission and distribution network remain natural monopolies). The existence of the network as a natural monopoly makes sense (1) if duplication of a certain infrastructure or service for which such infrastructure is a precondition is uneconomic (i.e., where this service would be more expensive when its performance required competition between two or more companies – the problem of so-called subadditivity of costs) (Petrov and Nunes, 2009, p. 3), and (2) if an activity performed by a monopoly is in the public interest.[4]

2. Infant Industry Argument – the term refers to an industry that could not be developed without assistance from the state (for more details, see Hamilton (2014)).

3. A natural monopoly arises when, due to particular structural (technological) features, the optimal number of operating firms in a market is one. Accordingly, competition is neither desirable nor efficient. A natural monopoly differs from a legal monopoly in that the condition preventing competition does not stem from technological features of the industry but is rather the result of legal provisions. Legal monopolies can be established, for instance, through the granting of public franchises, licences, patents, or copyrights. Available from: http://fsr-encyclopedia.eui.eu/natural-monopoly/ (accessed 18.09.2014).

4. Since a utility provides essential services for the wellbeing of society – both individuals and businesses – it is an industry "affected with the public interest" (Electricity Regulation, 2011, p. 3).

As a result, transmission and distribution are controlled by regulatory authorities. A good regulatory regime should provide companies with similar opportunities and incentives to those they would face in a competitive market. However, in mimicking market forces, the regulators should also balance this duty with the need to consider the interests of network owners in order to ensure that network businesses earn a reasonable rate of return on their (efficient incurred) investment (Energy Policy Regulation, 2014).

These circumstances helped identify the need for and objectives of the independent regulatory bodies to be set up. The major responsibilities of these bodies include the following:

- protecting consumers from monopolistic market forces;
- protecting investors from the prominent role of state administration (i.e., from its arbitrary decisions);
- improving the economic efficiency of the system; and
- managing externalities (particularly emphasized in the need to solve environmental problems).

7 OUTLINE OF INVESTMENT INCENTIVE SCHEMES

Finding investment incentive schemes is the responsibility of regulators. There is a lot of literature on theoretical or practical solutions in the EU. However, every energy system is unique and any "copy–paste" solution usually results in major and expensive mistakes. Originality in combining investment incentives should not be considered as a goal in itself, but rather as something necessary to reach an appropriate solution.

As already noted, the focus of this discussion is limited to mechanisms that can be used to encourage investment in an interconnection infrastructure that is strictly regional (i.e., analysis of mechanisms encouraging investment that affect the region as a whole).

In the remaining part of this chapter, we provide an overview of investment incentive schemes. This overview is based on studies developed or commissioned by the EU and its bodies. We then consider the importance of certain investment incentive mechanisms from a strictly regional perspective. We also look at the importance of harmonizing such mechanisms.

7.1 Tariff Methodologies

Tariff methodologies to determine the allowed revenue of regulated companies define a number of elements of the regulatory framework that are very significant for investors: the RAB, the weighted average cost of capital (WACC), the efficiency factor (*X* factor), network losses, bad debt, and so on. According to almost all the experts interviewed, regulatory issues are the most important factor in the financing of energy infrastructure projects. Key issues include regulatory remuneration (the foundation of all investment cases), stability of the regulatory regime, and related remuneration. These issues are equally important for both the TSO planning the investment and the financing institution providing the funds (Roland Berger Strategy Consultants, 2011, p. 7).

To analyze the tariff methodologies of SEE EC countries, we used "Status Review of Main Criteria for Allowed Revenue Determination," a report compiled in 2013 by the EC Regulatory Board (ECRB). In this report, the ECRB summarizes, among other things, all significant elements of regulated network tariffs applied in the countries of the region. The report covers Albania, Bosnia and Herzegovina, Croatia, FYROM, Georgia, Kosovo, Moldova, Montenegro, Serbia, Turkey, and Ukraine. The results for Bosnia and Herzegovina – which differ from those for the Federation of Bosnia and Herzegovina (FBIH), Republic of Srpska (RS), and Brcko Canton in FBIH – are presented separately in this report (hereinafter the Report). The Report was based on a questionnaire created by the ECRB and completed by the national regulatory authorities (NRAs). It has provided information and data that allow us to formulate a number of recommendations on how regulation can effectively encourage investment. The main findings from our recent survey can be summarized in 14 key messages:

1. The most commonly applied price control mechanisms are the cost plus (rate of return) regulation and revenue cap regulation; price cap regulation is occasionally used.
2. Regulators who applied incentive regulation (revenue cap or price cap) obviously had problems when determining the efficiency factor (X). While some regulators have $X = 0$ (Croatia and Albania), in Montenegro it is determined as ½ inflation, while in Kosovo it is 4%.
3. As a result of recognizing the operating and maintenance costs involved in allowed revenue, regulatory agencies are required to insure that tariffs for transmission and distribution are transparent and reflect the real costs (recital 18 in Directive 2009/72/EC, 2009, p. 56) (i.e., they should insure maintenance of appropriate levels of services, but avoid excessive costs being included in tariffs). The Report reveals that regulators generally have differing predefined matrices of "justified" and "unjustified" costs. They even restrict certain categories of costs. All of which indicates a high degree of arbitration.
4. All regulators calculate return on assets in the classical way by multiplying the value of the RAB by the rate of return. The exemption is Ukraine, where the rate of return is not calculated at all, but is replaced by "projected revenue."
5. The RAB, by definition, is the assets of the company used for performing its activities. Most regulators have introduced or intend to introduce regulatory accounting, which would enable harmonization in this area, by connecting it to international financial standards. As a result of the importance of this element in the calculation of total revenue, some regulators have the right to reevaluate the assets of the company.

6. Almost all regulators approve investment plans *ex ante* and monitor their implementation. Successful completion of investment plans affects the value of the RAB to the extent that allowed revenue includes some investments in advance, which is the case in Montenegro and FYROM. Accordingly, it is possible in multiyear regulatory periods to make a correction to the RAB at the end of each year if planned investments are not implemented as planned (the correction in Montenegro is made by excluding from the RAB all completed investments with a value less than 50% of the planned value).
7. The issue of working capital is viewed differently by regulators. It has been included in allowed revenue by regulators in Albania, Bosnia and Herzegovina, Croatia, and Montenegro.
8. Capital contributions in terms of grants from governments or other institutions are excluded from the RAB by all regulators.
9. WACC[5] is a commonly used means of calculating the rate of return in all regulated energy companies. However, there are wide variations in the results achieved as a result of using this method. For example, the value of WACC varies from 0.67% to 8% in Bosnia and Herzegovina, and as much as 95% in Moldova for transmission. It also varies from 3.5% to 14% in Bosnia and Herzegovina, and 12% in Moldova for distribution. Slight variations can be explained because no one considered whether a pretax or posttax model was applied or whether it was a nominal or real WACC when completing the questionnaire. However, these results certainly affect investor awareness and significantly affect the level of regulatory risk. We believe that the practice of employing specialized companies to calculate WACC could be applicable to countries in the SEE region as happens in some countries outside the region.
10. The regulatory agencies of the markets analyzed usually use straight line depreciation to calculate depreciation.
11. Network losses appear to be a very important element of regulation. Unfortunately, there seems to be little consistency in treating this problem on the part of regulators in the region (the ECRB study remarked that "Definition and criteria applied by the regulators for recognizing loses in the allowed revenue of all regulated activities are not differ significantly between analysed markets"). There are many ways of resolving this issue, starting from excluding losses from operating costs (Ukraine and part of Bosnia and Herzegovina) to recognizing complete costs including commercial ones (Kosovo –conditional on reducing costs every year in accordance with predetermined dynamics). In addition, there are different methods for assessing recognized technical losses (from realized to

5. Weighted average cost of capital (WACC), in general terms is a financial metric used to measure the cost of capital to a firm or project as the weighted average of the cost of debt and the cost of equity.

technically desirable) and for assessing the energy cost price to cover approved network losses (from the cost price of domestic generators to the real market price). Consequently, recognized costs range from 0.86% in Georgia to 3.66% in Montenegro for transmission and from 8% in Croatia to 27.40% in Kosovo for distribution.

12. Revenues from nonregulated activities are mostly excluded (except in Croatia, Georgia, and Moldova) from the regulated revenue of a company to avoiding double earnings from the same assets (a typical example of such revenue is that from transmission capacity congestion).
13. Regulators in all countries of the region typically use energy balances when determining quantities of energy. Differences may occur depending on who is determining the energy balance and whether or not it can be corrected according to changes in the market.
14. After expiration of the regulatory period, regulatory agencies calculate the corrections – in other words, the difference between allowed and actually earned revenue – so that the amount of the resulting over/underreturn can then be recovered in the following regulatory period.

According to the authors of this Report, regulators use similar principles and criteria when determining network tariffs.

We believe it is reasonable to conclude that the degree of consistency of tariff methodologies in the region shows a distinctly upward trend. All these countries effectively started regulation in the second half of the last decade when they established independent regulatory authorities and developed their first tariff methodologies. We can also conclude that an enormous step forward has been taken in the implementation of European regulative and best regulatory practice. A significant level of harmonization between regulators does not exist in terms of determining revenues for network operators, but it is imperative to impose a greater level of harmonization. The reason more needs to be done is that this overview includes countries that became members of the EC SEE later (i.e., Ukraine and Moldova) and countries that have observer status "only" (i.e., Georgia and Turkey). These types of incentives can be applied anywhere, at any time, and in any country, but it is not natural to expect countries in any region, no matter how complex they are, to implement these incentive schemes in the same way and with the same intensity. The level of harmonization between regulators is no greater even at the level of the EU. These differences mean a lot to investors and are very important for the development of internal markets (European University Institute, 2010, p. 25).

What is interesting is the effect these mechanisms have on the region as a whole, rather than on any one particular party or country. With these considerations in mind, we have come to the following conclusions:

- From the standpoint of investors, it would be very convenient if all countries in the region had identical incentive mechanisms (the same regulatory frameworks, tariff methodologies, tax policies, and so on).

- However, such incentive mechanisms are little more than utopia. This is due to different structures in different countries (particularly, the SEE countries). We are of the view that it would be irrational to assume something like that in practice. It is worth noting that the number of authors taking such a position is on the increase.
- The best way forward would appear to be consideration of the minimum measures and mechanisms that could realistically be achieved or even imposed on the whole region.

7.2 Mechanisms for Investment Incentives at the Regional Level

Projects of regional importance refer mainly to interconnectors or international transmission lines. While the need for substantial investment in increased interconnection capacity is widely recognized, the ways and means of promoting such investment are more controversial. Managing and financing transmission systems have become so complex because overburdening of transmission systems has effects on all neighboring networks. This is because electricity transmission flows utilize all available paths on the interconnected system in accordance with the laws of physics. This is why distance based or "direct path" charging is not appropriate for electricity transmission (Commission of the European Communities, 2008, p. 9). In modeling the interconnector, investors do not usually have a clear picture of when and whom they will charge their investment from (i.e., who will eventually bear the losses and who will have the benefits of the new interconnector).

In October 2012 the EC approved its own energy strategy (Energy Community, 2014). This was followed up by a process in which Projects of Energy Community Interest (PECI) (Energy Community PECI, 2014)[6] were initiated, mirroring European discussions on projects of common interest (PCI). This approach is in line with the recently adopted European Regulation on Guidelines for Trans-European Energy Infrastructure, defining the priority corridors of trans-European energy infrastructure, establishing the label PCI, and identifying possible measures to support them.

For PECI to be promoted by regulators, the ECRB recommended two studies. Its own "Cooperation of Regulators with Regard to Cross Border Investment Projects" (ECRB, 2010) which recommends a number of instruments to be implemented. The other study was "Development of Best Practice Recommendations on Regulatory Incentives Promoting Infrastructure Investments" put together by E-Bridge Consulting in 2011 on behalf of the Energy Community Secretariat; this too recommended a number of possible regulatory investment incentives practiced in the EU.

6. For details on the PECI process see Energy Community PECI (2014).

The major recommendations of this ECRB study from 2010 can be listed as follows:

- Full implementation of *acquis*; harmonization of regulatory market rules; introduction of the investment incentive scheme for the purpose of motivating national-level investment.
- The second recommendation relates to "nondomestic investment." National regulators should be authorized to recognize nonterritorial expenses (investment in the transmission network of the territory of a neighboring/other country) in their regulated assets in the event national/domestic consumers benefit from it (known as the "regulatory gap").
- Special significance is given to regional investment planning. TSOs require development of a 10-year investment plan for the whole EU, which would be approved by the Agency for the Cooperation of Energy Regulators (ACER).

Such a harmonized set of incentive schemes would be ideal for investors; however, we do not believe that this is achievable in practice due to the social, political, economic, and, above all, technical differences between energy systems. Bearing that in mind, all the incentive mechanisms given in the first bullet point should be put in place and carefully dimensioned in accordance with the specific circumstances and sensibility of national regulators. However, we do not consider them critical to the regional aspect. Accordingly, for the purpose of this chapter this analysis will be limited to regional investment plans and incentives for crossborder investments and so-called negative incentives as mechanisms with the strongest integration potential.

7.2.1 Regional Investment Plans

Every country (i.e., the energy systems of every country) has its own long-term development plans or investment plans. Often these plans are just too ambitious or represent a wish list. Consequently, it has already been accepted that these plans cannot be achieved in the SEE region. Another important reason the plans cannot be implemented is that projects within these plans are dimensioned solely within narrow national interests. Such investment plans in transmission networks in the context of a single EU market make little practical sense, unless they are considered from a regional or even wider standpoint. Moreover, regional coordination is the only feasible means of gaining synergy effects for the small and only partly connected emerging energy markets of the region (Energy Community Secretariat, 2011).

Having recognized this problem, the EU through its III Energy Legislative Package has imposed a number of requirements in terms of regional cooperation to promote the development of electricity infrastructure by any and all means, within member states, between them, and between the EU and the surrounding regions. Under Directive 2009/72 and Regulation No. 714/2009 (which amended Regulation EC 1228/2003), the TSOs are clearly required, *inter alia*, to build sufficient crossborder capacities to integrate the European transmission infrastructure; to develop 10-year development plans (TYDP) which should be updated every (or every second) year; and to ensure that these plans are mutually synchronized.

7.2.2 Nondomestic Investment (the Regulatory Gap)

Benefits aside, regional plans under certain circumstances can also create problems. For example, a regional plan that includes projects for two or three neighboring systems can cause negative effects on systems not included in this plan. When setting up regional plans, this is the reason special attention should be paid to the creation of special rules for harmonization by regulatory bodies and monitoring by the EC. First, this includes a provision on the harmonization of regulatory rules beyond national borders and issues relating to the harmonization of plans for crossborder projects. The III Energy Legislative Package stipulates that costs arising from foreign crossborder capacity flows will be taken into consideration when investing in new transmission infrastructure. National regulators typically do not accept burdening domestic consumers with tariffs for the investment needed for the construction of interconnectors since the consumers of neighboring systems would also benefit from them. Nondomestic investment is not covered in advance/*ex ante* (e.g., by recognition through regulated assets, tariffs, or guarantees given by the regulator to cover the cost of investment as part of long-term planning), but *ex post* through International Trade Center (ITC) compensation. However, it is not entirely certain whether ITC mechanisms are appropriate to allocate nondomestic costs of investments to end users, especially when the ITC fund is capped/limited (Stefanovic, 2011, p. 4).

7.2.3 Negative Incentives

National regulators are allowed to introduce so-called negative incentives by decreasing tariffs by the amount of revenue from transmission capacity congestion, which has not been used for its long-term resolution (i.e., for new investment to increase the capacity of interconnectors). Bearing in mind the responsibility the TSOs have to resolve transmission capacity congestion by new investment, negative investment incentives could be used as a possible regulatory tool.

8 A DIFFERENT APPROACH

Recent studies and analyses in both the EU and the EC clearly show that there is a huge need for investment in national transmission networks and interconnectors. The complexity of managing transmission networks, the unstable economic environment of countries in transition, and nonharmonized regulatory practices have resulted in investors lacking confidence in the feasibility of these projects or the security of their investment. Every analysis highlights the crucial influence of the regulatory framework in making investment decisions. Exit strategies from this situation already exist. The EU and EC have already come up with strategic documents – such as PCI, PECI, Ministerial Council (MC), and EC decisions – which relate to harmonization of regulatory systems both in EU member states and the EC CP.

Despite this, there still seems to be a dilemma as to whether there is a need to harmonize them? Similarly to European University Institute (2010, p. 25), we are also faced with a number of dilemmas relating to the scope and level of harmonization of regulatory mechanisms:

- The first dilemma is about the level of consistency we need. Investment is not a goal in itself, but should serve end users. Not all the countries in the region have the same investment needs because network development, demand satisfaction, technical aspect of the infrastructure, achieved efficiency, and much more is simply not at the same level. In addition, the levels of economic development and the social structures of end users are far from identical, which also implies a different intensity of investment incentives. Complete harmonization would mean disrespect for the specific features and needs of individual members of the region or of the EU.
- Do regulators have a clearly defined line of harmonization? In other words, to what extent should the regulatory framework be harmonized? We believe that we need a more precise definition of the magic words "the harmonization of regulatory systems." There are some indications of the dimensioning of harmonization in the authorizations and responsibilities of the ACER when it comes to the EU, whereas the ECRB still lacks such mechanisms. In our opinion the ACER is not a mechanism (legally or technically) that can articulate the needs of countries outside the EU and within the SEE, just as the ECRB with its current capacities is not a mechanism that can lead equally and successfully toward the achievement of the goals of countries that have EU candidate country status and those that are far from that position or perhaps even have no intention of becoming EU members.
- Would a significant (not to mention complete) harmonization tailor-made for investors lead to preregulation (i.e., to complete antagonism of economic regulation)? We believe this would be the case. Harmonization by definition leads toward equalizing rules or mechanisms. It is impossible to equalize and generalize incentive mechanisms (for anything) in different circumstances without hurting the free market and competition.
- It is reasonable to say that principles are harmonized rather than individual regulatory mechanisms. If this is the case, surely we already have established principles of economic regulation in Europe?
- By what means or mechanisms should we implement a high degree of harmonization – neither ACER nor an eventually empowered ECRB (if we are talking about urgently needed investment then we primarily need mechanisms that give quick results) will have enough power to harmonize national regulatory systems over the longer term.
- Apart from these regulatory doubts, we have a problem with broader economic principles. To be precise, any special stimulation or incentive is a

kind of subsidy and a violation of the law of the market. We cannot hide behind the phrase "monopoly regulation." On the contrary, economic regulation does not require us to abandon economic principles. For example, what does it mean to introduce a special rate of return on some investments? Usually, it comes down to a couple of situations. (1) The rate of return (cost of capital or the sum of risks) for standard investments is underestimated, and consequently the regulator tries to make amends by providing a special rate of return to new or selected projects. (2) The regulator follows political principles rather than economic ones and assigns a higher rate of return to a project of "special significance." In both cases, economic principles are violated.

9 CONCLUSIONS

Based on the previous discussions, we can summarize our conclusions into three broad areas. The first area relates to the need for new investment in the region, where incentive measures are being proposed which meet the necessary degree of mutual harmonization:

1. Investment in the transmission network is necessary to create the conditions for strengthening the single competitive European market for electricity.
2. National regulators, when creating their tariff methodologies and market rules, need to follow best European practice. Harmonization of these mechanisms is desirable but not necessary.
3. In our opinion, the mechanisms that should be harmonized are those relating to interconnection capacity and lines of international importance.

The second area relates to assessing whether or not a more or less firm model or concept developed by the EC for the purpose of investment incentives for the transmission networks of the region actually exists. Although we do not have a model, we propose something softer: an investment-friendly platform.

Finally, the third area related to the possibility of transferring this model/concept to another region. We are of the view that this is a predominantly political issue. The political roots of the EC have direct implications for the answer to our thesis and our conclusion related to possible transferability of this framework to another region. The countries that have signed the Athens Memorandum have all aspired to EU membership. Some have already become members and some have the status of candidate countries. Would contracting parties be equally committed to fulfilling their obligations to the EC if there was no political factor (acting both as a carrot and stick)? It is not easy to answer this question either way. We can only guess, but we believe it is quite obvious that the political commitments relating to integration into the EU and the carrot constantly offered by that integration are the anchor point of the reform process.

REFERENCES

Hamilton, A., 2014. Report on manufactures. The Annals of the Second Congress, December 5th 1791. http://nationalhumanitiescenter.org/pds/livingrev/politics/text2/hamilton.pdf (accessed 18.09. 2014.).

Haney, A.B., Pollitt, M., 2010. Exploring the Determinants of "Best Practice" in Network Regulation: The Case of the Electricity Sector. Working Paper, Judge Business School, University of Cambridge, Cambridge.

Helm, D., 2005. 'The Assessment: The New Energy Paradigm' in The Oxford Review of Economic Policy, vol. 1, no. 1. Oxford University Press, Oxford, pp. 1–18.

Kapff, L, Pelkmans, J., 2010. Interconnector Investment for a Well-Functioning Internal Market: What EU Regime of Regulatory Incentives? Bruges European Economic Research (BEER)no. 18College of Europe, Bruges.

Kuhlmann, J., 2014. 'Regulatory aspects of Energy Community' in Energy Community Law, Enforcement, Oil, Gas & Energy Law Intelligence, vol. 12, no. 2. http://www.ic-group.org/fileadmin/user_upload/Publikationen/ogel-_short_version.pdf (accessed 24.09.2014.).

Monastiriotis, V., 2008. Quo Vadis Southeast Europe? EU Accession, Regional Cooperation and the need for a Balkan Development Strategy in Hellenic Observatory Papers on Greece and Southeast Europe, no.10 LSE, London. http://www.lse.ac.uk/europeanInstitute/research/hellenicObservatory/pdf/GreeSE/GreeSE10.pdf (accessed 24.09.2014.).

Oettinger, G., 2012. Energy Infrastructure Priorities for 2020 and Beyond. Keynote Closing Speech at the Information Day on the Process of Identifying Projects of Common Interest. European Commission, Brussels.

Petrov, R., 2007. Legal and Political Expectations of the Neighbour Countries from the European Neighbourhood Policy. Workshop Paper, Florence, European University Institute.

Petrov, K., Nunes, R., 2009. Analysis of Insufficient Regulatory Incentives for Investments into Electric Networks: An Update. Final Report, Kema Consulting for the European Copper Institute, p. l.

Pollitt, M., 2007. Evaluating the Evidence on Electricity Reform: Lessons for the South East Europe (SEE) Market. CCP Working Paper 08-5, Judge Business School, Cambridge, University of Cambridge.

Roland Berger Strategy Consultants, 2011. The Structuring and Financing of Energy Infrastructure Projects, Financing Gaps and Recommendations Regarding the New TEN-E Financial Instrument. Final Report, Berlin.

Stefanovic, N., 2011. Regulatory Incentives for New Cross Border Transmission Capacity. Position Paper, Belgrade, CIGRE.

Uvalic, M., 2009. EU Policies Towards the Western Balkans: The role of Sticks and Carrots. Paper presented at AISSEC, Perugia, University of Perugia.

Varadarajan, A., Tovoularcas, S., Jovanovic, M., 2007. Development of Power Generation in the South East Europe, Update of Generation Investments Study, vol. 1. Summary Report, Final Report, the World Bank Group & South East Europe Consultants Ltd, Washington & Belgrade.

Zuokui, L., 2010. EU's conditionality and the Western Balkans' accession roads European perspectives. J. Eur. Perspect. Western Balkans 2 (1), 79–98.

Documents

Athens Memorandum, 2002. Memorandum of Understanding on the Regional Electricity Market in South East Europe and its Integration into the European Union Internal Electricity Market. http://www.energy-community.org/pls/portal/docs/36296.PDF (accessed 24.09.2014.).

Commission of the European Communities, 2008. Impact Assessment, Accompanying Document to the Inter Transmission System Operator Compensation Mechanism and Harmonization of Transmission Tariff for Electricity, Brussels.

Directive 2009/72/EC, 2009. Directive 2009/72/EC of the European Parliament and of the Council, Official Journal of the European Union.

Energy Community, 2014. Report from the Commission to the European Parliament and the Council. https://www.energy-community.org/pls/portal/docs/1810178.PDF (accessed 24.09.2014.).

Energy Community PECI, 2014. Regional energy strategy. http://www.energy-community.org/portal/page/portal/ENC_HOME/AREAS_OF_WORK/Regional_Energy_Strategy/PECIs (accessed 25.09.2014.).

Energy Community Regulatory Board, 2010. Cooperation of Regulators with Regard to Cross Border Investment Projects.

Energy Community Secretariat, 2011. Development of energy community strategy, proposal on further steps. Annex 17/19.07.2011, Vienna, Energy Community, 2011. http://www.energy-community.org/portal/page/portal/ENC_HOME/AREAS_OF_WORK/Investments/Energy_Community_Strategy (accessed 22.09.2014.).

Energy Regulation, 2011. Electricity Regulation in the US: A Guide. RAP, Montpelier.

European University Institute, 2010. The Overview of the European Regulatory Framework in Energy Transport. EU Working Paper, Robert Schuman Centre for Advanced Studies, Florence.

Websites

http://www.munkdebates.com/debates/europe

Energy Policy Regulation, 2014. Price regulation. http://www.dnvkema.com/services/advisory/mar/energy-policy-regulation/price-regulation.aspx (accessed 22.09.2014.).

Chapter 2

Defining Euro-Mediterranean Energy Relations

Francesca Pia Vantaggiato
School of Politics, Philosophy, Language and Communication Studies (PPL), University of East Anglia (UEA), Norwich, UK

This chapter investigates the process of issue redefinition of Euro-Mediterranean energy relations operated by the European Commission (EC) over time. Section 1 reviews several themes, characterizing the literature on the European Union (EU) external energy policy, insofar as it is applied to Euro-Mediterranean energy cooperation. It then proposes to use policy formation literature to understand how priorities emerged and/or shifted in EU external energy policy and how they were discursively framed. Section 2 reconstructs the stages in the history of Euro-Mediterranean energy cooperation. The launch of the European Neighbourhood Policy (ENP) and the failed endorsement of the Mediterranean Solar Plan (MSP) are highlighted as key turning points in the discursive definition of Euro-Mediterranean energy cooperation. Section 3 examines the most recent energy policy debate in light of the insights gained from the previous two sections. Section 4 is a conclusion. The analysis highlights three facts:

- Euro-Mediterranean regional energy cooperation was located within the realm of the EU's energy policy; since 2004, the EC moved it to the "external relations" framework, to recently embed it into its Foreign and Security Policy;
- Since the enlargement and in particular since Russia–Ukraine gas disruptions, the political salience of import dependency on Russia increased, causing the EC to predicate its policy interest in Euro-Mediterranean energy cooperation on drawbacks in its relations with Russia;
- The failed endorsement of the MSP brought prospects of regional electricity market integration to a standstill, pushing the EC to fall back on existing policy templates such as the Union for the Mediterranean (UfM), the ENP, and the concept of a "pan-European Energy Community."

Regulation and Investments in Energy Markets. http://dx.doi.org/10.1016/B978-0-12-804436-0.00002-3

1 THE EU EXTERNAL ENERGY POLICY: FRAMEWORKS OF ANALYSIS

Few topics of European integration have triggered as much debate as the EU external energy policy. This is probably due to its evolving nature, its uncertain competence boundaries, its shifting policy priorities, and the challenges therein. Contributions in the literature have clustered around a few key themes. These themes constituted frameworks, serving as lenses through which policy debates have been analyzed.

A particularly prolific framework is the "markets versus empires" metaphor (Correljé and van der Linde, 2006; Youngs, 2009; Escribano, 2011). In this framework, markets represent an organization of energy issues around markets, regulation, and supranational multilateral institutions charged with achieving consensus among members and dealing with rule enforcement. In this scenario, energy matters are solved through market mechanisms, inspired by efficiency and competitiveness. In the empires scenario, by contrast, the world is split into blocks, segmented along energy-based political alliances. Rivalry and quest for resources characterize the framework, which does not exclude military intervention to secure resource access. In the literature, the EU is usually seen as a committed supporter of the markets framework.

A second very successful notion is the one of the EU as a "normative power" (Manners, 2002). In this vision, the EU boasts a magnetic force for countries surrounding it, where it seeks to export values and norms through the means of soft power and suasion. Thanks to its success as a regional model, the EU is able to draw countries toward its socioeconomic model based on democracy and the rule of law. The concept of "market power" Europe counterbalances this normative vision (Damro, 2012; Pollack, 2010) and underlines the usage the EU makes of its market size in order to impose its hegemony in its neighborhood (Haukkala, 2008)). In energy issues, the concept of normative (or market) power is called upon in contributions underlining the EU's attempt to extend its rules (the energy *acquis*) onto neighboring countries.

A further, although less thoroughly explored, topic in analyses of the EU external dimension (including energy) is "regionalism." The EU has been said to "play it regional" (Bicchi, 2006) by regionalizing areas in its neighborhood, which do not perceive themselves as regions, and by applying onto them its own institutional templates. The influence of sociological isomorphism on the EU's institutional practice would explain this approach (Bicchi, 2006). Energy is a sector often tackled on a regional basis. Integrating markets across borders is considered to bring advantages both in terms of efficiency and in terms of economic growth, security, and political stability.[1] This conception can explain the

1. This circumstance is often mentioned in contributions on the importance of energy policy for the EU, itself born on the basis of energy-related agreements (see Kanellakis et al., 2013, for a thorough review).

EU-specific understanding of the "Euro-Mediterranean region," which includes the EU as well as southern and eastern Mediterranean countries (SEMCs). This concept differs from the notion of the Middle East and North African (MENA) region,[2] which includes countries belonging to the Cooperation Council for the Arab States of the Gulf (often referred to as the GCC).

More recently, the literature has relied on frameworks based on ideas and discourse to analyze the EU external energy policy. In this case, the EU's upholding of its energy, legal, and regulatory framework in third countries is seen as premised on deeply rooted and path-dependent frameworks of understanding, which are communicated through specific discourses. The energy policy discourse revolves around key concepts, such as market framework, liberalization, and regulation, which identify the EU as a supporter of markets and norms, or as "a liberal actor in a realist world" (Goldthau and Sitter, 2014). Nevertheless, observers have diagnosed inconsistency across EU institutions in their external energy policy discourse toward Russia (Kuzemko, 2013), revealing a mixture of neoliberal and geopolitical themes underpinning it. This view appears to be confirmed by the recent European Commission (2014), which adopts "unprecedented geopolitical tones" (Youngs, 2014).

Overall, the strenuous attempt of the EU to replicate, or project, itself onto its neighborhood pervades this whole literature, representing its common denominator.

The policy literature also fed into this debate. Contributors have understood the EC as a determined institutional entrepreneur, who appropriated the energy issue over time by exploiting political opportunity windows as they presented themselves (Maltby, 2013). In this way, the EC managed to frame internal and external energy issues so that member states acknowledged the need of supranational, EU action. The centrality of the EC to EU energy policy analyses is due to the key role it played in a matter, which was initially placed out of its institutional reach, such as energy. The process whereby the EC consolidated its grip on the rules of the internal market proceeded virtually simultaneously with its quest for a role in external energy matters. Observers have remarked on the ability of the EC to lock energy issues in overlapping policy areas falling under its legislative competence (Solorio and Morata, 2012), such as competition policy and external trade relations.

It is worth noting how on the EU side, the whole external and internal energy policy discourse has always been premised on security of supply, given its scarce fossil fuel endowment. The EU external action in energy has thus also been interpreted as an attempt to securitize its borders through extending its internal dynamics to its neighborhood (Lavenex, 2004). Specifically, the need to

2. The MENA region concept is widely adopted by international organizations such as the IMF (see El-Erian and Fischer, 1996), the Organization of Economic Cooperation and Development (OECD), and the World Bank (WB) as well as in policy papers such as El-Katiri (2014) and Romdhane et al. (2013).

ensure security of supply underlies the notion of interdependence between the EU and its neighborhood in energy issues. However,

> *The perception of interdependence is not a fixed entity, and varies with the conjuncture of security concerns within the Union. (...) Such a perspective explains not only why specific issues of 'domestic politics' gain priority in relations with neighbouring countries, but also why these priorities fluctuate over time, such as manifested in EU–Mediterranean relations. Securitization from this perspective does not directly derive from objective external threats but is the outcome of framing processes within an evolving institutional environment (Lavenex, 2004).*

Further insights can be gained, I argue, by relying on the policy formation and agenda-setting literature. This literature argues that policy issues are framed and reframed dynamically in time in response to endogenous or exogenous shocks. At the macro level, collective issue definition determines which issues make it to the policy agenda. At the micro level, policy actors compete to frame an issue recognized as salient according to their own policy interests and preferences. The combination of the micro level and the macro level helps put items on the policy agenda and, after a competitive process of issue definition, gives rise to the policy itself (see Kingdon, 1984; Stone, 1989; and also Nowlin, 2015). The various facets of policy may change in salience or prominence as conditions change. Conditions create windows of opportunity for policy actors to put forward their preferences and shape policy issues accordingly. Hence, policy issues are defined and redefined in accordance with variations in their political salience. The concept of issue definition is part and parcel of the policy literature on agenda setting and institutional entrepreneurialism. Maltby (2013) explores the growing role the EC carved for itself within the EU energy policy in gas, managing to frame it in security terms and thus requiring a European solution.

This contribution underlines how the security aspect of energy policy has been exacerbated by recent events involving Russia and Ukraine, allowing for stronger EC policy entrepreneurialism. Most importantly, this contribution outlines how these events impacted on the salience and the definition of Euro-Mediterranean energy cooperation at the EU level, through an analysis of the EC framing of the issue. In order to grasp the magnitude of the change, I delineate the key episodes in the EC framing of Euro-Mediterranean energy relations since the beginning of the Euro-Mediterranean Partnership (EMP) in 1995.

The picture emerging from examination of the EC's definitions of the Euro-Mediterranean energy relations over time is that of a movable feast: the EC shifted the Euro-Mediterranean energy relations' issue from energy to the regional cooperation policy framework in the early 2000s; most recently, it moved it again, from the regional cooperation basket to a foreign policy and security heading, in a crescendo of securitization of energy issues.

2 EURO-MEDITERRANEAN ENERGY RELATIONS

This section dwells on milestone events for Euro-Mediterranean energy relations. Section 2.1 briefly reconstructs the history of Euro-Mediterranean energy relations, from 1995 to the present time. Section 2.2 illustrates the reasons for the failure of the MSP, the first electricity market integration project conceived for the Euro-Mediterranean area, which transcended policy documents to come close to be implemented. Section 2.3 examines the most recent policy events, concerning EU energy policy and Euro-Mediterranean energy cooperation within it. Throughout the section, the main purpose is to highlight the shifts in the EC's definition of Euro-Mediterranean energy relations.[3]

2.1 A Short History of Euro-Mediterranean Energy Relations

Talks regarding regional cooperation between the EU and the southern shore of the Mediterranean Sea started as early as 1995. A few years earlier, Algeria had plunged into civil war, prompting EU policy action. In 1995, the Barcelona Conference marked the start of the so-called Barcelona Process and the establishment of the EMP. The EMP was a broad policy program, covering items from democracy to trade. However, the centrality of energy issues was acknowledged, among others, in the only EC Communication entirely dedicated to energy cooperation in the so-called "Euro-Mediterranean region" (European Commission, 1996). The EMP energy vector was characterized by a twofold regional approach: political dialog between energy ministers from both shores in the Inter-Ministerial Conferences and expert dialog between representatives of the partner countries and the EC in the Euro-Mediterranean Energy Forum. Both the conferences and the forum began convening soon after the Barcelona Conference, but fell short of regular meetings thereafter.[4] Cooperation did not progress as expected.

In 2004, the EC took a major policy decision by establishing the ENP, a program dedicated to cooperation between the EU and its composite southern and

---

3. The EC can communicate its vision over any matter of relevance through nonbinding documents such as Communications, White Papers, and Green Papers. The contents of these documents normally outline EC policy goals and prelude legislative or policy initiatives. Hence, EC Communications often point to the direction policy will take. This is why relevant EC Communications and Green Papers are often referred to in this contribution.

4. The first Inter-Ministerial Conference was held in Trieste in 1996. Other meetings followed in Rome and Athens in 2003. They were then discontinued until 2007 (Limassol, Cyprus) and 2014 (Rome, Italy). The Euro-Mediterranean Energy Forum convened for the first time in Brussels in 1997, then in Granada in 2000. It proved extremely difficult to retrace the chronology of forum meetings thereafter. The Athens meeting of 2003 appears to mix Forum and Inter-Ministerial Conferences together. Following the Athens meeting, the Energy Forum as conceived within the EMP seems to have been discontinued.

eastern neighborhood over a number of areas. The EMP was fitted to the ENP.[5] The ENP novelty was that it combined regional aspects with a pronounced bilateral dimension: the EC proposes bilateral action plans to each of its "neighbors," framing cooperation around a series of issues, among which is energy. The plans encourage convergence toward the EU legal and regulatory framework insofar as each concerned country is available to commit itself (Escribano, 2010). It has been argued that the bilateral dimension of the ENP was meant to bypass the political obstacles hampering regional cooperation among SEMCs (Tholens, 2014).

The ENP did not only represent a reframing of the Euro-Mediterranean energy cooperation (see Vantaggiato, 2013 for an overview of the regional energy vectors of the ENP), but of the whole of the EU external dimension. Three main events occurred in or around the year 2004: (1) the Second Energy Package entered into force framing the EU internal energy policy in much more ambitious terms than ever; (2) the eastward enlargement of the EU took place; and (3) the Energy Community Treaty (ECT), extending the energy *acquis* to countries in South Eastern Europe (SEE), holding a membership perspective, was signed in 2005 and entered into force in 2006 – protracted and intense policy activity preceded it. This bundle of significant political steps encompassed both internal and external matters, resulting in a default extension of the EU energy market approach to a sizable area in eastern Europe.

The same could not be hoped for in SEMCs.[6] Political divisiveness impeded regional cooperation among SEMCs. In the absence of a membership perspective, the EC had no coercive strength. The presence, among SEMCs, of net exporting countries (NECs) and net importing countries (NICs) that also represented important transit countries unveiled steep asymmetries of interests and bargaining power (Padgett, 2011) both among them and between them and the EU.

The ENP thus came to perform a key policy function: it marked the boundaries of the EU, separating members (and potential members such as the ECT signatories) from countries without a membership perspective (Cardwell, 2011). It also juxtaposed bilateral relations over an initiative initially premised on a regional approach. The regional component of the cooperation remained part of the ENP. However, the discourse framing regional cooperation was significantly altered.

In the 1990s the EC discourse on Euro-Mediterranean energy cooperation was constructed in straightforward economic terms. In EC words (European Commission, 1996), regional energy cooperation should aim:

> *to develop energy planning tools based on the highly complementary nature of the Northern and Southern Mediterranean markets and supply networks; to increase*

---

5. The ENP financial instrument, called ENPI, financed various voluntary stakeholder cooperation initiatives in the area, such as the Association of Mediterranean Energy Regulators (MedReg), the Association of Mediterranean Transmission System Operators (Med-TSO), and others (see Hafner and Tagliapietra, 2013; Carafa, 2015).

6. Comparison between the EU approach (and results) to external energy policy in the case of the establishment of the Energy Community Treaty (ECT) and Euro-Mediterranean relations are often made in the literature (see Padgett, 2009 for a short and clear overview).

trade in energy products; developing and linking up the energy networks in the various regions around the Mediterranean; promoting RTD and investment with the aid of partnerships on renewable energy sources and energy efficiency; to create a favourable environment in order to promote investment by starting or continuing reforms of the Mediterranean partners' energy industries.

Moreover, the EC discourse left room for coownership of the legal and regulatory aspect of energy cooperation ("An appropriate legal framework should be devised to encourage and promote regional and transregional trade") and embedded regulatory convergence in a proinvestment discourse ("Uniform regulatory and contractual conditions must therefore be defined to encourage investments by foreign firms. The Mediterranean partners must not only open up their markets but also adopt rules which are as uniform as possible so that undertakings do not have to adapt to different regulatory frameworks in each country.")

With the ENP, the discourse surrounding Euro-Mediterranean regional energy cooperation transformed into the idea to: "Extend the EU's internal market, through expansion of the Energy Community Treaty to include relevant EEA and ENP countries" (Joint Paper, 2006). The idea of expanding the coverage of the ECT was first advanced in the 2006 Green Paper and subsequently reiterated in the 2011 Communication, meant as a response to the Arab Spring.

The approach underpinning proposals of extension of the ECT, however, never gathered SEMCs support: the discourse framing them is couched in a top–down policy approach, which is not supported by sufficient political clout to entice SEMCs. Moreover, cooperation and coordination among SEMCs is mostly absent. Hence their collective commitment to a regional initiative is improbable. It is thus unsurprising that energy cooperation gathered more momentum in the bilateral component than in the regional one (Tholens, 2014).

Crucially, with the launch of the ENP, the EC made Mediterranean regional energy issues exit the realm of the "EU Energy Policy" and enter the realm of "EU External Relations."[7] Before the launch of the ENP, "regional cooperation" was one item of the Community's "energy policy." Thereafter, "energy" became one item under the "regional cooperation" umbrella.

1. Since 2004, the EC used the heading "energy policy" to essentially refer to the internal energy market (IEM). All other energy matters became functional to its achievement. This shift in policy framing demarcates the discursive change mentioned earlier. In practice, the EC outsourced regional cooperation matters to parallel, umbrella policy frameworks in order to concentrate on its core business: achieving the IEM. Moreover, the IEM became its main negotiation tool in relation to SEMCs, which are offered a stake in the IEM in exchange for regulatory convergence.
2. In this perspective, policy frameworks such as the ENP do not appear as demonstrations of reinforced EC interest in regional energy cooperation

7. This is also evidenced by the fact that the EUR-Lex database does not include ENP-related documents under the Energy subject matter. The ENP is in the scope of activity of the EEAS.

with the SEMCs. To the contrary, they suggest a scaling down of regional energy cooperation priorities and a reinforced focus on strengthening the IEM, which by itself would increase the bargaining power of the EC in external energy matters.[8]

Furthermore, the enlargement altered the balance of energy priorities among the member states and for the EC. The membership of central and eastern European member states worsened figures of EU energy dependency on Russian imports. New member states advocated for the diversification of supply routes embedding their claims into a discourse of threatened national security, which transferred to EC level (Maltby, 2013). External energy policy priorities thus refocused away from SEMCs toward Russia, whose image as a dependable supplier[9] was turned upside down. More precisely, energy policy interest toward SEMCs became predicated on the ebbs and flows of relations with Russia.

In the years following the launch of the ENP, slow progress in the achievement of the IEM and deterioration of energy relations with Russia added an element of urgency and a perception of threat to the EU's energy security. Gas disruption episodes in 2006 and 2009 prompted renewed EC interest in the Mediterranean, in the form of bilateral strategic partnerships with ENP countries (such as Algeria, Morocco, and Egypt), as well as in the relaunch of regional cooperation in both gas and electricity.[10] Therefore, initially EU–Russia problems played in favor of the Euro-Mediterranean energy dimension, which gained bolstered importance in 2008.

2.2 The Mediterranean Solar Plan and Its Demise

In 2008, the EMP was repackaged into the UfM, a program similar to the EMP but with a sharper focus on energy issues. Its importance resided in constituting the policy container of the first concrete attempt to approach the regional dimension of the Euro-Mediterranean Energy Partnership: the MSP. The MSP was an industrial initiative promoted in the same year by a consortium of energy companies and transmission system operators (TSOs) from various member states. It was aimed at creating a vast integrated regional electricity market across the northern and southern shores of the Mediterranean, fueled by renewable energy sources (RES). The MSP followed the wake of the DESERTEC initiative: an ambitious German-led industrial initiative proposing the development of

8. In this context, the proliferation of stakeholder initiatives in the Mediterranean area can be seen as an attempt to rebalance the policy focus: in the face of scant ENP progress, the stakeholders took it upon themselves to steer the process of regional energy cooperation in the Euro-Mediterranean area. As a matter of fact, MEDREG, Med-TSO, and the MSP are all voluntary initiatives.

9. Expressed, for instance, in the 2000 Green Paper.

10. At the 5th EuroMed Energy Ministers Conference, held in Limassol (Cyprus) in 2007. On that occasion, the then EC Commissioner for energy launched a renewed Euro-Mediterranean Energy Partnership. An Action Plan for the region was also launched on that occasion covering the 2008–2013 time span.

RES-generated electricity in the Middle East and North Africa (MENA) region (thus including Gulf countries) and interconnection with the EU.

The rationale for supporting RES deployment in SEMCs has been amply discussed in the literature.[11] Briefly, it can be summarized as follows: the potential of RES-generated electricity in the area is such, that it could cover a substantial part of the internal (steadily growing) demand of SEMCs. This would free up gas resources to sell on international markets (for the NECs, such as Algeria and Egypt). Also, it would reduce the deficits of NICs (such as Morocco, Jordan, and Tunisia), who feed their fossil fuel-based energy mix through imports from their neighbors. In addition, electricity generated through RES technology in the SEMCs could be exported to Europe. The EU would benefit from importing electricity from SEMCs in that it would achieve its ambitious climate targets and bolster its supply security. A bigger and integrated electricity market would ensure gains from trade for both sides. When the MSP was launched, it gained EC support as a "European" initiative. The EC sponsored the "Paving the Way for the Mediterranean Solar Plan (PWMSP)" initiative, launched in 2010 and gathering experts in charge of studying the feasibility of the plan.

The 2009 dispute between Russia and Ukraine caused grave gas supply disruption to several member states. The image of Russia further deteriorated in the eyes of the EU. The most affected member states, in Eastern and Central Europe, strongly advocated reducing the EU dependency on Russian gas supplies. The MSP came to be framed in these terms as well: RES-generated electricity from the SEMCs would both meet the EU's climate objectives and reduce its supply vulnerability.

The PWMSP program lasted for three years, thus ending in 2013. The UfM had been tasked with drafting an MSP master plan, outlining the stages of its implementation, by the end of the program. The master plan was due to be approved at the Inter-Ministerial Conference purposefully convened in Brussels in December 2013. The ministers' endorsement would put a European seal on the project that would set in motion investments. Unexpectedly, the plan was not endorsed, principally due to Spain's opposition (see Carafa, 2015 for their motivation).

Lack of endorsement of the MSP is regrettable for various reasons. One of the most important from a policy perspective is its perceived ability to bring about regulatory convergence in the Euro-Mediterranean by convincingly framing it in terms of facilitated market outcomes for both shores (Escribano et al., 2013). In other words, stakeholders involved in the MSP had managed to reframe the discourse of regulatory convergence in terms more similar to those the EC used in the 1990s than to those it adopted in the early 2000s. By approximating their institutional and policy framework to that of the EU, SEMCs would lower their country and policy risk, besides lowering the costs of entry and eventually being

11. See Mason and Kumetat (2011), Brand and Zingerle (2011), Jablonski et al. (2012), Trieb et al. (2012), Carafa (2015), Khalfallah (2015) to mention just some of the most recent contributions.

able to participate in the EU Internal Energy Market. All stakeholders thus eyed the MSP as the gateway to the realization of several important objectives in Euro-Mediterranean regional energy relations.

The reasons for the MSP upheaval precede the Brussels Conference of December 2013. Whereas policy statements regarding this failure are rather generic,[12] industry stakeholders are adamant in recognizing that:

> *...the rapid and sudden evolution of the Energy Reference Scenario was manifested in its entirety, caused by new phenomena: i) the structural change in the international fuel market (nonconventional fuels), ii) the interference between the production of electricity from renewable and conventional sources, iii) the reduction of electricity consumption on the Northern shore of the Mediterranean. This change of scenario impacts the assumptions of grid planning: no longer a strong production of electricity from renewable sources (RES) in the South for its export to the North, but a trading system much more articulated and complex, aimed at the integration of electricity and energy systems of the two shores of the Mediterranean (Med-TSO[13] – emphasis added).*

---

12. "Question for written answer P-014428/13 to the Commission from Gaston Franco (PPE) – (23 December 2013). Subject: Failure to adopt the Master Plan for the implementation of the Mediterranean Solar Plan. (...) The Master Plan for the implementation of the Mediterranean Solar Plan (MSP) was intended to be a strategic reference document to promote renewable energies and energy efficiency, strengthen electricity interconnections between the two shores of the Mediterranean and develop integrated regional markets. In the end the long-awaited adoption of this Master Plan following the meeting of the Energy Ministers of the Union for the Mediterranean, held on 11–13 December 2013 in Brussels, did not take place, owing to Spain's veto in particular.

1. Could the Commission provide its analysis of the failure to adopt the MSP Master Plan at the meeting of the Energy Ministers?Could the Commission provide its analysis of the failure to adopt the MSP Master Plan at the meeting of the Energy Ministers?
2. Does it believe that the European energy market needs to be perfectly integrated prior to adopting the MSP Master Plan?
3. When will this Master Plan be tabled for adoption again?
4. Does the Commission intend to support electricity interconnections in the Mediterranean by building underwater cable links?

Answer given by Mr Oettinger on behalf of the Commission – (27 January 2014). The Mediterranean Solar Master Plan is a non-legally binding, policy orientation document. It suggests guidelines for the development of consistent and effective renewable energy and energy efficiency policies in the Mediterranean; it does not identify concrete projects nor provide for any type of financing. The Commission regrets that consensus could not be reached on the Master Plan. It considers, however, that lack of formal endorsement by the Union for the Mediterranean Ministerial (UfM) meeting does not prevent UfM Members from continuing and intensifying efforts for promoting renewable energy and energy efficiency, and more generally energy cooperation, in the Mediterranean. The Commission seeks to facilitate the integration of electricity markets around and across the Mediterranean. This requires, inter alia, the construction of electricity interconnectors between the south and the north rims of the Mediterranean. In this sense, the Commission supports and works with all relevant public and private stakeholders committed to the development of trans-Mediterranean electricity interconnections." (http://eur-lex.europa.eu/legal-content/EN/TXT/?qid=1436089288587&uri=CELEX:92013E014428 (accessed 05.07.2015.).

13. http://www.med-tso.com/mediterranean.aspx?f= (accessed 05.07.2015.).

Opportunities and benefits for a massive renewable deployment in Southern Med countries were identified by institutional and industrial stakeholders, embedding a consistent import of energy to Europe for their bankability. When RES4MED was created, the vision of South Med as a 'green energy reservoir for EU' started to be challenged and today does not hold anymore (RES4MED Annual Conference[14] – emphasis added).

Due to the economic crisis, energy demand in the EU has been shrinking for a few years now. Further, a quicker rate of emissions reduction was achieved than was expected, due to energy efficiency measures but also due to reduction of economic activity. At the same time, the RES support schemes enacted in several member states meant the take up of RES proceeded quicker than expected. The fact that the marginal cost of RES-generated electricity is zero allowed it to flood the wholesale markets and displace all other fuels. Furthermore, additional RES capacity has not been matched by equivalent plant decommissioning. This state of things would require a growing demand to be efficient, but demand is not there. In short, many member states suffer from overcapacity on their electricity markets. Lack of adequate interconnection infrastructure traps overcapacity within national boundaries. Hence, the failure of the MSP is premised also on the failure of coordination of member state national energy policies.

Demand in northern Mediterranean countries (NMCs) may recover to levels sufficient to alleviate these concerns. This is however not expected to happen in the short term. The situation is so critical that the idea is emerging of investigating the possibility to export electricity from NMCs to North Africa in the short term, rather than the contrary (Dii Report, 2013, p. 2; OME Electricity Committee Meeting Minutes, 2013, p. 5, 2014, p. 2). Demand in SEMCs, in fact, is projected to keep growing steadily in the next two decades.

Following the December 2013 Ministerial Conference, the MSP evaporated and silence fell over the Euro-Mediterranean electricity market integration debate until November 2014, when the Italian Presidency of the European Council gathered energy ministers from the Euro-Mediterranean region for the relaunch of regional energy cooperation in the area.

2.3 The Recent Relaunch of Euro-Mediterranean Regional Energy Cooperation

During its Presidency of the European Council, the Italian Government hosted an Inter-Ministerial Conference of Euro-Mediterranean energy ministers. The outcome of the conference was the establishment of three platforms for

14. http://www.res4med.org/uploads/activities/AnnualConference2015/VIGOTTI_RES4MED_20%20APRIL%202015.pdf (accessed 05.07.2015.).

Euro-Mediterranean energy cooperation: gas, electricity, and RES. A chronology of the main steps taken toward this result is given.

In March 2014, European Council Conclusions called on the EC to "conduct an in-depth study of EU energy security and to present by June 2014 a comprehensive plan for the reduction of EU energy dependence (...) including through the development of interconnections. Such interconnections should also include the Iberian peninsula and the Mediterranean area."

In response, in May 2014, the EC released its Communication on the Energy Security Strategy, which places energy among the topics of EU Foreign Policy, and argues for the creation of a Mediterranean gas hub.

In July 2014, a conference entitled "Security of gas supply: the role of gas developments in the Mediterranean region" was held in Malta. It was attended by energy ministers from the EU and the SEMCs, industry representatives, and key stakeholders in the energy sector, including the EC. The ministers agreed on the establishment of a Mediterranean gas platform of cooperation.

In November 2014, the conference held by the Italian Government during its Presidency of the Council took place. On this occasion, regional cooperation in electricity and RES, largely absent from EC communications, is put back on the agenda.

In February 2015, the EC issued a further communication, launching the European Energy Union. It strongly advocated the completion of the internal market and for enhanced interconnection among member states. The discourse framing these topics was based on geopolitical worries and energy security. The Mediterranean entered the picture of the Energy Union due to gas supplies present in the region. The concept of the Mediterranean gas hub was reiterated. Steering committee meetings for the gas platform began in March 2015.

In May 2015, the new EC Commissioner for Energy undertook an energy diplomacy trip to Algeria and then to Morocco, so as to deepen energy ties. A political dialog on energy was established between the EC and Algeria. In Morocco, the three platforms were relaunched.

The platforms are inscribed into the framework of the UfM. However, the Observatoire Méditerranéen de l'Energie (OME) industry association has been nominated as secretariat for the gas platform. The UfM secretariat, instead, will coordinate the work of the electricity platform together with Mediterranean Energy Regulators (MEDREGs) and Med–TSO associations, as well as the RES platform.

3 ISSUE (RE)DEFINITION IN THE MEDITERRANEAN: THE SECURITIZATION OF ENERGY MATTERS

Recent events may suggest that Euro-Mediterranean energy cooperation is back among EC agenda priorities. In order to understand what caused this revived interest, one should take a broader view and bring member states back into the picture.

Escribano (2011) clustered EU member states according to their main energy suppliers to gauge their expected policy preferences in external energy matters.[15] He distinguished four groups: European, Euro-Russian, Russian, and Mediterranean. One of the countries in the "Russian" group identified by Escribano (2011) is Poland, a long-standing advocate for reducing reliance on Russian gas imports at the EU level. On April 10, 2014 the Polish Prime Minister Donald Tusk submitted a "nonpaper" to the EC. The document called for an energy union to increase the EU's bargaining power toward its key gas and oil suppliers, first and foremost, Russia. Weeks before this move, a referendum made Crimea part of the Russian Federation. Soon afterward, the outbreak of unrest in the separatist provinces of Ukraine triggered EU sanctions toward Russia.

It seems that the tensions in Ukraine made eastern European calls for import diversification echo louder at the EU level. In other words, the circumstances of the Russia–Ukraine dispute heightened the political salience of diversifying gas suppliers and routes. The EC seized the occasion to borrow the concept of the Energy Union expressed in Tusk's paper and reframe it to encompass the whole range of its policy priorities. Importantly, the EC maintained both the tones of urgency and the perception of threat contained in Tusk's paper.

Within the new discourse, the "Euro-Mediterranean region" is functional to ensuring the EU's gas supply. Given this reformulation of the EU energy policy, the failed endorsement of the MSP, and the crisis on EU wholesale electricity markets, the issue of electricity market integration with SEMCs lost its prominence on the policy agenda. A further element confirming this is the EC choice to rely on an industry association to coordinate the works of the gas platform, while placing electricity and RES within old, enfeebled policy frameworks such as the UfM.

Moreover, the recent EC/European External Action Service (EEAS) joint consultation paper (2015) on the future ENP does not make reference to gas. The EC/EEAS joint communication on the ENP implementation in 2014 does not make any reference to energy cooperation within the Euro-Mediterranean area. Hence, recent policy documents suggest that the EC has pushed Euro-Mediterranean energy (particularly gas) relations out of a regional framework such as the ENP, which is premised on the notion of "cooperation," to firmly inscribe them into foreign policy and security items.

In this respect, it is worth mentioning that in 2003, the European Council issued a document entitled "A Secure Europe in a Better World," where it outlined the European security strategy by identifying five key threats: terrorism, proliferation

15. Security of supply is certainly not the only aspect that matters in energy policy. For instance, although Denmark belongs to the European cluster and Germany to the Euro-Russian cluster in Escribano's classification, these countries both have considerable industrial policy interests in RES technology. This explains their interest in RES technology deployment for electricity generation in the Mediterranean and also why they figure as the only European sponsors of the RCREEE initiative (www.rcreee.org).

of weapons of mass destruction, regional conflicts, state failure, and organized crime. The document "provides the conceptual framework for the Common Foreign and Security Policy (CFSP)."[16] In that document, energy dependence was only mentioned as a concern of the EU, without further specification being given.

On July 20, 2015 the Council of the EU adopted its conclusions on the EU Energy Diplomacy Action Plan proposed by the EC and the EEAS within its CFSP framework, arguing that "The full range of foreign policy instruments should be used to provide support in promoting common messages and 'narratives' on the top priorities and challenges for EU energy diplomacy."[17]

In the process of securitizing energy issues, the EC has radically redefined its attitude and demands toward the Euro-Mediterranean area. The energy relations discourse with countries in the area (in accordance with its overall change) shifted farther away from the tones it had on its inception, following a discursive evolution pattern bearing striking similarities to the process of securitization of public issues outlined in Buzan et al. (1998):

> *'Security' is the move that takes politics beyond the established rules of the game and frames the issue either as a special kind of politics or as above politics. (…) In theory, any public issue can be located on the spectrum ranging from nonpoliticized (…) through politicized (…) to securitized (meaning the issue is presented as an existential threat, requiring emergency measures and justifying actions outside the normal bounds of political procedure).*

The urgency injected into the EU energy policy discourse by the Energy Union concept justifies the unprecedented geopolitical tones (Youngs, 2014) retrieved in EC communications. It also explains the current EC Commissioner's energy diplomacy trips (i.e., the establishment of openly political dialogs for an actor, normally depicted as supporting market frameworks and whose energy policy discourse had been premised on the necessity to depoliticize energy issues). Most importantly, it justifies the centralization of external energy policy issues at European level and a strong push toward completing the IEM as an energy security measure.

4 CONCLUSIONS

This chapter examined the evolution of Euro-Mediterranean energy relations by emphasizing changes in the discursive frameworks adopted by the EC to define them. The analysis recognized a distinct shift in discursive approach at the time

---

16. http://www.eeas.europa.eu/csdp/about-csdp/european-security-strategy/ (accessed 23.07.2015.).

17. The EU Energy Diplomacy Action Plan (annexed to the Council Conclusions) appears to complement the previously established Climate Change Diplomacy. The EC and the EEAS defined "climate change" as a threat multiplier in 2008 (Climate Change and International Security – Paper from the High Representative and the European Commission to the European Council) and in 2011 (Joint Reflection Paper – "Towards a renewed and strengthened EU climate diplomacy"), arguing for the framing of climate change targets within the European Security Strategy outlined in the Council document of 2003.

of the establishment of the ENP, when Euro-Mediterranean energy cooperation exited the "energy policy" box to become one of the items in the basket of a broad external relations strategy, thus losing specificity and policy focus.

The EC policy dialog, initially rooted in straightforward economic terms and mutual gains from cooperation with the ENP, transformed into a discourse bearing hegemonic tones and expecting compliance in return for access to an as yet incomplete IEM. Furthermore, the enlargement increased the political salience of dependency on Russian gas supplies, bringing the EC to reorient its external energy policy attention eastward.

The launch of the MSP in 2013 breathed new life into a stagnating Euro-Mediterranean energy dialog. The failure of its endorsement froze prospects of large-scale electricity market integration between the two shores of the Mediterranean for the short to the medium term.

Most recently, worsening of the Russia–Ukraine crisis represented a window of opportunity to speed up the pace of completion of the IEM and diversification of gas import routes. The EC seized this occasion, reframing its energy policy discourse once again through a strategy of securitization, whereby energy issues entered the realm of EC Foreign Policy. This meant its discursive framing of Euro-Mediterranean energy cooperation changed as well: it is now premised on an explicitly geopolitical dialog.

Electricity and RES issues have lost prominence on the policy agenda, and have been inscribed into existing policy frameworks, such as the UfM, devoid of purpose after the failure of the MSP.

Important actors within this context, stakeholder associations are instead already working on reframing cooperation in electricity and RES, including by considering the option to export electricity from NMCs to SEMCs in order to obviate to the overcapacity problem plaguing them. Stakeholder work on regional electricity market integration between the two shores will likely continue in the background of the current developments, waiting for more propitious political times.

REFERENCES

Bicchi, F., 2006. 'Our size fits all': normative power Europe and the Mediterranean. J. Eur. Public Policy 13 (2), 286–303.

Brand, B., Zingerle, J., 2011. The renewable energy targets of the Maghreb countries: impact on electricity supply and conventional power markets. Energy Policy 39 (8), 4411–4419.

Buzan, B., Wæver, O., De Wilde, J., 1998. Security: A New Framework for Analysis. Lynne Rienner Publishers, Boulder, CO.

Carafa, L., 2015. Policy and markets in the MENA: the nexus between governance and renewable energy finance. Energy Procedia 69, 1696–1703.

Cardwell, P.J., 2011. EuroMed, European Neighbourhood Policy and the Union for the Mediterranean: overlapping policy frames in the EU's governance of the Mediterranean. J. Common Market Stud. 49 (2), 219–241.

Correljé, A., van der Linde, C., 2006. Energy supply security and geopolitics: a European perspective. Energy Policy 34 (5), 532–543.

Damro, C., 2012. Market power Europe. J. Eur. Public Policy 19 (5), 682–699.

Desert Power: Getting Started – The Manual for Renewable Electricity in MENA – Policy Report, 2013. http://www.desertenergy.org/fileadmin/Daten/Desert_Power/Desert%20Power%20Getting%20Started-Policy%20Report%20English.pdf (accessed 23.07.2015.).

El-Erian, M.A., Fischer, S., 1996. Is MENA a Region? The Scope for Regional Integration. IMF Working Papers, IMF.

El-Katiri, L., 2014. A Roadmap for Renewable Energy in the Middle East and North Africa. Oxford Institute for Energy Studies, Oxford.

Escribano, G., 2010. Convergence towards differentiation: the case of Mediterranean energy corridors. Mediterr. Polit. 15 (2), 211–229.

Escribano, G., 2011. Markets or geopolitics? The Europeanization of EU's energy corridors. Int. J. Energy Sector Manag. 5 (1), 39–59.

Escribano, G., Marín-Quemada, J.M., San Martín González, E., 2013. RES and risk: renewable energy's contribution to energy security. A portfolio-based approach. Renew. Sustain. Energy Rev. 26, 549–559.

Goldthau, A., Sitter, N., 2014. A liberal actor in a realist world? The Commission and the external dimension of the single market for energy. J. Eur. Public Policy 21 (10), 1452–1472.

Hafner, M., Tagliapietra, S., 2013. A New Euro-Mediterranean Energy Roadmap for a Sustainable Energy Transition in the Region. MEDPRO Policy Paper.

Haukkala, H., 2008. The European Union as a regional normative hegemon: the case of European Neighbourhood Policy. Eur. Asia Stud. 60 (9), 1601–1622.

Jablonski, S., Tarhini, M., Touati, M., Gonzalez Garcia, D., Alario, J., 2012. The Mediterranean Solar Plan: project proposals for renewable energy in the Mediterranean Partner Countries region. Energy Policy 44, 291–300.

Kanellakis, M., Martinopoulos, G., Zachariadis, T., 2013. European energy policy – a review. Energy Policy 62, 1020–1030.

Khalfallah, H., 2015. Connecting Mediterranean countries through electricity corridors: new institutional economic and regulatory analysis. Utilities Policy 32, 45–54.

Kingdon, J.W., 1984. Agendas, Alternatives, and Public Policies, Little, Brown.

Kuzemko, C., 2013. Ideas, power and change: explaining EU–Russia energy relations. J. Eur. Public Policy 21 (1), 58–75.

Lavenex, S., 2004. EU external governance in 'wider Europe'. J. Eur. Public Policy 11 (4), 680–700.

Maltby, T., 2013. European Union energy policy integration: a case of European Commission policy entrepreneurship and increasing supranationalism. Energy Policy 55, 435–444.

Manners, I., 2002. Normative power Europe: a contradiction in terms? J. Common Market Stud. 40, 235–258.

Mason, M., Kumetat, D., 2011. At the crossroads: energy futures for North Africa. Energy Policy 39 (8), 4407–4410.

Nowlin, M.C., 2015. Modeling issue definitions using quantitative text analysis. Policy Stud. J., doi: 10.1111/psj.12110.

OME, Electricity Committee Meeting September 19, 2013 Minutes. http://www.ome.org/files/live/sites/ome/files/docs/Technical%20Committees/Electricity%20Committee/Meetings/Minutes/eleccommminutestunissept2013pdf (accessed 23.07.2015.).

OME, Electricity Committee Meeting June 25, 2014. http://www.ome.org/files/live/sites/ome/files/docs/Technical%20Committees/Electricity%20Committee/Meetings/Minutes/eleccommminutesromejune2014pdf (accessed 23.07.2015.).

Padgett, S., 2009. External European Union Governance in Energy. Full Research Report, ESRC.

Padgett, S., 2011. Energy Co-operation in the Wider Europe: Institutionalizing Interdependence. JCMS: Journal of Common Market Studies 49(5), 1065–1087.

Pollack, M., 2010. Living in a material world: a critique of "Normative Power Europe". Available from: http://ssrn.com/abstract=1623002 or http://dx.doi.org/10.2139/ssrn.1623002

Romdhane, S.B., Razavi, H., Santi, E., 2013. Prospects for an integrated North Africa energy market: opportunities and lessons. Energy Strategy Rev. 2 (1), 100–107.

Solorio, I., Morata, F., 2012. Introduction: The Re-evolution of Energy Policy in Europe. Edward Elgar, Cheltenham.

Stone, D.A., 1989. Causal stories and the formation of policy agendas. Polit. Sci. Quart. 104 (2), 281–300.

Tholens, S., 2014. An EU–South Mediterranean Energy Community: the right policy for the right region? Int. Spect. 49 (2), 34–49.

Trieb, F., Schillings, C., Pregger, T., O'Sullivan, M., 2012. Solar electricity imports from the Middle East and North Africa to Europe. Energy Policy 42, 341–353.

Vantaggiato, F.P., 2013. Mechanism and outcomes of the EU external energy policy: an alternative approach. In: Cambini, C., Rubino, A. (Eds.), Regional Energy Initiatives. Routledge, London.

Youngs, R., 2009. Energy Security: Europe's New Foreign Policy Challenge. Routledge, London.

Youngs, R., 2014. A new geopolitics of EU energy security. http://carnegieeurope.eu/publications/?fa=56705 (accessed 19.07.2015.).

EC POLICY DOCUMENTS

Council of the European Union, 2015. Council Conclusions on Energy Diplomacy (Annex) and EU Energy Diplomacy Action Plan (Annex to the Annex). http://data.consilium.europa.eu/doc/document/ST-10995-2015-INIT/en/pdf (accessed 24.07.2015.).

European Commission, 1996. Communication from the Commission to the European Parliament and the Council Concerning the Euro-Mediterranean Partnership in the Energy Sector.

European Commission, 2000. Towards a European Strategy for the Security of Energy Supply. Green Paper.

European Commission, 2006. A European Strategy for Sustainable, Competitive and Secure Energy. Green Paper.

European Commission, 2011. A Partnership for Democracy and Shared Prosperity with the Southern Mediterranean. Joint Communication to the European Council, the European Parliament, the Council, the European Economic and Social Committee and the Committee of the Regions.

European Commission, 2014. European Energy Security Strategy. Communication from the Commission to the European Parliament and the Council.

European Commission, 2015. Energy Union Package. A Framework Strategy for a Resilient Energy Union with a Forward-Looking Climate Change Policy. Communication from the Commission to the European Parliament, the Council, the European Economic and Social Committee, the Committee of the Regions and the European Investment Bank.

European Commission and High Representative of the European Union for Foreign Affairs and Security Policy, 2015. Towards a New European Neighbourhood Policy. Joint Consultation Paper.

European Commission and High Representative of the European Union for Foreign Affairs and Security Policy, 2015. Implementation of the European Neighbourhood Policy in 2014. Joint

Communication to the European Parliament, the Council, the European Economic and Social Committee and the Committee of the Regions.

European Council, 2003. A Secure Europe in a Better World.

European Council, 2014. European Council Conclusions, March 20/21, 2014.

Joint Paper by the Commission and the Secretary-General/High Representative, 2006. An External Policy to Serve Europe's Energy Interests

Tusk, D., 2014. Roadmap towards an Energy Union for Europe. Non-paper addressing the EU's energy dependency challenges. https://www.msz.gov.pl/resource/34efc44a-3b67-4f5e-b360-ad7c71082604:JCR (accessed 24.07.2015.).

Chapter 3

Renewable Energy in the Southern and Eastern Mediterranean: Current Trends and Future Developments

Simone Tagliapietra
Fondazione Eni Enrico Mattei, Milan, Italy

1 BOOMING ENERGY DEMAND IN SEMCs

Economic and demographic trends are the key driving forces of energy demand and for this reason it is always worthwhile to take into account the macroeconomic background of a given area before analyzing its energy landscape.

As an overall trend, over the last few decades the southern and eastern Mediterranean region[1] experienced a general trend of economic growth. This trend, which was relatively smooth during the 1980s and the 1990s, considerably accelerated since the early 2000s, and is expected to further increase over the next years.[2]

In the recent past southern and eastern Mediterranean countries (SEMCs) have also expanded in terms of population and this upward trend is expected to further increase in the future.[3] Egypt and Turkey are the two most populous countries in the region and together they count for more than a half of the overall regional population.

But the population of SEMCs is not only growing; it is also young and urbanizing: two elements that also have a great impact on energy demand.

In fact, these economic and demographic trends are at the basis of the region's booming energy demand. In fact, the primary energy demand in SEMCs – calculated by the International Energy Agency (IEA) as total primary

1. Algeria, Egypt, Israel, Jordan, Lebanon, Libya, Morocco, the Palestinian Territories, Syria, Tunisia, and Turkey.

2. For an in-depth analysis of the determinants of economic growth in SEMCs (Coutinho, 2012; Dabrowski and De Wulf, 2013).

3. For a comprehensive discussion of demographic developments in SEMCs (Groenewold and De Beer, 2013).

Regulation and Investments in Energy Markets. http://dx.doi.org/10.1016/B978-0-12-804436-0.00003-5

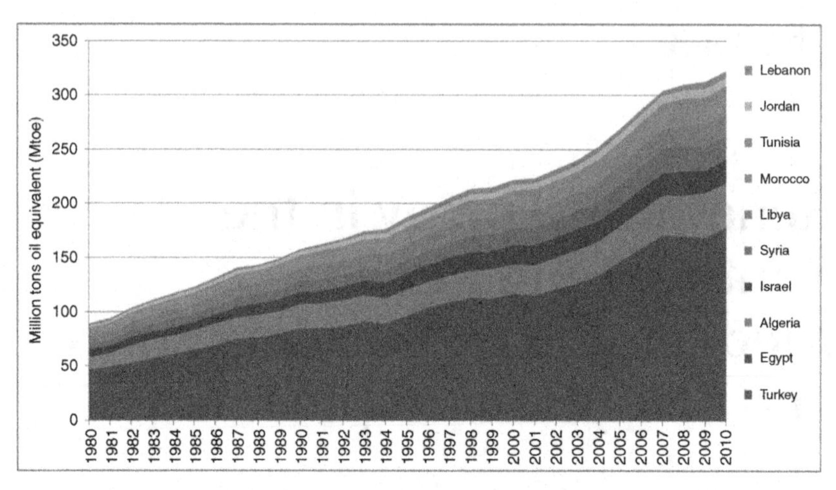

FIGURE 3.1 Total primary energy demand in SEMCs (1980–2010). *(Source: Author elaboration on International Energy Agency, Extended World Energy Balances Database (accessed May 2014).)*

energy supply (TPES)[4] – has progressively increased over the last few decades, and most notably after the early 2000s (Fig. 3.1).

In absolute numbers, the primary energy demand in SEMCs grew from 89 million tons oil equivalent (Mtoe) in 1980 to 157 Mtoe in 1990, to 221 Mtoe in 2000, and to 322 Mtoe in 2010. Presented in this way, these figures might not be very expressive. However, they can become more useful if translated into growth rates. Such a perspective reveals that the primary energy demand in SEMCs rose by 77% between 1980 and 1990, by 41% between 1990 and 2000, and by 45% between 2000 and 2010. These growth rates are even more interesting if compared with other regions in the world.

In fact, over the last few decades the growth rates of primary energy demand in SEMCs well surpassed – in comparative terms – the ones of OECD Asia Oceania, OECD Americas, and OECD Europe: the world's three key economic areas (Fig. 3.2).

This trend, before significantly accentuated in the 1980s and then less pronounced in the 1990s, is particularly interesting in relation to the first decade of the 2000s. In fact, in this period of time the growth rate of the primary energy demand in SEMCs strongly exceeded that of OECD Asia Oceania, an area generally considered as the key driver of energy demand worldwide. Figure 3.2 clearly exemplifies how rapid and consistent the evolution of primary energy demand in SEMCs currently is.

But how is the primary energy demand in SEMCs actually composed? Today the fuel mix of SEMCs is composed as follow: 44% oil, 36% natural gas, 13% coal, 2% hydro, 5% other renewable energy sources (biofuels, geothermal,

4. TPES is calculated by the IEA as production of fuels + inputs from other sources + imports – exports – international marine bunkers + stock changes.

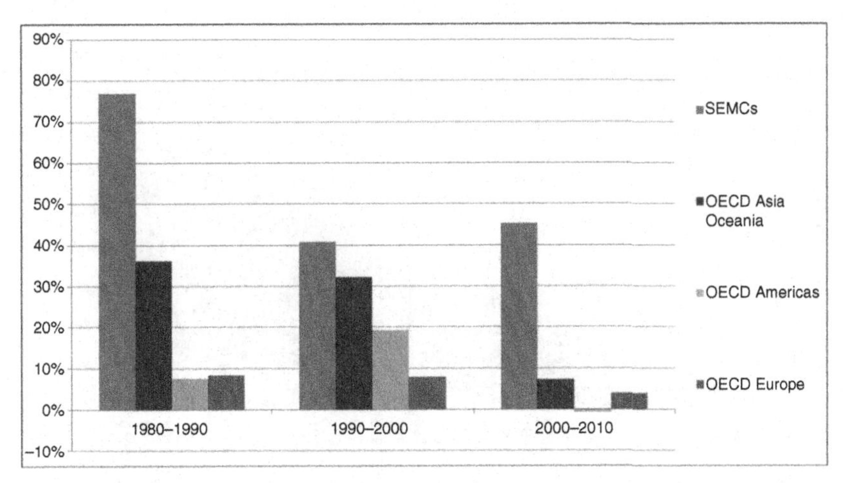

FIGURE 3.2 Growth rates of total primary energy demand in SEMCs and other regions. *(Source: Author elaboration on International Energy Agency, Extended World Energy Balances Database (accessed May 2014).)*

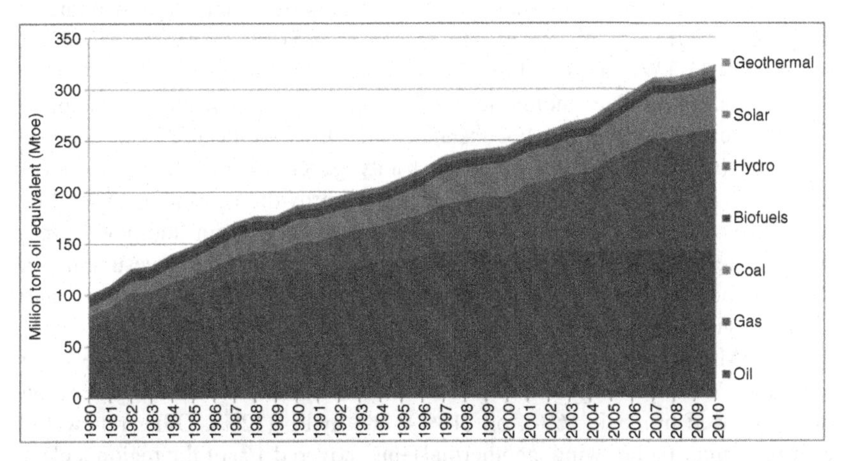

FIGURE 3.3 Total primary energy demand by fuel in SEMCs (1980–2010). *(Source: Author elaboration on International Energy Agency, World Energy Balances Database (accessed May 2014).)*

solar, wind). The composition of this fuel mix considerably changed over the last few decades, in particular as far as the shares of oil and natural gas are concerned (Fig. 3.3).

2 THE CRUCIAL ROLE OF ELECTRICITY

Electricity plays a crucial role in SEMC energy systems. Between 1990 and 2010 electricity consumption in SEMCs grew by an annual growth rate of about 6%. Just to provide a quick comparison, in the same period of time electricity

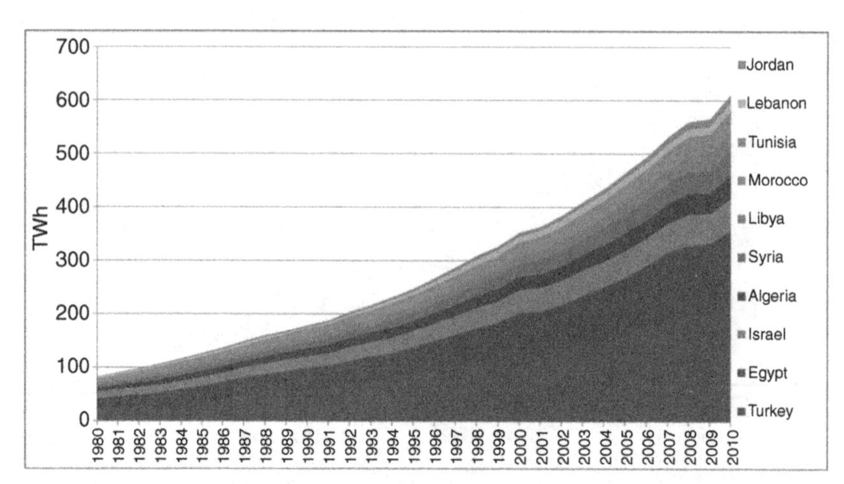

FIGURE 3.4 Electricity generation in SEMCs by country (1980–2010). *(Source: Author elaboration on International Energy Agency, World Energy Balances Database (accessed May 2014).)*

consumption in north Mediterranean countries grew by an annual growth rate of 1.8% OME (2011a, p. 58). Electricity generation in SEMCs grew from 82 TWh in 1980 to 611 TWh in 2010 (Fig. 3.4).

A higher level of economic growth and population will push up demand for electricity in the future. For instance, the Observatoire Méditerranéen de l'Energie (OME, 2011a) forecasts electricity generation in SEMCs to reach 1534 TWh in 2030, implying an average annual growth rate of about 5% OME (2011a, p. 58). Such a rapid and consistent growth will put additional pressure on the existing electricity infrastructure, requiring major investments on the construction of new electricity generation facilities, transmission lines, and distribution networks.

In SEMCs the electricity generation mix is mainly based on fossil fuels and hydro. In 2010 natural gas covered 55% of the region's electricity generation mix, followed by coal (16%), oil (16%), and hydro (12%). Other renewable energy sources (solar, wind, geothermal) only covered 1% of the region's electricity generation mix (Fig. 3.5): a number that notably clashes with the region's huge renewable energy potential that will now be described.

3 RENEWABLE ENERGY POTENTIAL OF THE REGION

SEMCs are endowed with massive renewable energy resources, most notably solar (concentrated solar power (CSP) and photovoltaic (PV)) and wind.

3.1 CSP Potential

Solar energy is generally considered as the potentially most important renewable energy resource in the southern and eastern Mediterranean region, particularly as far as solar thermal electricity generated with CSP systems is concerned.

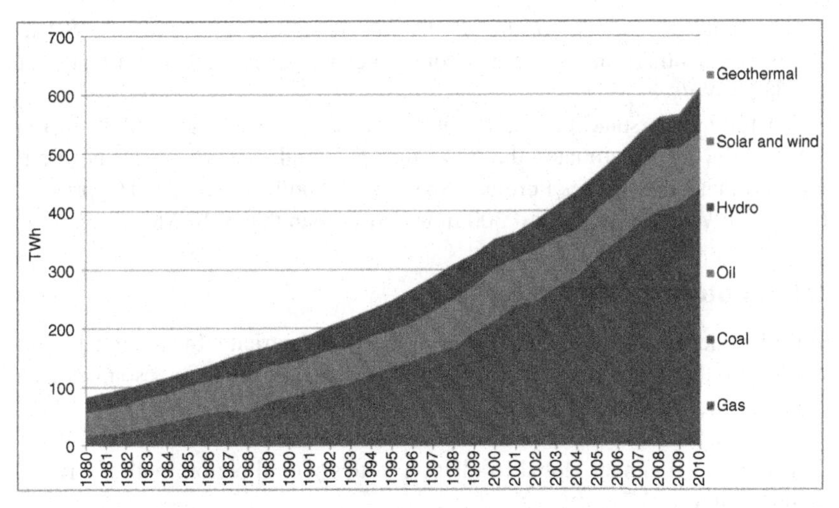

FIGURE 3.5 Electricity generation in SEMCs by fuel (1980–2010). *(Source: Author elaboration on International Energy Agency, World Energy Balances Database (accessed May 2014).)*

FIGURE 3.6 Annual DNI in 2002. *(Source: DLR (2005, p. 59).)*

The study "MED-CSP – Concentrating Solar Power for the Mediterranean Region," (DLR, 2005) commissioned by the German Federal Ministry for Environment, Nature Conservation and Nuclear Safety, and published in 2005 by the German Aerospace Center (DLR),[5] revealed that in SEMCs the sunshine duration ranges between 2650 and 3600 h/year, while direct normal irradiance (DNI) ranges between 1300 kWh/m^2/year on the coast and 3200 kWh/m^2/year in the Sahara desert (DLR, 2005, p. 59) (Fig. 3.6).

5. This consortium also carried out two additional studies: the TRANS-CSP study, which focused on the interconnection of the electricity grids of Europe, the Middle East, and North Africa; and the AQUA-CSP study, which focused on the potential of concentrating solar thermal power technology for large-scale seawater desalination for urban centers in the Middle East and North Africa.

This analysis has demonstrated that the economic potential of CSP in the overall southern and eastern Mediterranean region could be estimated at 431,382 TWh/year.

To fully understand this figure, it might be useful to remind that, in the same study, DLR estimates the economic potential of CSP in the northern Mediterranean region (i.e., Portugal, Spain, Italy, Malta, Greece, and Cyprus) at 1450 TWh/year: a level almost 300 times lower than that of SEMCs.[6]

3.2 Photovoltaic Potential

But solar energy is not just all about solar thermal electricity. In fact, in addition to CSP another technology that could be used to convert solar resources into electricity is, of course, PV.

In order to assess the regional resources for PV, in the "MED-CSP" study DLR calculated annual global irradiation in the overall region, with the result presented in Fig. 3.7.

On this basis, DLR estimates the economic potential of PV in SEMCs at 122 TWh/year.

To fully understand this figure, it might be helpful to consider that, in the same study, DLR estimates the economic potential of PV in the northern Mediterranean region (i.e., Portugal, Spain, Italy, Malta, Greece, and Cyprus)

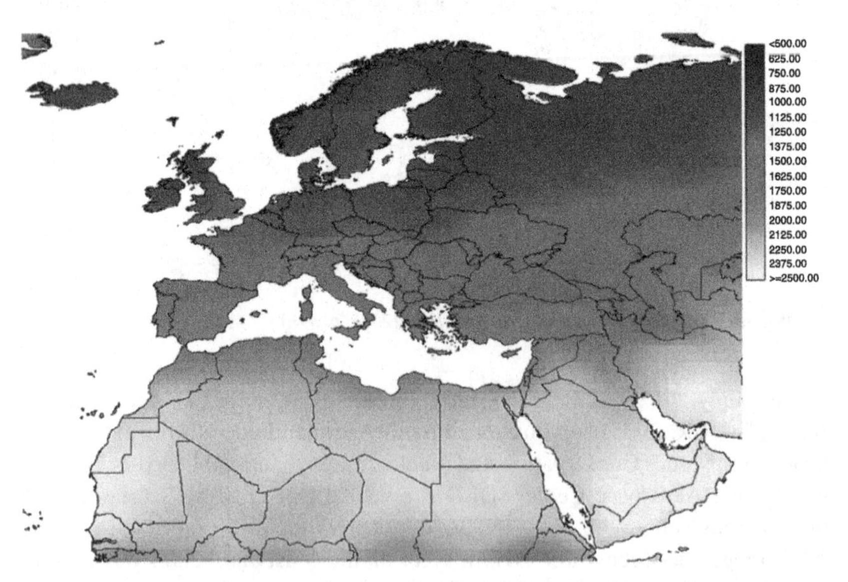

FIGURE 3.7 Annual global irradiation on surfaces tilted south with latitude angle in kWh/m²/year. *(Source: DLR (2015, p. 69).)*

6. These calculations are based on the MED-CSP Databank. Available from: http://www.dlr.de/tt/desktopdefault.aspx/tabid-2885/4422_read-6575/

at 22 TWh/year: a level almost five times lower than that of the southern and eastern Mediterranean region.[7]

3.3 Wind Power Potential

Less generally recognized, but no less important, the southern and eastern Mediterranean region also has favorable wind conditions. In the "MED-CSP" study, DLR estimates that some areas in the region belong to the world's top sites in terms of wind potential, with several sites exceeding a mean wind velocity of 7 m/s and, in some cases, even 11 m/s (Fig. 3.8).

These favorable wind conditions are particularly exceptional on Morocco's Atlantic coast and the Red Sea but attractive sites for wind power generation might also be found in Turkey, Algeria, Libya, and Syria, as indicated in Fig. 3.9.

According to the analysis of DLR, the technical potential of wind power in SEMCs amounts to 21,967 TWh/year. Following our usual comparison, the technical potential of wind power in northern Mediterranean countries amounts to 648 TWh/year, a figure 34 times lower than that of SEMCs.

The wind power potential of SEMCs is considerably lower if translated into economic potential. In its model, DLR assumes a maximum installed capacity of 10 MW per square kilometer of land area and considers the economic

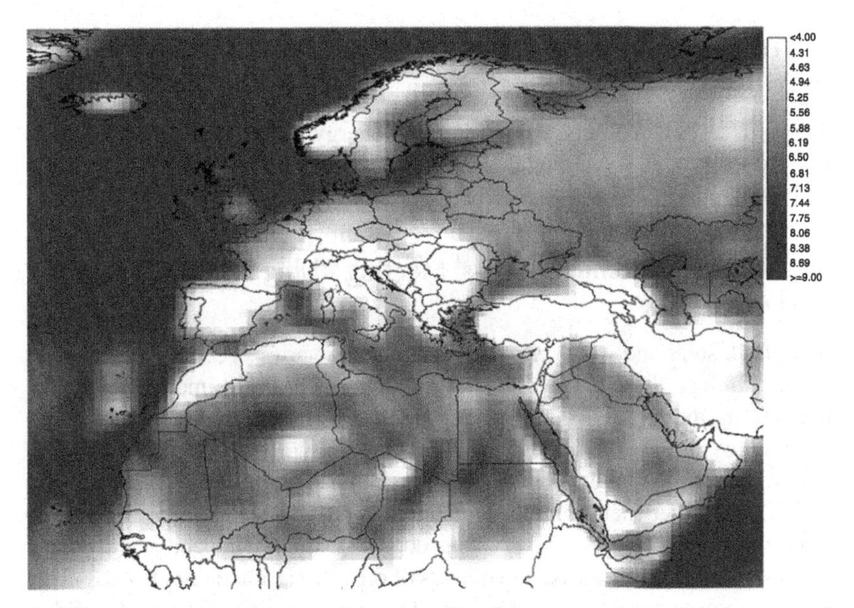

FIGURE 3.8 Annual average wind speed at 80-m above ground level in meter/second. *(Source: DLR (2015, p. 68).)*

7. These calculations are based on the MED-CSP Databank. Available from: http://www.dlr.de/tt/desktopdefault.aspx/tabid-2885/4422_read-6575/

	1980	1990	2000	2008	2000-2008 change (%)
Algeria	0.055	0.080	0.081	0.100	2.7
Egypt	0.112	0.115	0.105	0.112	0.8
Israel	0.080	0.082	0.078	0.069	−1.5
Jordan	0.121	0.207	0.195	0.154	−2.9
Lebanon	0.094	0.090	0.110	0.060	−7.3
Libya	0.134	0.274	0.121	0.106	−1.6
Morocco	0.083	0.072	0.085	0.086	0.2
Syria	0.134	0.215	0.165	0.121	−3.8
Tunisia	0.101	0.113	0.104	0.085	−2.5
Turkey	0.132	0.118	0.119	0.109	−1.1
EU-27	0.135	0.109	0.091	0.080	−1.6

FIGURE 3.9 **Economic potential of wind power in selected SEMCs.** *(Source: DLR (2005).)*

potential of "the areas with annual full load hours over 1400 h/year equivalent to a capacity factor of 16%" (DLR, 2005, p. 68).

However, technical potential seems to represent here the real key to understanding the region's wind power landscape, as economic potential needs to be progressively adapted to the new economics and technology realities of wind power, which are rapidly evolving toward better performances, largely unexpected in the past.

4 THE POTENTIAL BENEFITS OF RENEWABLE ENERGY IN THE REGION

Having described the huge solar and wind energy potential of SEMCs, let us now try to provide a comprehensive answer to the following question: What kind of benefits could renewable energy bring to the region?

4.1 Macroeconomic Benefits: Meeting Rising Energy Demand at a Lower Price

First of all, solar and wind energy could generate various economic benefits in SEMCs. In fact, these renewable energy sources could be primarily devoted to diversifying the region's energy mix, which still remains today largely dominated by fossil fuels. This could contribute to satisfying the region's rapidly increasing energy demand at a lower cost, both in net energy–exporting countries (NECs) and in net energy–importing countries (NICs).

In the case of NECs, a consistent deployment of solar and wind energy sources might free up consistent volumes of oil and gas, alternatively used in the domestic power generation sector, for additional exports to Europe. Considering that an important gas infrastructure connecting North Africa with Europe is already in place, this choice would involve an immediate, significant economic return for the region's gas-exporting countries just because of the growth in the export value of gas stocks. The same dynamic could be applied to oil exports, which even has a higher netback value than gas.

In the case of NICs, the potential economic benefit of renewable energy is even more obvious. In these countries the energy bill due to the import of oil and gas represents a dramatic burden to public finances, particularly because of the universal fossil fuel consumption subsidies regime currently in place.

4.2 Socioeconomic Benefits: Creating New Jobs and Alleviating Energy Poverty

In addition to these macroeconomic benefits, the exploitation of solar and wind energy could also generate wider socioeconomic benefits in SEMCs, such as job creation and alleviation of energy poverty.

Job creation is generally considered one of the key benefits of renewable energy. For instance, the United Nations Environment Programme (UNEP) and the International Labour Organization (ILO) estimate that for the production of the same quantity of MW/h, PV plants would use on average seven times more labor than coal-fired plants, and wind power would use on average 1.8 times more labor than natural gas (ILO, 2008, p. 102).

In the case of SEMCs, these data must be carefully interpreted. In particular, it must be recognized that SEMCs generally do not have, at the moment, the technological know-how to directly manufacture high-tech renewable energy devices such as, for instance, receivers for CSP or PV cells. However, many other components, such as mounting structures, could well be manufactured locally. Furthermore, in addition to direct job opportunities, a consistent number of jobs could also derive from indirect job opportunities such as installing, operating, and maintaining renewable energy generation facilities.[8]

The second socioeconomic benefit that the exploitation of solar and wind energy could potentially generate in SEMCs is the alleviation of energy poverty. In fact, situations of energy poverty still persist in many SEMCs and renewable energy could well play a role to alleviate them.

4.3 Environmental Benefits: Lowering the Energy Intensity of SEMCs

In parallel with the macroeconomic and socioeconomic benefits just described, renewable energy could generate important environmental benefits in SEMCs. This element is crucially important, particularly considering the regional reliance on fossil fuels.

8. For an assessment of the local manufacturing potential for CSP in the MENA region (World Bank, 2011a,b, 2013).

The carbon footprint of SEMCs has grown dramatically over the last few decades, mainly because of rapid urbanization, rising living standards, and rapid energy intensive industrialization. Albeit often underevaluated by the region's governments, climate change represents a serious threat to SEMCs. According to the Intergovernmental Panel on Climate Change (IPCC), SEMCs are particularly vulnerable to global warming because of their geographical position and their dependence on climate-sensitive economic sectors. Sea level rise endangers the living conditions of millions of people living on the Mediterranean coast. Important economic sectors like tourism and agriculture depend heavily on weather conditions, and the increase in temperature and the frequency of extreme events calls for comprehensive preventive measures to avoid future economic costs. In this overall situation the deployment of solar and wind energy would clearly represent an important contribution to the much-needed sustainability path of SEMCs.

4.4 Cooperation Benefits: Enhancing Both Intra-SEMC and EU–SEMC Cooperation

To conclude, it might be interesting to outline the potential positive spillovers eventually generated by the deployment of solar and wind energy in the region, both in terms of intra-SEMC cooperation and EU–SEMCs cooperation.

An extensive deployment of renewable energy in SEMCs would require the development of a regional electricity market able to permit the exchange of electricity in substantial volumes. In particular, the basic requirement for an efficient large-scale deployment of renewable energy sources in SEMCs would be the development of a regional system of electricity interconnections, able to physically link SEMCs into a unique network. This element is very important, particularly considering that notwithstanding the geographical proximity and common heritage, trade links between SEMCs remain at a very low level.

Moreover, a large-scale deployment of renewable energy sources in SEMCs is often expected to enclose a relevant potential for cooperation between the two shores of the Mediterranean. In fact, over the last decade various organizations have claimed that part of the electricity potentially produced via solar and wind energy sources in SEMCs might eventually be exported to Europe via high-voltage direct current (HVDC) electricity interconnections.

This idea has represented the foundation of several large-scale renewable energy projects, aimed at enhancing Euro-Mediterranean cooperation through renewable energy, such as Desertec, the Mediterranean Solar Plan (MSP), MEDGRID, and RES4MED.

In addition to these projects, the abundant renewable energy resources previously described also stimulated almost all SEMC governments to adopt their own national renewable energy plans, as described in the following section.

5 SEMC NATIONAL RENEWABLE ENERGY PLANS

As an overall trend, SEMC national renewable energy plans mainly target solar energy, most notably PV and CSP, and wind. These plans generally include targets for shares of electricity generation from renewable energy, typically 10–20%, shares of total final energy supply from renewable energy, total amount of energy production from renewable energy, or installed electrical capacities of specific technologies. The timeframe of these targets is generally between 2020 and 2030 (Fig. 3.10).

These targets are generally enclosed in dedicated national renewable energy plans such as, for instance, Morocco's "National Renewable Energy and Efficiency Plan" launched in 2008, Algeria's "Renewable Energy and Energy Efficiency Program" launched in 2011, and Tunisia's "Solar Plan" launched in 2009. These plans do not just provide overall targets such as the ones just presented, but also specific renewable energy capacity targets by technology. According to these plans, if all targets were met, the installed solar and wind energy capacity of SEMCs could approximately – reach 75,000 MW by 2030 – an

Country	Targets
Morocco	o 42% of electricity generation by 2020 of which 14% is solar, 14% is wind, and 14% is hydro
Algeria	o 6% of electricity generation by 2015; 15% by 2020; 40% by 2030
Tunisia	o 11% of electricity generation by 2016; 25% by 2030 o 16% of installed power capacity by 2016; 40% by 2030
Libya	o 7% of electricity generation by 2020 and 10% by 2025
Egypt	o 20% of electricity generation by 2020
Israel	o 10% of electricity generation by 2020
Palestinian territories	o 25% of primary energy by 2020 o 10% of electricity generation by 2020
Jordan	o 7% of primary energy by 2015; 10% by 2020
Lebanon	o 12% of electrical and thermal energy by 2020
Syria	o Specific capacity targets for PV, solar heat, and wind by 2030
Turkey	o 30% of electricity generation by 2023

FIGURE 3.10 Overall renewable energy targets in SEMCs. *(Source: Author elaboration on UAE/IRENA/REN21 (2013) and national renewable energy plans.)*

impressive figure, almost 40 times higher than the 1700 MW estimated for 2010 (Hafner et al., 2012a,b). However, these targets should be handled very carefully, as they are generally not legally binding. Furthermore, it might also be considered that these targets often represent a means used by governments to show their country's commitment to renewable energy, particularly to international investors.

Having said that, the targets just presented at least demonstrate how policymakers in SEMCs are increasingly aware of their countries' abundant solar and wind energy resources and of the potential benefits of their exploitation. This growing awareness is also demonstrated by the fact that many SEMCs have put in place dedicated agencies to support their renewable energy plans. These agencies complement the activities of the ministries responsible for the energy sector and of the energy regulatory authorities in the promotion of renewable energy (Fig. 3.11).

Considering all these elements (i.e., the development of Euro-Mediterranean projects supporting renewable energy in the region, the rise of national renewable energy plans, and the establishment of dedicated agencies), it might be natural to think that the situation of renewable energy in the region is well on track. However, solar and wind energy continue to represent less than 1% of the regional electricity generation mix.

This paradox represents a clear signal that the efforts made so far to promote renewable energy in the region have not been completely efficient and

Country	Agency	Year	Functions
Morocco	Moroccan Agency for Solar Energy (MASEN)	2010	Public–private agency created for the implementation of the Moroccan Solar Plan and the promotion of solar resources in every aspect.
Algeria	New Energy Algeria (NEAL)	2003	Agency established by the Algerian government and Algeria's national energy companies to encourage domestic production, use, and export of renewable energy.
Tunisia	National Agency for the Promotion of Renewable Energy (ANME)	2009	Agency established by the Tunisian government to encourage domestic production, use, and export of renewable energy.
Libya	Renewable Energy Authority (REAOL)	2007	Governmental institution created to implement renewable energy projects, increase the contribution of renewable energy in the mix, and propose legislation to support renewable energy.
Egypt	New & Renewable Energy Authority (NREA)	1986	Agency established to act as the national focal point for expanding efforts to develop and introduce renewable energy technologies on a commercial scale.
Jordan	National Energy Research Centre (NERC)	1998	Dedicated to research, development, and training in the fields of new and renewable energy.
Syria	National Energy Research Centre (NERC)	2003	Established by the Ministry of Electricity to conduct studies on renewable energy and implement experimental pilot projects.

FIGURE 3.11 Renewable energy agencies in SEMCs. *(Source: Author elaboration on MASEN, NEAL, ANME, REAOL, NREA, and NERC official websites.)*

successful. It is thus essential to investigate the reasons for this paradox, in order to understand how the regional deployment of solar and wind energy might be better promoted in the future.

For this reason, the next section will provide an analysis of the barriers to the development of renewable energy in the region, with the aim to elaborate a new approach able to translate the regional renewable energy potential into reality.

6 BARRIERS TO THE DEVELOPMENT OF RENEWABLE ENERGY IN THE REGION

Generally speaking, the grade of success of the deployment of a new technology is mainly dependent on its comparative cost advantage in relation to existing technologies. This overall concept is useful to understand part of the difficulties encountered by solar and wind technologies in the past, not only in SEMCs, but all around the world.

In fact, the typical structure of a solar or wind energy business plan often makes such projects less competitive than the conventional ones. This is due to the fact that solar and wind energy projects are characterized by high upfront costs, which implies a long payback period, given the marginal profit derived from each unit of electricity sold.

Notwithstanding the fact that, due to its abundant renewable energy resources, lifecycle costs for solar and wind energy in SEMCs could lie well below those in most other regions in the world, SEMCs lag far behind most other countries in Europe, North America, and Asia in terms of solar and wind energy deployment. This is due to the fact that the development of renewable energy in SEMCs faces additional barriers, related to some specific features of the regional market.

In particular, these additional regional barriers could be categorized into four areas: (1) commercial; (2) infrastructural; (3) regulatory; and (4) financial. Each one of these barriers will be discussed hereafter, with the underlying conviction that only by fully addressing them will the region be finally able to translate its renewable energy potential into reality.

6.1 The Commercial Barrier: The Need to Reform Energy Subsidies

Energy subsidies represent the cornerstone of the political economy of energy in many SEMCs and – among many other things – they represent a key barrier to the development of renewable energy in the region.

Energy subsidies have long been used by governments all over the world to reach specific political, economic, social, or environmental targets. Energy subsidies can assume different forms and modalities with a direct or indirect outcome on energy production costs and/or final prices.

As outlined by Bergasse and Paczynski (2012, p. 16), "[in SEMCs energy subsidies] mostly consist of universal direct energy price reductions focused on

LPG (used for cooking and heating), diesel (local transport and agriculture) and electricity (general and agriculture). Subsidies to fuels (LPG, diesel) account for the largest share, followed by electricity and natural gas."

As an overall trend, these universal energy subsidies (i.e., fossil fuel consumption subsidies) have been employed by SEMCs with the official target of alleviating energy poverty. However, the situation slightly differs from country to country, and particularly from energy-producing countries and nonenergy-producing countries.

Two key indicators of the overall SEMCs' policy of providing low-cost energy supplies are represented by pump price for diesel fuel and electricity prices.

As indicated in Fig. 3.12, the pump price for diesel fuel in Libya, Algeria, Egypt, Syria, Tunisia, Lebanon, Morocco, and Jordan are among the lowest in the world. Only the Palestinian Territories, Israel, and Turkey match European levels.

Furthermore, as indicated in Fig. 3.13, electricity prices in Syria, Libya, Egypt, Lebanon, Algeria, and Jordan also are among the lowest in the world.

Only Tunisia, Morocco, the Palestinian Territories, Turkey, and Israel match European levels in terms of electricity prices, albeit if, as El-Katiri (2014, p.10) outlines, "both Moroccan government and the Palestinian authorities subsidize electricity prices, suggesting that actual generation costs may well exceed these already high price levels."

In short, energy prices are subsidized in all SEMCs, with the only exception of Turkey and Israel, which are in line with European energy prices.

In theory, the employment of energy subsidies might appear beneficial for the socioeconomic development of a region – such as the southern and eastern Mediterranean – where an important share of the population still lacks access to modern energy services. However, the concrete implementation of this principle turns out to be rather inefficient. In fact, energy subsidies lead to an inefficient

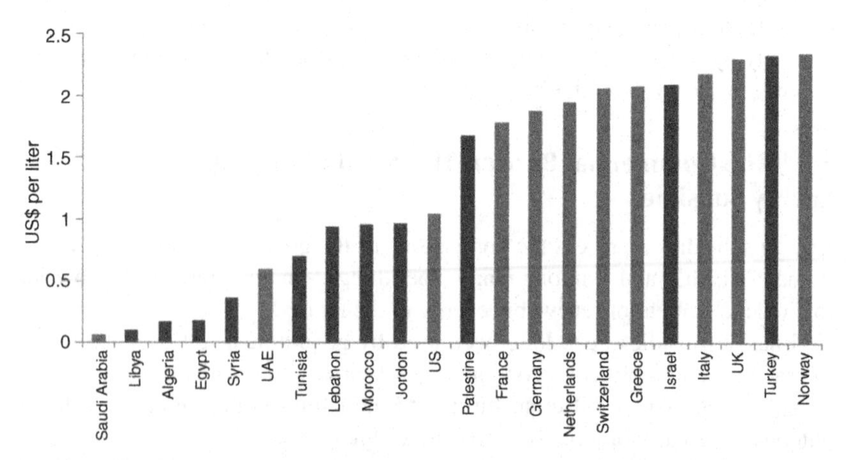

FIGURE 3.12 Comparison of average pump price for diesel fuel (2012). *(Source: Author elaboration on World Bank, World Development Indicators (accessed May 2014).)*

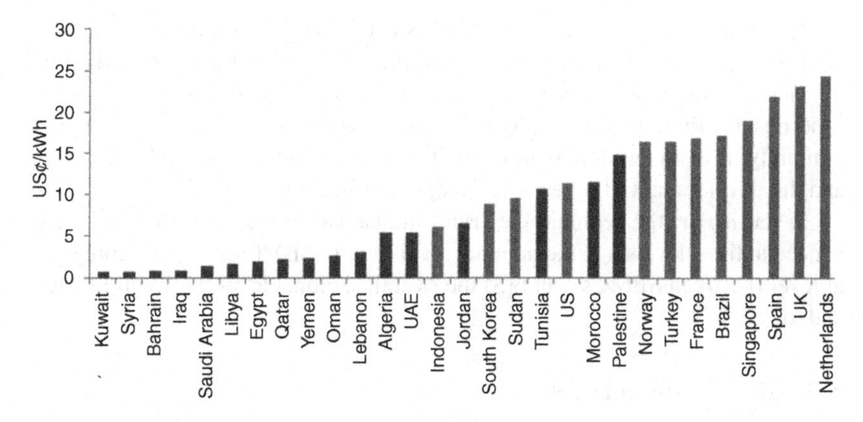

FIGURE 3.13 Comparison of average residential electricity prices in selected countries (2008). *(Source: El-Katiri (2014, p. 9).)*

allocation of resources and market distortions by encouraging a rent-seeking behavior and thus an excessive production or consumption. Furthermore, they generate an unsustainable financial burden in energy-importing countries that purchase energy at world prices and sell it domestically at lower prices, and also huge economic losses in net-energy exporting countries, as an important share of their resources is sold domestically at a fraction of its international market value. Energy subsidies also appear inefficient in addressing their social targets, as on average only 8% of fossil fuel subsidies go to the poorest income group and in some cases they may not even reach the poor at all (IEA, 2011, p. 519). Finally, energy subsidies distort the economics of energy and the price signals of energy resources – particularly in SEMCs that are oil and gas producers – also limiting the competitiveness of renewable energy sources (El-Katiri, 2014, p. 11).

The long list of negative consequences generated by universal energy subsidies just presented might raise the question why such economically – detrimental and socially – inefficient schemes continue to be adopted throughout the southern and eastern Mediterranean region. The answer to this question must be sought in the political sphere rather than in the economic one. In fact, energy subsidies represent a key element of the "social contract " in many SEMCs, notably the energy-producing ones. A reform of energy subsidies (e.g., aimed at phasing out universal fossil fuel consumption subsidies to replace them by individualized/targeted subsidies) would be essential to improve the energy, economic, and social sustainability of the overall southern and eastern Mediterranean region. Such a move would also favor the deployment of solar and wind energy by removing a key market distortion.

6.2 The Infrastructural Barrier: The Key Role of Med-TSO

Another key barrier to the deployment of solar and wind energy in SEMCs is represented by the lack of an adequate electricity infrastructure in the region.

This fact is strictly connected to the issue of energy subsidies, as in fact the development of a regional electricity transmission system has been limited by the lack of a regional market, largely due to energy subsidies themselves. The rigidities that this situation imposes mean that existing infrastructure is not used optimally, investment in new infrastructure is distorted and probably hindered, and the development of renewable energy is ultimately delayed.

In particular, the regional electricity transmission systems need to be enhanced at three levels: (1) the national level (within SEMCs); (2) the subregional level (between SEMCs); and (3) the overall regional level (between SEMCs and the EU).

6.3 The National Level

Electricity transmission lines within the respective SEMCs need to be reinforced. These lines are often weak and characterized by considerable technical distribution losses. Nontechnical (commercial) distribution losses also remain at very high levels (up to 40% in Lebanon and 20% in Algeria) at the expense of paying customers and distributors.

The weakness of the existing electricity transmission lines, combined with the rapid and consistent growth of electricity demand in SEMCs, will generate a serious pressure on the existing infrastructure, requiring major investments in the future not only on the construction of new electricity generation facilities, but also of new electricity transmission lines and distribution networks.[9]

6.4 The Subregional Level

Electricity transmission lines between SEMCs need to be constructed and, in certain cases, reinforced. Over the last decade SEMCs have pledged to connect their electricity networks but up to this date the southern and eastern Mediterranean region is not fully interconnected (Fig. 3.14).

Electricity interconnections do exist between Morocco, Algeria, and Tunisia, as well as between Libya, Egypt, Israel, Palestinian Territories, Lebanon, Jordan, and Syria. However, connections currently do not exist between Algeria/Tunisia and Libya or between Syria and Turkey. Such a fragmented situation has actually resulted in the creation of three not as yet interconnected blocks: the western block (from Morocco to Tunisia), the eastern block (from Libya to Syria), and Turkey.

Many electricity interconnection projects have been on the table for many years. These projects involve the connection of the three blocks to one another as well as connecting the overall southern and eastern Mediterranean region with Europe, as described later.

9. A detailed analysis of the respective SEMC electricity networks is provided in PWMSP (2011, 2012a,b).

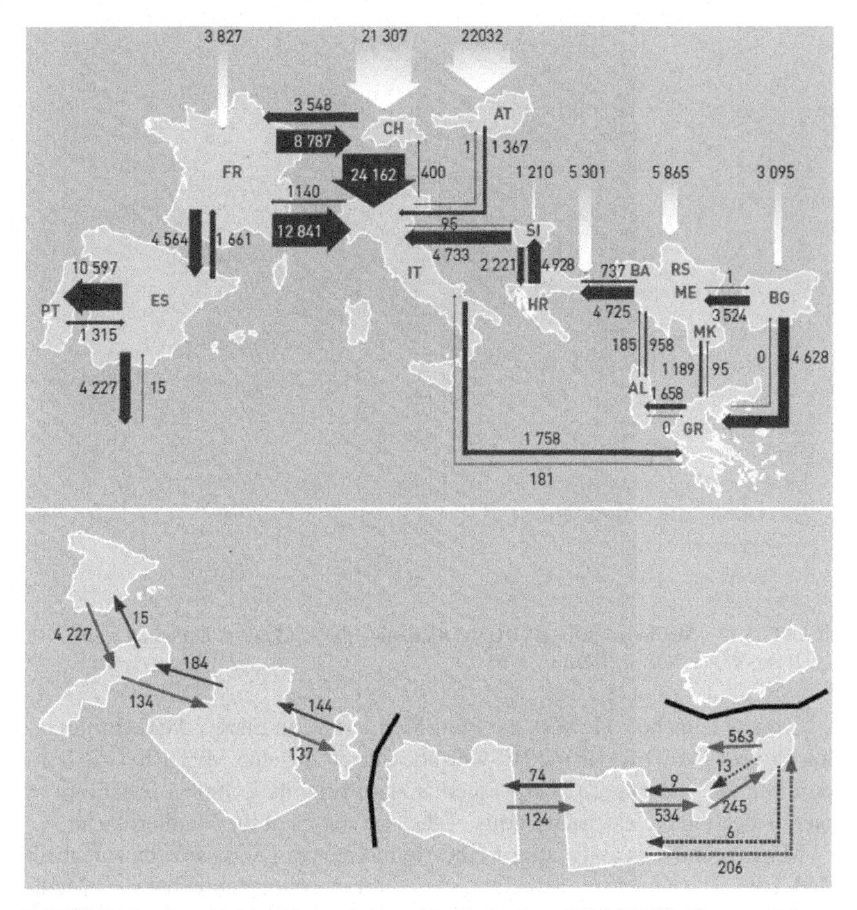

FIGURE 3.14 Annual electricity exchanges, physical values, 2008 (GWh). *(Source: Elaboration of OME (2011b, p. 139) in MED-EMIP (2010).)*

6.5 The Regional Level

The two shores of the Mediterranean have been connected since 1997 through an interconnector linking Morocco and Spain across the Strait of Gibraltar. This line, doubled in 2006, is set to be tripled in the near future to reach 2100 MW.

In addition to this, many SEMC–EU electricity interconnection projects have been under evaluation for many years, the most advanced of which is the Tunisia–Italy HVDC cable (Fig. 3.15).

Other, far less advanced, projects involve the construction of HVDC cables connecting respectively Algeria and Spain, Algeria and Sardinia, and Libya and Sicily.

These south–north interconnections have always be considered as crucial not only for the deployment of renewable energy sources in SEMCs, but also for the overall Euro-Mediterranean energy cooperation. This is the reason the EU

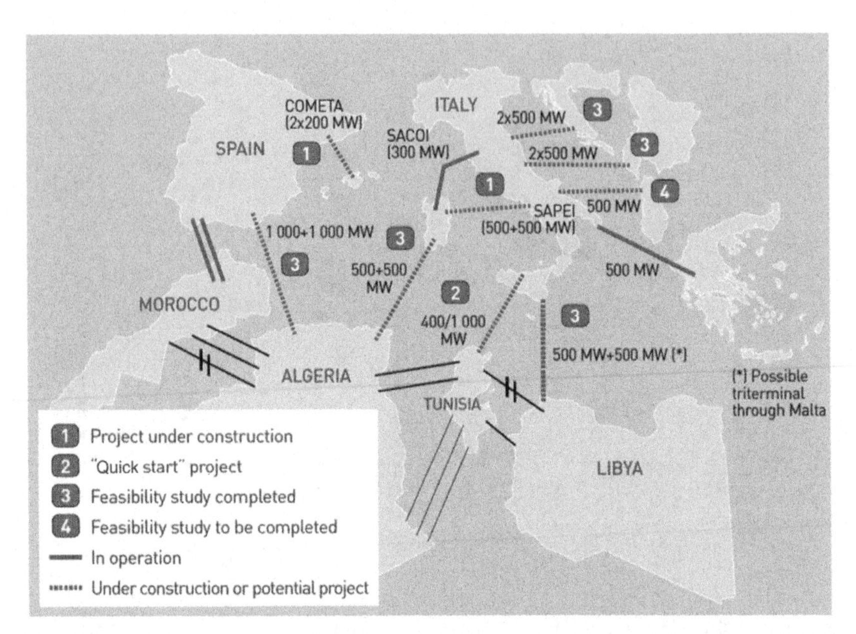

FIGURE 3.15 Mediterranean HVDC transmission links. *(Source: Elaboration of OME (2011b, p. 143) in CESI/MEDELEC (2010).)*

has already launched, in 2000, the Euro-Mediterranean Energy Market Integration Project (MED-EMIP), 2010 with the aim to develop MEDRING (Mediterranean electric ring), a study aimed at analyzing the technical feasibility of interconnecting the electricity grids of the two shores of the Mediterranean.

More recently, an association of the Mediterranean TSOs has been launched: Med-TSO. The aim of this association is to promote cooperation between Mediterranean TSOs[10] and to encourage the implementation of a regulatory framework aiming at the integration of Mediterranean electricity systems, through the development of necessary transmission infrastructures. Med-TSO elaborated in 2013 the First Master Plan of the Mediterranean electricity interconnections, aimed at sharing criteria among the Mediterranean TSOs and analyzing projects of interconnections and related reinforcements of internal grids also eligible for European projects of common interest (Fig. 3.16).

To conclude, it is important to underline that, as the OME (2011b) correctly points out, "to promote market interdependence, the involvement of both transmission operators and regulators needs to be officially encouraged and the relevant associations need to contribute. This can foster development of appropriate

10. The funding members and future associates of Med-TSO are: OST (Albania), SONELGAZ, GRTE, OS (Algeria), Cyprus Transmission System Operator (Cyprus), EETC (Egypt), RTE (France), ADMIE (Greece), IEC (Israel), TERNA (Italy), NEPCO (Jordan), GECOL (Libya), CGES (Montenegro), ONE (Morocco), REN (Portugal), ELEKTRO-SLOVENIJA (Slovenia), REE (Spain), STEG (Tunisia), TEIAS (Turkey).

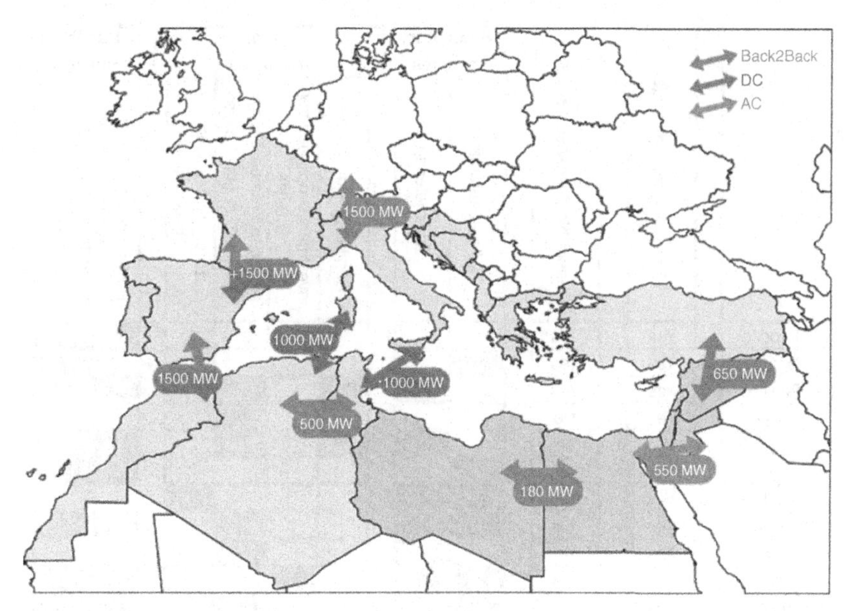

FIGURE 3.16 **Med-TSO Master Plan.** *(Source: Med-TSO (2013).)*

regulations (competition, transparency, unbundling) in SEMCs that will facilitate the convergence of norms and pricing," (OME, 2011b, p. 137) in order to achieve a unique southeast Mediterranean network, which can be integrated into the European network.

6.6 The Regulatory Barrier: The Key Role of MEDREG

As an overall trend, a stable regulatory framework represents a fundamental prerequisite for the large-scale deployment of solar and wind energy. This is particularly due to the fact that, as mentioned before, solar and wind energy projects are characterized by high upfront costs, which implies a long payback period. In this period of time the "rules of the game" should not change, as any regulatory alteration might represent a major threat to the long-term business plan on which any solar or wind energy project is developed.

It has been said that policymakers in SEMCs are increasingly aware of their countries' abundant renewable energy resources and of the potential benefits of their exploitation. This growing awareness is also demonstrated by the fact that almost all SEMCs have progressively put in place dedicated regulatory policies aimed at promoting renewable energy.

The regulatory measures commonly implemented in SEMCs are feed-in tariffs (FITs) and net metering. Many types of fiscal incentives are also deployed among SEMCs, including capital subsidies and tax production credits or reductions (Fig. 3.17). However, the most utilized renewable energy support policy in

	Renewable energy targets	Renewable energy strategy/plan	Regulatory policies		Fiscal incentives				Public financing	
			Feed-in-tariffs	Net-metering	Capital subsidy, grants	Investment/production tax credits	Reduction in sales, energy, CO_2, VAT, or other taxes	Energy production payment	Public investment, loans, or grants	Public competitive bidding
Morocco	Yes	Yes							■	■
Algeria	Yes	Yes	■							■
Tunisia	Yes	Yes		■	■		■		■	
Libya	Yes	No					■			
Egypt	Yes	Yes		■	■		■		■	■
Israel	Yes	Yes	■						■	■
Palestine	Yes	Yes	■	■				■		
Jordan	Yes	Yes	■	■			■		■	■
Lebanon	Yes	Yes		■			■			
Syria	Yes	No	■	■		■			■	■
Turkey	Yes	Yes	■				■		■	■

FIGURE 3.17 Renewable energy support policies in SEMCs. *(Source: Author elaboration on REN21 (2012) and Mondaq (2013).)*

SEMCs is public competitive bidding for fixed quantities of renewable energy and public financing policies, including grants and subsidies.

As this brief overview might have suggested, the current SEMC regulatory landscape in the field of renewable energy continues to be characterized by a high level of fragmentation. This feature represents a key barrier to the deployment of renewable energy in the region, as a fragmented (and unstable) regulatory framework does not allow investors to be fully committed in developing projects in the region. Progressive harmonization and stabilization of the regional regulatory framework would thus be crucial for the future development of solar and wind energy in SEMCs.

The removal of this regulatory barrier needs to be at the national level and in particular through institutions in charge of regulating the renewable energy market in each country of the region. Only such a bottom-up and inclusive approach has the potential to effectively pave the way for a harmonized and stable regulatory framework both at national and, consequently, regional level.

Being the only institution grouping the Mediterranean energy regulators, MEDREG represents the best platform to promote a clear, stable, and harmonized regulatory energy framework in the Mediterranean.

As stated before, in order to promote the common development of the Mediterranean renewable energy market, a strong cooperation between MEDREG (an association of Mediterranean energy regulators) and Med-TSO (an association of Mediterranean TSOs) would be crucially important. In fact, if correctly pursued, such a cooperation scheme might effectively have the potential to abate the infrastructural and regulatory barriers that still characterize the renewable energy market in the southern and eastern Mediterranean region, also paving the way for new investments in the area.

6.7 The Financial Barrier: The Key Role of Institutional Investors

The three barriers just described (e.g., commercial, infrastructural, and regulatory) individually represent an obstacle to the deployment of renewable energy in SEMCs. However, when combined, they also generate – in one fell swoop – a fourth barrier to the deployment of renewable energy in the region: the financial barrier.

In fact, the combination of a distorted energy market (due to the use of universal fossil fuel consumption subsidies), a lack of an adequate electricity infrastructure, and a lack of a stable and harmonized regulatory framework, ultimately prevent private and institutional investors from financing renewable energy projects in SEMCs.

As a result, government investment and finance from various international institutions (most notably the EU) continue to act as the cornerstone of the regional renewable energy-financing scheme.

In particular, over the last decade the following European financing mechanisms have contributed to the financing of renewable energy projects in SEMCs: the Facility for Euro-Mediterranean Investment and Partnership (FEMIP), the Neighborhood Investment Facility (NIF), and the InfraMed Fund.

These three financing mechanisms have surely provided a considerable contribution to the development of certain renewable energy projects that, as in the case of the Ouarzazate Solar Plant in Morocco, might also serve as pilot projects for the overall region. However, this financial framework is not sustainable. In fact, a large-scale deployment of solar and wind energy in the region requires a much more solid financing scheme that only capital markets can provide.

Solar and wind energy projects are characterized by high upfront costs, which implies a long payback period, given the marginal profit derived from each unit of electricity sold. In financial terms, this means that only investors with a long-term investment horizon can potentially find financing this segment of the energy sector attractive. This is even truer in the aftermath of the economic and financial crisis, with the traditional sources of renewable energy investments – governments, commercial banks, and utilities – facing significant constraints. For instance, commercial banks are drastically reducing infrastructure investments as their ability to provide long-term financing is being negatively affected by new banking regulations (e.g., Basel III), generating a trend toward short-term allocation.

In this overall situation the only sustainable option for the financing of renewable energy projects in SEMCs seems to be represented by institutional investors such as pension funds, insurance companies, mutual funds, and sovereign wealth funds (SWFs). In fact, the long duration of their liabilities allows institutional investors to make buy-and-hold investments in long-dated productive assets, achieving higher yields to offset longer-term risks and lower liquidity inherent in many of these assets. In other words, their longer time horizons enable institutional investors to behave in a patient, countercyclical manner, also restraining "short-termism" (European Commission, 2013).

In OECD countries institutional investors held over US$83 trillion in assets in 2012 (Kaminker et al., 2013) and in nonOECD countries SWFs are key sources of capital, with US$6 trillion in assets in 2012.[11] In the framework of a broader research activity on institutional investors and long-term financing, the OECD has analyzed the potential role of institutional investors in green infrastructure investments. In particular, a seminal study carried out by Kaminker et al. (2013) explores the current barriers and the future opportunities for institutional investor investment in green infrastructure, such as renewable energy projects.

According to this study, institutional investors could well play a role in this investment area, particularly considering that "given the current low-interest rate environment and weak economic growth prospects in many OECD countries, institutional investors are increasingly looking for tangible asset classes that can deliver diversification benefits and steady, preferably inflation-linked, income streams with low correlations to the returns of other investments." (Kaminker et al., 2013, p. 6).

In practice, institutional investments in green infrastructure could be channelled in various ways (Fig. 3.18).

However, institutional investors should not be considered as the unconditional "saviors" of the renewable energy sector, as their primary concern is to have a proper risk-adjusted financial performance.

This is a crucial point to bear in mind, also considering that institutional investors "have to invest in accordance with the 'prudent person principle.' Assets have to be invested in the best interest of member and beneficiaries and policyholders and in such a manner as to ensure their security, profitability, liquidity and quality." (Kaminker et al., 2013, p. 8).

In particular, Kaminker et al. identify three kinds of barriers to institutional investor investment in green infrastructure: the first is general and concerns infrastructure investments overall, the second is specific to green investments, and the third relates to the lack of suitable investment vehicles (Fig. 3.19).

On the basis of a comprehensive analysis of recent policy and investment trends in the renewable energy sector, Kaminker et al. recommend seven actions necessary to address these barriers in order to facilitate institutional investments in renewable energy infrastructure projects (Kaminker et al., 2013, p. 13):

---

11. Sovereign Wealth Fund Institute.

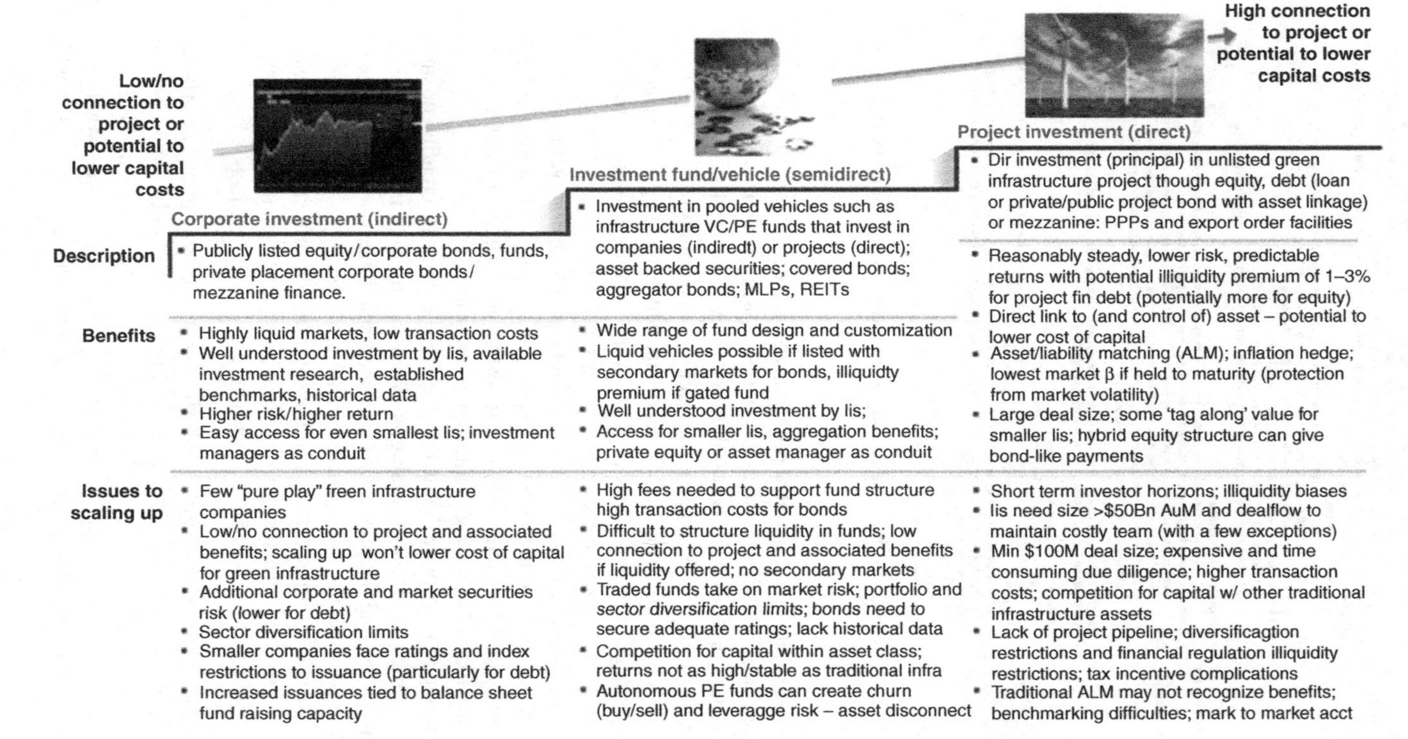

FIGURE 3.18 Channel for institutional investments in green infrastructure. *(Source: Kaminker et al. (2013, p. 6).)*

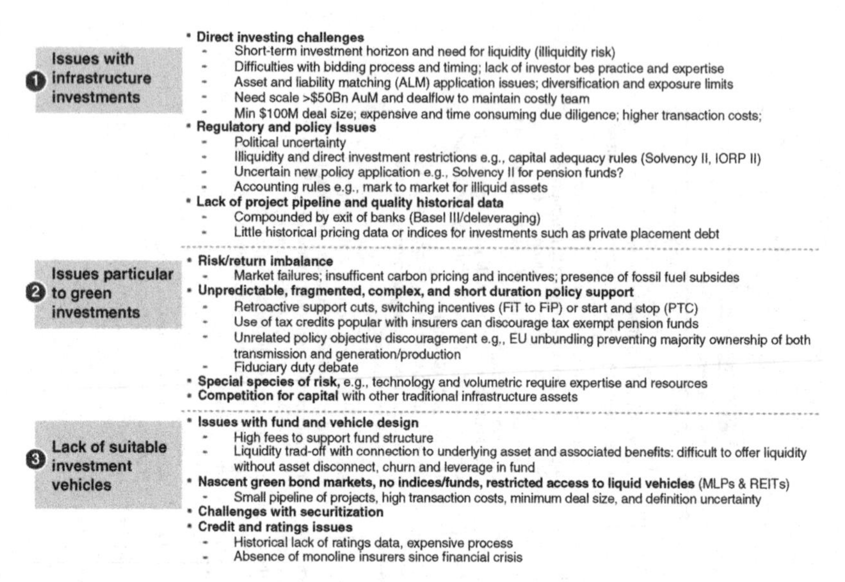

FIGURE 3.19 Barriers to institutional investment in green infrastructure. *(Source: Kaminker et al. (2013, p. 11).)*

1. Ensure a stable and integrated policy environment.
2. Address market failures (such as energy subsidies) that create risk–return investment profiles that favor polluting or environmentally damaging infrastructure projects over green infrastructure investments.
3. Provide a national infrastructure roadmap, which would give investors confidence in government commitments to the sector and demonstrate that a pipeline of investable projects will be forthcoming.
4. Facilitate the development of appropriate green financing vehicles.
5. Reduce the transaction costs of green investment by fostering collaborative investment vehicles between investors, so helping to build scale and in-house expertise among institutional investors.
6. Promote public–private dialog on green investments by creating or supporting existing platforms for dialog between institutional investors, the financial industry, and the public sector to understand the barriers and opportunities to investment in green infrastructure projects.
7. Promote market transparency and improve data on infrastructure investment by strengthening formal requirements to provide information on investments by institutional investors in infrastructure and green projects.

If transposed into the reality of SEMCs, some of these points clearly reflect what has previously been said with regard to the potentially crucial role of institutions such as Med-TSO and MEDREG in creating a stable and harmonized regulatory environment and in enhancing SEMC national renewable energy plans. However, such an action at the regulatory and policy level would not be sufficient to open the southern and eastern Mediterranean renewable energy sector to capital markets and particularly to institutional investors.

In fact, as the action plan delineated by Kaminker et al. outlines, an additional effort is needed to create appropriate, collaborative, financing vehicles, and risk-mitigating tools. Neither SEMC governments nor institutions such as Med-TSO or MEDREG can deliver such financing mechanisms. The only way to provide such a toolkit seems to be represented by a new coordinated effort of the European Investment Bank (EIB) (i.e., the Facility for Euro-Mediterranean Investments and Partnership, FEMIP) with a key financial player that has only recently extended its action to the region: the European Bank for Reconstruction and Development (EBRD).

6.8 The EIB's FEMIP

The EIB has traditionally played an important role in supporting the deployment of renewable energy in SEMCs. In particular, this action has been carried out by the EIB through FEMIP.

As an overall trend, FEMIP represents the main financial instrument ever designed by the EU for the southern and eastern Mediterranean region.[12] FEMIP brings together the whole range of services provided by the EIB to assist the economic development and the integration of SEMCs, with the aim of accompanying the region along a path of sustainable and socially inclusive development. FEMIP does not focus only on renewable energy but also on pollution reduction in the Mediterranean Sea and the development of maritime and land highways.[13]

Since its establishment in 2002, FEMIP invested €13 billion in the region, 40% of which targeted the energy sector (EIB, 2012, p. 7). These investments have mainly targeted the exploitation of the renewable energy potential, including solar energy, wind energy, and hydropower, the upgrading of national energy infrastructures, and the strengthening of regional energy interconnections. For instance, FEMIP has contributed to the construction of new power plants (in Syria, Egypt, and Tunisia), new gas pipelines (in Egypt, Tunisia, and Jordan), new LNG plants (Egypt), new hydropower plants (Morocco), and new wind farms (Morocco and Egypt).[14]

FEMIP essentially offers three types of products (EIB, 2010b):[15]

1. *Loans*. Most FEMIP operations consist of long-term loans, extended in one of two forms. (1) Individual loans made directly to private or public sector promoters of projects where the investment cost exceeds €20–25 million.

---

12. Specifically, FEMIP is devoted to the so-called "Mediterranean Partner Countries": Algeria, Egypt, Gaza/West Bank, Israel, Jordan, Lebanon, Morocco, Syria, and Tunisia.

13. These priorities were identified at the Paris Summit in July 2008. For a brief overview of FEMIP operations on each priority area (EIB, 2010a).

14. For more information about FEMIP activity on the energy sector (EIB, 2010b).

15. List here reported as published online. Available from: http://www.eib.org/

The financing terms (maturity and grace period) are tailored to the type of investment, and appropriate security is required. (2) Credit lines cater for projects with an investment cost below €20–25 million. They are advanced to local intermediary banks and leasing bodies, which on-lend the funds to finance investment undertaken by small or medium-sized enterprises (SMEs) or local authorities. Credit lines can also be used to meet the needs of micro-enterprises, in partnership with specialized associations.

2. *Private equity*. FEMIP finances private companies by providing equity or quasiequity funding (participating or mezzanine loans) from its budgetary resources. In terms of potential beneficiaries, it targets mainly but not exclusively, SMEs. As part of its support activity, FEMIP also deploys risk capital resources via participating loans to certain microcredit associations. Private equity operations must present a number of features: (1) governance – the EIB's hands-on approach to individual investment leads to the application of best market practice (often EU standards) in the areas of financial discipline and governance; (2) leverage – the bank is always a minority investor, committing at a very early stage to the projects that it finances. Through its early involvement, the bank often allows the company to attract other sources of funding; and (3) sustainability – private equity operations are designed to support profitable ventures creating value added and jobs. The achievement of an expected return commensurate with the risk taken is a *sine qua non* for each investment decision by the EIB.
3. *Technical assistance*. Technical assistance is a key instrument for improving the quality of lending operations and enhancing their development impact. It is an integral part of the appraisal and implementation process for FEMIP loans, and provides the analytical data needed to understand the economic and financial development issues of Mediterranean partner countries. The technical assistance offered by FEMIP is based on two funds. (1) The FEMIP Support Package supports existing projects that have already been identified. This assistance can be delivered at various stages of the project, from preparation through to completion. (2) The FEMIP Trust Fund provides upstream technical assistance for loan operations, which is not necessarily linked to an EIB investment.
4. *Guarantees*. In order to stimulate the regional capital market, to mobilize additional resources to augment the volume of public capital, to reduce exchange rate risks, and to support the development of substate entities, FEMIP also provides guarantees to SMEs, local banks, public sector promoters, and substate entities.

6.9 The New Role of the EBRD in the Southern and Eastern Mediterranean Region

The EBRD was established in 1991 in response to major changes in the political and economic climate in central and eastern Europe, with the aim to support the development of market economies in the region. Considering the bank's experience with accompanying countries in political and economic transition, in May 2012 the EBRD's shareholders gave unanimous backing to the expansion of the bank's mandate, allowing future activities in the southern and eastern Mediterranean region.

In November 2012 Egypt, Jordan, Morocco, and Tunisia were granted the status of potential recipient countries, enabling the bank to conclude six investment transactions in the region by the end of that year, representing a total commitment of €181 million.

Looking forward, the EBRD is estimated to have the capacity to invest up to €2.5 billion in SEMCs by 2015 (EBRD, 2013a, pp. 7–13). The bank's overall goal in SEMCs is to improve financing of the private sector via investments in loans and equities, while providing support and expertise through policy dialog, capacity building, and other forms of technical assistance.

More specifically, the bank has identified a limited number of initial priorities for the region, among which is the renewable energy sector. The EBRD will thus support renewable energy and energy efficiency investments in SEMCs and it will assist governments in unbundling and gradually liberalizing the sector while strengthening regulatory agencies (EBRD, 2013a, p. 13).

This effort would fit into the wider activity framework of the EBRD in the field of renewable energy: the Sustainable Energy Initiative (SEI). This initiative was launched in 2006 to address the twin challenges of energy efficiency and climate change in the EBRD region, which is among the most energy intensive areas in the world. SEI market segments include renewable energy, energy efficiency, sustainable energy financing facilities through financial intermediaries, and carbon market support. From 2006 to 2012, the EBRD invested €11 billion in 602 sustainable energy projects in 33 countries. In 2012, SEI investments accounted for 26% of the EBRD's activities (EBRD, 2013b).

Considering the experience and the expertise of the EBRD, this *nouvelle vague* represents a unique opportunity for the southern and eastern Mediterranean region to tackle the financing barrier to the deployment of solar and wind energy at the regional level.

Considering the complex interdependence of the various issues on the table (e.g., commercial, regulatory, infrastructural, and financial) the action of the EIB and the EBRD in the region might turn out to be more effective if framed into a cooperation framework with Med-TSO and MEDREG. This cooperation might be promoted with an innovative, dedicated, platform designed as proposed in the conclusions of this paper.

7 CONCLUSIONS: TOWARD A NEW EURO-MEDITERRANEAN RENEWABLE ENERGY PLATFORM

The chapter has outlined that – notwithstanding the huge renewable energy potential and the numerous benefits potentially related to its exploitation, and the various actions that have been undertaken at both national and international levels to promote the deployment of such technologies over the last decade – SEMCs continue to lag far behind most other regions in the world in terms of solar and wind energy deployment.

The chapter has argued that this paradox is mainly due to the fact that the deployment of renewable energy in the region faces four key barriers: commercial, infrastructural, regulatory, and financial.

In order to effectively tackle these barriers, a "double-track" approach seems to be essential. In other words, these barriers are so resilient that they should be faced both singly and globally, in one fell swoop. Let us see how this might be done, and what role European institutions might play in this chessboard.

As illustrated in the chapter, reforming universal energy subsidies is not an easy game. Subsidies act as a cornerstone of the social contract in many SEMCs, particularly in oil and gas-producing countries. Over recent decades, some governments tried to reform universal energy subsidies, but did not accomplish the target. In the aftermath of the Arab uprising, such a reform process seems to be even more difficult, particularly in the countries where the political situation is not fully stabilized. For this reason it is not possible to expect substantial changes on this crucial point across the region. However, in the medium term all SEMCs should advance an energy subsidy reform process, phasing out universal fossil fuel consumption subsidies in favor of targeted subsidies aimed at effectively addressing the problem of energy poverty. In this field little support can be provided by European institutions. In fact, the regional governments are well aware of the major economic burden represented by universal energy subsidies; however, economic rationalities currently do not match with political requirements that only an overall process of political and economic stabilization in the region might eventually change.

As far as the development of an adequate regional electricity infrastructure is concerned, Med-TSO – in quality of association of the Mediterranean transmission system operators – might play a potentially crucial role in coordinating the various players in the field, in promoting a clear regional transport code, and in developing operational tools for the coordinated planning process of regional interconnection. Such an inclusive and bottom-up approach seems to represent the best way to promote the development of an adequate electricity infrastructure in SEMCs, among SEMCs, and between SEMCs and the EU.

The same rationality could well be applied to energy regulation. In fact, as illustrated in the chapter, MEDREG – -in quality of association of the Mediterranean energy regulators – - might play a key role in promoting a clear, stable, and harmonized regulatory energy framework in the Mediterranean region.

Concerning the financial dimension of renewable energy in the region, a "great leap forward" is urgently needed. As illustrated in the chapter, strong cooperation between the EIB and the EBRD seems to represent the best way to develop a set of proper financing mechanisms aimed at attracting institutional investors such as pension funds, mutual funds, insurance companies, and sovereign wealth funds into the regional renewable energy market. On the basis of the principle that institutional investors will jump into this area only if a proper risk-adjusted return is considered as guaranteed, the EIB and the EBRD might develop a new "Euro-Mediterranean Renewable Energy Infrastructure Fund" aimed at channelling financial resources from institutional investors into renewable energy companies acting in the region, solar and wind energy projects, asset-backed securities, or bonds. Such a mechanism, already applied in the region –albeit on a smaller scale – by the InfraMed Fund, could have many important advantages: (1) by making use of the multiyear experience of the EIB and the EBRD in financing renewable energy projects, such a mechanism might allow institutional investors to play a role in a sector with which they are often not familiar; (2) considering the financial capability of the EIB and the EBRD, risk-mitigating and credit-enhancing tools might also be included in the mechanism; and (3) considering the high reputation of both the EIB and the EBRD, such a mechanism might progressively attract a number of institutional investors, thus allowing the exploitation of economies of scale in the determination of the projects to be financed.

Such a new "Euro-Mediterranean Renewable Energy Infrastructure Fund" might be launched in synergy with the "Euro-Mediterranean Platform on Regional Electricity Markets" and the "Euro-Mediterranean Platform on Renewable Energy and Energy Efficiency" to be developed by the UfM Secretariat also with the support of MEDREG and Med-TSO.[16] In fact, considering the complex interdependence of all these infrastructural, regulatory, and financial issues, a strong coordination between Med-TSO, MEDREG, and the EIB–EBRD tandem seems to be essential.

These new entities could well have the potential to emerge as the focal point for the development of renewable energy in the southern and eastern Mediterranean region. As a matter of fact, the Euro-Mediterranean region is currently characterized by the presence of countless large-, medium-, and small-scale organizations dealing with renewable energy. In such an intricate – and often redundant – situation, it is necessary "to put the house in order" and focus on the key players in the field to develop a new renewable energy mechanism able to break with convention.

TSOs, energy regulators, and the key financial institutions in the region are the most important players to proceed in this direction. On the basis of an inclusive, pragmatic, and bottom-up approach, they have the potential to boost the regional renewable energy market, reversing the trend of disillusionment that has progressively characterized the southern and eastern Mediterranean renewable energy sector over recent years, ultimately translating regional renewable energy potential into a reality.

16. http://www.medreg-regulators.org/Portals/45/immagini_home/Rome_Final_statement_on_the_HighLevel_Conference.pdf

REFERENCES

Bergasse, E., Paczynski, W., 2012. The Relationship between Energy and Economic and Social Development in the Southern Mediterranean. MEDPRO Technical Report No. 15, Center for European Policy Studies, Brussels.

Coutinho, L., 2012. "Determinants of Growth and Inflation in Southern Mediterranean Countries", MEDPRO Technical Report No. 10, Center for European Policy Studies, Brussels.

Dabrowski, M., De Wulf, L., 2013. Economic Development, Trade and Investment in Southern and Eastern Mediterranean Countries: An Agenda towards a Sustainable Transition. MEDPRO Technical Report No.4, Center for European Policy Studies, Brussels.

DLR, 2005. MED-CSP – Concentrating Solar Power for the Mediterranean Region. Deutsches Zentrum fur Luft- und Raumfahrt (study commissioned by the Federal Ministry for the Environment, Nature Conservation and Nuclear Safety of Germany), Stuttgart.

http://www.dlr.de

EBRD, 2013a. EBRD Annual Report 2012. London.

EBRD, 2013b. Sustainable Energy Initiative: Financing sustainable energy. London.

EIB, 2010a. Union for the Mediterranean: Role and Vision of the EIB. Luxembourg.

EIB, 2010b. FEMIP and the Energy Challenge in the Mediterranean. Luxembourg.

EIB, 2012. FEMIP Annual Report 2011. Luxembourg.

El-Katiri, L., 2014. A Roadmap for Renewable Energy in the Middle East and North Africa. OIES Paper MEP 6, Oxford Institute for Energy Studies, Oxford.

European Commission, 2013. Long-Term Financing of the European Economy. COM(2013) 150 final.

Euro-Mediterranean Energy Market Integration Project (MED-EMIP), 2010. MEDRING Update, Vols. 1–4, MED-EMIP, Heliopolis.

Groenewold, G., De Beer, J., 2013. Population Scenarios and Policy Implications for Southern Mediterranean Countries, 2010–2050. MEDPRO Policy Paper No. 5, Center for European Policy Studies, Brussels.

Hafner, M., Tagliapietra, S., El Andaloussi, H., 2012a. Outlook for Electricity and Renewable Energy in Southern and Eastern Mediterranean Countries. MEDPRO Report No. 16, Center for European Policy Studies, Brussels.

Hafner, M., Tagliapietra, S., El Andaloussi, H., 2012b. Outlook for Oil and Gas in Southern and Eastern Mediterranean Countries. MEDPRO Report No. 18, Center for European Policy Studies, Brussels.

Kaminker, C., et al., 2013. Institutional Investors and Green Infrastructure Investments: Selected Case Studies. OECD Working Papers, OECD, Paris.

Med-TSO, 2013. Med-TSO: A Mediterranean Project. Presentation made by Noureddine Boutarfa (Med-TSO Chairman) at the Euro-Mediterranean Ministerial Meeting on Energy, 11 December 2013, Brussels.

http://www.medelec.org

Mondaq, 2013. Incentives on Renewable Energy in Turkey. December 16.

OME, 2011a. Mediterranean Energy Perspectives 2011. Nanterre.

OME, 2011b. Mediterranean Energy Perspectives – Egypt. Nanterre.

PWMSP, 2011. Benchmarking of Existing Practice Against EU Norms – Country Reports. Paving the Way for the Mediterranean Solar Plan ENPI 2010/248-486.

PWMSP, 2012a, "Power Systems at 2020: State of Play of the Existing Infrastructures". Paving the Way for the Mediterranean Solar Plan ENPI 2010/248-486.

PWMSP, 2012b, "Regional Road Map for Regulatory and Legislative Convergence". Paving the Way for the Mediterranean Solar Plan ENPI 2010/248-486.

REN21, 2012. Renewables – Global Status Report. REN21, Paris.

REN21, 2013. Renewable – Global Futures Report 2013. REN21/ISEP, Paris.

UAE/IRENA/REN21, 2013. MENA Renewables Status Report 2013. Abu Dhabi.

World Bank, 2011a. MENA Economic Developments and Prospects: Investing for Growth and Jobs. World Bank Middle East and North Africa Region, Economic Developments & Prospects, World Bank, Washington, DC.

World Bank, 2011b. Middle East and North Africa Region Assessment of the Local Manufacturing Potential for Concentrated Solar Power (CSP) Projects. Energy Sector Management Assistance Program (ESMAP), Washington, DC.

World Bank, 2013. MENA Economic Developments and Prospects: Investing in Turbulent Times. World Bank Middle East and North Africa Region, Economic Developments & Prospects, World Bank, Washington, DC.

Chapter 4

Scaling Up Renewable Energy Deployment in North Africa

Georgeta Vidican
Department of Sustainable Economic and Social Development, Deutsches Institut für Entwicklungspolitik/German Development Institute, Bonn, Germany

1 ENERGY SYSTEMS IN NEED OF TRANSFORMATION

Nobody can dispute anymore that countries across the Middle East and North Africa (MENA) need to transform the way they consume and produce energy. Countries rich in oil and gas resources must not only improve their environmental footprint by reducing their consumption of energy and water resources, but also find new opportunities for exporting these natural resources and for diversifying their energy sector. Those countries that depend on energy imports must find ways to improve their energy security and to reduce their vulnerability to price shocks. Thus, increasingly, national governments and donors alike have started to view renewable energy as an important solution for the energy challenge in the MENA region. Renewable energy technologies can also be viewed to open the path to inclusive and green growth in the region, through job creation, private sector development, greening the industrial sector, and environmental sustainability (UNEP, 2011; World Bank, 2012).

In spite of wide economic and political differences across the region, the energy question is high on all governments' agendas, as, in light of future growth, energy demand is expected to grow by 2% yearly well into the 2030s (IEA, 2012, p. 57). While renewable energy is seen as part of the solution, emphasis on its role varies across countries. The immediate concern for decision makers is, of course, finding quick solutions to satisfy increasing needs for energy. Even if renewable energy sources are abundant across the region, generation costs and different operation models (as compared with fossil fuels) prevent, time and again, governments from making transformative decisions regarding the energy sector. Vested interests in the conventional energy sector act as strong barriers to renewable energy deployment even where renewables have reached grid parity with fossil fuel-based generation. Fossil fuel subsidies, the highest worldwide, create a real cost disadvantage

Regulation and Investments in Energy Markets. http://dx.doi.org/10.1016/B978-0-12-804436-0.00004-7

for renewable energy technologies, by lowering the market price for conventional electricity and fossil fuel products (such as kerosene, butane gas, gasoline). Reducing subsidies is a politically complex process that has been attempted many times before, but rarely succeeded in the MENA region. Vested interests prevent real transformation also because the existing energy system is characterized by deep path dependencies created not only by an existing infrastructure that supports large fossil-fueled power plants and centralized electricity generation. Rather, entire systems of power and patronage networks are at the core of the energy sector, defining relations between the state, energy-intensive industries, and energy producers (World Bank, 2014).

Thus, a shift toward a more diversified energy mix where renewables play a much more important role will not come easy in the region, in spite of the widely demonstrated benefits and opportunities. To enable such a process of transforming the energy system change needs to happen at different levels. Critically, a new social contract is needed, one that is based on more inclusive forms of governance and supports growth closely aligned with sustainability goals. Such transformation would be enabled by new ways of managing rents (withdrawing rents from fossil fuels and allocating them to sustainable development goals) in a politically and socially acceptable way, by fostering inherent interlinkages between policy interventions, and coordinating renewable energy agendas at the regional level. As such, a new narrative that discredits the old (fossil fuel) regime and sheds light on the cobenefits of pursuing a more diversified energy mix and scaling up renewables is needed.

So far important preparatory steps have been taken with respect to setting renewable energy targets, implementing policies, and regulatory frameworks, and the first few projects have been rolled out. Yet, without these wider (societal/governance) changes, large-scale deployment of renewable energy technologies and the energy system transformation are likely to fall behind.

With a focus on North African countries (in particular, Egypt, Tunisia, Algeria, and Morocco), this chapter explores these challenges and opportunities in greater detail. While other chapters in this book treat some of these issues in much greater detail, here the focus is on presenting the dynamic interaction between these factors and on embedding the analysis in larger societal transformation aspirations. The chapter starts with a general discussion on the important steps taken so far by North African countries toward renewable energy deployment, with a focus on policies, projects, and strategies. It then turns to elaborate several challenges to large-scale deployment and the transformation of the energy sector. Once these aspects are underlined, the chapter considers in more detail the larger, higher-level transformations, which are needed (with respect to changing the social contract and managing rents) to trigger a shift to a more sustainable energy system.

2 INITIAL STEPS TO SUPPORT DEPLOYMENT

While large-scale transformations have not yet materialized, important initial steps have already been taken across the MENA region. Supported by various Euro-Mediterranean energy initiatives[1] and the advancing climate change mitigation agenda, national governments from North Africa, and international donors channeled their attention to harnessing the wide potential for renewable energy generation in the region. Depending on the perceived urgency to diversify the energy mix, governments set renewable energy targets, implemented policies to support deployment, identified potential projects, and attracted much needed investments. Countries such as Morocco, highly dependent on energy imports, have been much more proactive; but also fossil fuel–rich countries, such as Algeria, followed course seeing opportunities in maintaining their comparative advantage in the energy sector. Table 4.1 illustrates some outcomes from these developments and prospects for future plans in terms of project development.

These developments follow previously agreed targets for renewable energy across all countries in the region (Table 4.2 for the North African region). The setup of these objectives has been followed by more concrete actions with respect to setting up the necessary institutional framework, attracting investments, and developing more long-term strategies. These advances will be briefly discussed in the next section, followed by a critical analysis of challenges to large-scale deployment in Section 3.[2]

2.1 Egypt

In the context of turbulent times following the Arab Spring uprisings in 2011 and the prolonged political transition, the energy sector in Egypt is facing various challenges. To satisfy increasing demand, Egypt needs to double its generation capacity by adding about 3000 MW annually in the next 10 years with investment of about US$32.5 billion (US$3 billion per year) (EgyptERA, 2014). In response to rising demand and dwindling natural resources, in 2008 the Ministry of Electricity and Energy (which later became Ministry of Electricity and Renewable Energy (MoERE)), set the 2020 target of 20% electricity generation from renewable energy (7.2 GW from wind by 2020 and 3.5 GW from solar by 2027) and also planned to add 4 GW of nuclear power by 2020/2022 (EgyptERA, 2014). In this process, wind energy has been prioritized over solar energy due to its lower generation cost, which is already at grid parity with conventional energy (Vidican, 2012).

1. Examples of such initiatives are the Union for the Mediterranean (UfM), DESERTEC Industrial Initiative (DII), and Association of Mediteranean Energy Regulators (MedReg).

2. In this analysis, we only focus on four North African countries, excluding Libya because of its volatile political and economic situation.

TABLE 4.1 Fossil Fuel Reserves and Renewable Energy Deployment in North Africa

	Net oil balance* (1000 b/d), 2010	Net natural gas balance (BCM), 2009	Wind installed capacity (MW)	Solar water heating installed capacity (MW)	Solar installed capacity (MW)		Capacity in the pipeline** (MW)	
					PV	CSP	Solar	Wind
Egypt	–77.4	18.3	550	525.0	15	20	106	1070
Libya	1500.2	9.9	0	0.02	4.8	0	–	610
Tunisia	–0.3	–1.3	154	437.5	4	0	5	100
Algeria	1765.7	52.7	0	0.2	7.1	25	175	20
Morocco	–205.1	–0.5	291	245.0	15	20	173	1553

**Net balance is the difference between production and consumption.*
***These projects have been approved but not yet developed.*
Source: REN21 (2013). Data are for year 2012; Fattouh and El-Katiri (2012).

TABLE 4.2 Renewable Energy Targets in North Africa

Countries	Renewable energy targets and target dates
Egypt	20% of electricity generation by 2020, of which 12% is wind.
Libya	3% of electricity generation by 2020; 10% by 2025.
Tunisia	11% of electricity generation by 2016; 25% by 2030; 16% of installed power capacity by 2016; 40% by 2030.
Algeria	6% of electricity generation by 2015; 15% by 2020; 40% by 2030, of which 37% is solar (PV and CSP) and 3% is wind.
Morocco	42% of installed power capacity by 2020 (including hydropower).

Source: REN21 (2013).

To roll out the renewable energy plans, additional responsibilities have been given to the New and Renewable Energy Authority (NREA),[3] which operates under the auspices of the MoERE, to implement the agreed upon targets for solar and wind energy. For medium- and large-scale projects NREA organizes a competitive bidding process for build-operate-own (BOO) projects with long-term power purchase agreements (PPA) (Vidican, 2012). A feed-in-tariff (FiT) has also been agreed on in 2014, aimed especially at smaller scale solar photovoltaic (PV) projects. To cover the investment cost for these projects Egypt relies on concessional funds, FiT mechanisms, competitive bidding, and consumers' own investment using the grid permitted under third-party access regulations (EgyptERA, 2014).

While the potential for renewable energy in Egypt is one of the highest in the region, deployment has been rather limited until now (Table 4.1). Nevertheless, the urge to find answers to the energy crisis has fostered entrepreneurial initiatives in the clean energy sector, especially focused on off-grid solutions (e.g., KarmSolar, RodoSol, Shamsina) (Vidican, 2012; Williamson, 2015). These initiatives, however, do not currently receive large financial and institutional support, but rather they are the result of grass root efforts.

While there is no doubt that renewable energy will be part of the energy sector in Egypt, several challenges could act to undermine its development process. First, an increasing focus on boosting coal and natural gas–based electricity generation and strong commitment to nuclear energy,[4] communicated recently as part of the revised energy agenda under the new leadership, could limit the institutional space for renewable energy (EIPR, 2015). Second, fossil

3. NREA was created in 1985, but until recently, its expertise has been mostly in the area of hydropower.

4. Egypt recently signed an agreement with Russia to build its first nuclear energy plant (MESP, 2015).

fuel subsidies are among the highest in the MENA region, even if Egypt's net energy balance is increasingly negative (Table 4.1). The high rate of subsidization not only drains the government budget and detracts investments from socioeconomic goals, but also creates a strong price disadvantage for renewable energy technologies (Vidican, 2014).

While the government has started the reform process already (at the end of 2014), it remains to be seen whether drastic measures will be taken and suitable compensation measures implemented (Vidican, 2015a). Finally, given the high level of unemployment in Egypt (but also across the region), investments in developing a new sector should ensure that employment opportunities are being created. However, the participation of local companies in the value chain of these technologies remains limited (Vidican, 2012).

2.2 Tunisia

Like Morocco, Tunisia has limited natural resources. As a result of economic growth and high levels of energy subsidies, energy demand increased significantly, leading to a rising energy deficit (GIZ, 2012). The Tunisian Solar Plan initiated in 2010, and earlier programs since 2005, had the objective to promote renewable energy as an alternative to the strong reliance on natural gas. Within this context, the programs focused on reducing energy intensity, maintaining moderate growth, and increasing the share of renewable energy in primary energy consumption (GIZ, 2012).

The Tunisian Solar Plan for the 2010–2016 period has been developed within the larger framework of integration into the Mediterranean area. The plan's intention is to promote large-scale deployment of renewable energy (480 MW by 2016), enhance energy efficiency (energy savings of 23% of primary energy demand), establish interconnection lines to export power to Europe, and develop a cluster of expertise in the solar equipment industry (GIZ, 2012).

The most important achievement for Tunisia has been with respect to the deployment of solar water heaters. At the end of the PROSOL program in 2010, the total installed collector surface area reached around 490,000 m^2. This development made it possible to create a market for 50 local suppliers, including 7 manufacturers, and over 1200 small installation businesses (GIZ, 2012). Solar PV has been primarily developed in the context of rural electrification, while wind energy received a stronger focus only more recently.

The regulatory framework for solar water heaters is currently one of the most effective in the region. Several innovative incentives and instruments are used to encourage deployment, such as financial and fiscal support, customs duty reduction, VAT exemptions, and bank loans with reduced interest rates (UNEP, 2015). Further, repayment of loans for solar water heaters is organized through regular utility bills from state electricity utilities, while local banks receive support for providing reduced interest rate loans. In addition, the government provides a subsidy of 20% of the system cost, while customers are expected to finance a minimum of 10% of the purchase and installation costs (UNEP, 2015).

In spite of this early success for rural electrification and solar water heater programs, the scale-up of renewable energy deployment is lagging behind.The high level of fossil fuel subsidies plays a role in this outcome, as well as limited grid infrastructure and a lack of investment to support such a scale-up.

2.3 Morocco

In the North Africa region, Morocco experiences the highest dependency on energy imports. Currently approximately 95% of its energy needs are satisfied by imports (MEMEE, 2011), making Morocco highly vulnerable to changes in energy prices on the international market, fluctuations in supply, and budget constraints. To remedy this situation, in 2008 the government elaborated on its National Energy Strategy aiming to supply 42% of electricity generation from solar (2 GW), wind (2 GW), and hydro sources (2 GW), thus harnessing the vast renewable energy resources present in Morocco. The Plan Solaire was developed to provide a road map for achieving solar energy targets, by constructing five large solar power plants.

To ensure implementation of the Plan Solaire the government committed US$9 billion and created in 2010 an agency/company, the Moroccan Agency for Solar Energy (MASEN) tasked with project development. MASEN also organizes the competitive bidding projects for the solar plants and coordinates the combination of grants and loans by multilateral and bilateral donors (Vidican et al., 2013). The government has also taken its first steps in creating a legal and institutional framework to support the formation of the domestic market for renewable energy. Law 13-09 allows medium- and high-voltage projects to access the electricity network and open up competition in this new market. The law also permits projects with capacities of a maximum 50 MW to be built and operated by private enterprises (Vidican et al., 2013). Wind energy developed much faster, due to its lower cost of generation.

While these developments enabled the first projects, such as the landmark 160 MW Ouarzazate I, lack of incentives and regulatory framework for distributed generation and small-scale projects (such as a rooftop PV), as well as not clearly specifying the type of technologies targeted by the Plan Solaire, prevented market expansion. For instance, a FiT is lacking and net-metering schemes have been under discussion for a very long time.

A particularly important trend is the Moroccan authorities' focus on linking solar energy applications to other sectors, such as housing (i.e., rooftop PV systems and solar water heaters) and agriculture (i.e., solar water pumps). The Agence Nationale pour le Développement des Energies Renouvelables et de l'Efficacité Energétique (ADEREE) initiated collaborations with the Ministry of Agriculture and Ministry of Housing to integrate renewable energy technologies in new projects. Such initiatives are critical to integrating clean energy technologies in Moroccan society more broadly. To scale up such programs, however, proceeding with the reform of fossil fuel subsidies is crucial in order eliminate energy price distortions.

Yet, as in the case of Egypt, the involvement of private companies manufacturing parts and components and providing services for the emerging renewable energy sector has been limited. Thus, job creation has also lagged behind.

2.4 Algeria

Algeria has very different energy conditions compared with the rest of the North Africa region. With large reserves of fossil fuels, Algeria is an important oil and gas exporter. Yet, especially since 2010, Algerian authorities have developed a strategy for diversifying its energy mix, realizing that conventional fuels are limited and that it can gain much more from exporting these resources than using them for electricity generation domestically. Nevertheless, with a strong fossil fuel interest group and centralized energy generation infrastructure, lock-in to conventional fuels is strong.

The Renewable Energy and Energy Program was developed in 2011, with a perspective spanning to 2030. The program consists of installing up to 22,000 MW of power generation, aiming to achieve 40% of electricity generation by 2030 (Table 4.2). For the first few years the program seeks to test the suitability of different technologies to local conditions; the deployment phase will start in 2015 and will be scaled up by 2020 (CDER, 2011). To support this process, a FiT was been set up in 2014 offering roughly 16 DZD/KWh (US$0.17 at current exchange rates) for projects between 1 MW and 5 MW (Ayre, 2015).[5] The FiT offers a guaranteed flat rate for the first 5 years of a project, with a performance-based rate being used for the next 15 years (Ayre, 2015). While it is too early to assess the effectiveness of the FiT scheme, several solar energy projects are currently in the pipeline (Table 4.1).[6]

Not surprisingly as a large producer of fossil fuels, Algeria provides generous subsidies to conventional energy products and electricity (approximately 11% of GDP, while only 6% of the budget is spent on education) (IMF, 2014). Without a change in the subsidy scheme investments in renewable energy are likely to increase the country's energy bill to unsustainable levels. Due to price distortions, large-scale deployment of solar energy technology is also likely to be hampered. Moreover, as the supply of energy is not a concern for Algeria at the moment, it is to be expected that renewable energy deployment will proceed at a much slower rate.

5. Projects larger than 5 MW benefit from variable FiT rates based on the predicted energy output of the specific installation (Ayre, 2015).

6. More recent sources estimate projects under development at a much higher level (i.e., 350 MW of solar PV projects) (Ayre, 2015).

3 SCALING UP IS CHALLENGING

In spite of ambitious plans and efforts to increase deployment of alternative energy technologies, the share of renewable energy in the energy mix in North Africa remains limited. The factors that contribute to slow progress are varied and point to the systemic nature of energy sector reform. A systemic approach is essential to scaling up renewable energy deployment, as piecemeal measures are not likely to trigger expected outcomes. Policy interventions must coalesce to not only derisk renewable energy investments (Carafa et al., 2015), but to also conceive an energy system that supports green and inclusive growth in the region, sensitive to its specific governance structure (Vidican, 2015b). Moreover, I argue that harmonization across national strategies can speed up the transition toward renewables across the Mediterranean, augmenting the benefits across the region.

To this end, several factors are considered important, relating to supportive institutions, infrastructure,[7] access to finance, private sector development, strategic coordination, and regional integration. In this process of scaling up renewables, politics is at the core, as it implies a redistribution of rents and realignment of interests between stakeholders who currently have conflicting interests (Lütkenhorst et al., 2014). Integrating these different factors into a unified and comprehensive strategy, while at the same time addressing the political challenges by forging alliances across diverse sets of interests, remains a challenging task. In essence, it is the integration and synergies between these factors (or interventions) that is likely to contribute to scaling up renewable energy deployment and the subsequent transformation in the energy system.

As a first step in responding to this challenge, I briefly examine some of these factors, followed by a short discussion (Section 4) on the interlinkages at the national and regional level necessary to achieve large-scale deployment and a transformation in the way energy is generated and consumed in the region.

3.1 Commitment to a "Green Growth" Agenda

Scaling up renewable energy cannot be achieved without a comprehensive strategy geared to achieving a transition in the energy sector, driven by a strong commitment to a green growth agenda. Such a vision and strong leadership has not yet been manifested in North Africa, in spite of it being the most dynamic region with respect to renewable energy project development. While ambitious, renewable energy targets are nonbinding and significant delays have been recorded in attracting investment. Moreover, a limited integration of renewables within various sectors (i.e., industry, energy, water, housing, agriculture) is likely to limit opportunities for scaling up these technologies and pursuing a green growth

7. The most relevant aspects here relate to grid access and extension of the grid, which are discussed in more detail in other chapters of this book.

agenda. Finally, a "single lens" focus on energy security drives political choices to favor cheap fossil fuels, to the detriment of renewables and the cobenefits they could deliver in the medium to long term. A case in point is the recent announcement by the Egyptian government regarding extensive investments in coal and nuclear power plants, as part of its revised energy strategy (Vidican, 2015a).

3.2 Regulatory Framework

Aside from short- to medium-term targets for renewable energy, most governments have also implemented laws and regulations to support deployment (Section 2). Yet, these institutions fall short of providing long-term guarantees and incentives to stimulate private sector investment. In particular, decentralized electricity generation has been neglected by existing institutions. Strong vested interests of key players in the dominance of centralized electricity generation (e.g., state utilities, and the *régies* in Morocco) have contributed to weakening the appeal of distributed solar PV generation, in spite of favorable economic calculations (Vidican et al., 2013). Nonexistent or weakly implemented FiT and net metering schemes (OECD, 2013) are likely to discourage investors. Competitive bidding has been the most widely used instrument for supporting large-scale renewable energy plants in the region. Yet, lack of transparency regarding associated incentives and the type of solar energy technologies envisioned for future projects is likely to weaken competition (Vidican et al., 2013). Moreover, the absence of state guarantees could undermine long-term infrastructure projects in the region (OECD, 2013).

3.3 Energy Prices

High fossil fuel subsidies and locking in to conventional energy generation infrastructures after decades of reliance on fossil fuels has created a real price disadvantages for renewable energy technologies. The reform of subsidies has often been attempted in the past, but due to high political risk, it was unsuccessful. Being highly embedded in the social contract, the removal of fossil fuel subsidies faces opposition not only from vulnerable population groups, but also from powerful interest groups, such as energy-intensive industries and energy producers (Vidican, 2014). Yet, a more recent experience in Morocco, where the government engaged in a gradual reform of the subsidy scheme (for electricity and the main fuel products), increased hopes that the process can be sustained over time. The first steps taken in Egypt, however, raise questions as to whether the government's commitment to reform is strong enough to advance the process and to systematically introduce the necessary mitigation measures (James, 2015).

3.4 Access to Financing

While financing large-scale energy generation in North Africa has been easier to accomplish due to assistance from multilateral donors, small-scale projects (on- and off-grid) have been severely constrained by lack of financial instruments.

While liquidity is not viewed to be a problem in the financial sector of most North African countries, the main reason quoted by financiers in the region is the weak bankability of such projects (Vidican et al., 2013). To partly alleviate financing constraints, state guarantees have been widely suggested (OECD, 2013). The experience of Tunisia with using innovative financing instruments for their new solar water heaters program could serve as an example for other countries, such as Morocco and Egypt (Vidican et al., 2013).

3.5 Private Sector Engagement

Limited private sector engagement in the renewable energy programs raises serious concerns not only regarding the sustainability of these projects, but also the degree to which these strategies and investments can deliver the much needed socioeconomic benefits. Given the development challenges that North African countries are confronted with, a limited focus on addressing energy security without emphasizing local value creation is likely to compete with other investments in socioeconomic goals.

A stronger private sector involvement should be envisioned along the entire value chain of renewable energy technologies, especially for solar photovoltaics and wind power, from manufacturing parts and components to service-related activities (e.g., feasibility studies, operation, and maintenance). While domestic private sector engagement has been much stronger in the wind energy sector (especially in Morocco), serious gaps are visible in solar energy projects. Achieving a higher local content in the tenders for solar energy power plants has been strongly articulated in tenders for projects in Ouarzazate, a city and province in Morocco. Yet, although nonnegligible quotas of 20–30% local content have been achieved, most will be satisfied through civil works, thus contributing trivial amounts to the local economy (Vidican et al., 2013).

Thus, to achieve long-lasting effects, a comprehensive long-term strategy of integrating private sector development goals with energy policy goals is necessary. Specifically, measures to increase private sector competitiveness and improve investment framework conditions (e.g., market size) would contribute to not only scaling up renewables but also to local value creation.

3.6 Regional Perspective

The regional perspective, the main focus of this book, is absolutely critical for scaling up renewables in the region. First of all, no national market in the North Africa region is large enough to deliver sufficient benefits from renewable energy investments. Thus, an integration of energy markets between and across the north and south of the Mediterranean can magnify the benefits (UfM, 2013). Yet, a high degree of fragmentation across the MENA region (regarding visions of development and reinforced by historical realities and political agendas) has led to weak levels of coordination and cooperation with respect to transitions in the energy sector.

4 CAPITALIZING ON EARLY STEPS TO TRANSFORM THE ENERGY SECTOR AND SCALE UP RENEWABLES

The North African region has an enormous potential to diversify its energy mix toward more sustainable forms of energy, as several chapters in this book illustrate. But, while various targets for renewable energy have been set, scaling up projects and initiatives has been challenging. Capitalizing on these early steps to transform the energy sector requires a systemic approach that not only integrates different objectives (e.g., finance, institutions, private sector development) but also embeds these efforts into a larger coherent and long-term vision for "green growth" and economic transformation in the region. The emergence of such a vision is partly impaired by a lack of experience regarding how such transformations occur (in both developed and developing countries), the need to experiment locally with different policy alternatives (i.e., systematic policy learning), and the consequent willingness to reconsider overall policy objectives (Johnson et al., 2014).

For these reasons, "one-dimensional" solutions, such as setting renewable energy targets or implementing FiTs, most likely will not result in large-scale outcomes regarding deployment, and thus will not deliver the much needed socioeconomic benefits for these countries. Thus, a focus on building alliances across diverse interests and improving coordination between different development agendas is needed.

I argue that a main problem in this regard is the significant degree of fragmentation in energy strategies (at national and regional level), both north and south of the Mediterranean. Specifically, north of the Mediterranean divergence across countries in terms of the role that renewable energy plays in the energy system can be easily observed (e.g., between Germany, France, and Spain). Fragmentation can also be observed within nations. Taking Germany as an example, although a high level of commitment has been made to renewables, vested interests from competing energy sources (e.g., coal) and diverging opinions on how energy needs in a postnuclear energy future are to be satisfied and on the degree of integration across energy markets (Lütkenhorst and Pegels, 2014), hinder the development of an integrated and coherent energy strategy. At the same time, a similar type of fragmentation (with respect to misalignment of interests and coordination) can be observed south of the Mediterranean, where energy security is high on the agenda in most countries. A lack of policy and strategy harmonization (within nations and across the North Africa region), weak commitments to a "green growth" agenda, and vested interests in the *status quo*, prevent the emergence of a momentum to scale up important initial steps toward renewable energy deployment.[8]

8. One manifestation of such fragmentation is lost momentum on the UfM's Master Plan for renewable energy and energy efficiency.

While a regional approach to the energy transition is not the only effective solution, lack of it can be problematic for several reasons. As mentioned earlier, interlinkages between policy interventions and strategies are critical. The reality is that no national market in North Africa (and the Mediterranean more generally) is large enough to enable, and to fully capture, the benefits of "scaling up." Alignment of regulatory frameworks (i.e., to phase out fossil fuels and phase in renewables) within and across countries would result in the creation of a greater market that could send the right signals to the financial sector and to private investors to orient resources toward renewable energy projects. Further, common "green growth" agendas could result in exploiting the comparative advantage in industrial/manufacturing capabilities across the region, which could further contribute to scaling up. Specifically, specialization in the production of parts and components for renewable energy technologies can result in higher socioeconomic benefits (due to scale effects) in terms of jobs and know-how, as opposed to each country making similar industrial investments.

Enabling these interlinkages between policy interventions and achieving harmonization of energy strategies is, however, the "Achille's heel" of energy transition in the Mediterranean region. To overcome tendencies to lock in and to follow one-sided interests (at national or actor level), a deeper comprehension of the political economy (e.g., constellation of interests, drivers for decision-making) is needed. Systematic learning should be at the core of policy-making with a view to capturing long-term benefits, as well as an effort to coordinate across development agendas and foster alliances between stakeholders with diverse interests in the energy sector and beyond. Ultimately, a new narrative that discredits the "old" energy regime and sheds light on the cobenefits from pursuing an integrated (and inclusive) approach to green growth and energy sector transformation is needed.

REFERENCES

Ayre, J., 2015. Algeria doubling renewable energy target, now 25 GW by 2030. CleanTechnica. http://cleantechnica.com/2015/01/30/algeria-doubling-renewable-energy-target-now-25-gw-2030/ (accessed 01.07.2015).

Carafa, L., Frisari, G., Vidican, G., 2015. Electricity transition in the MENA: a de-risking governance approach. J. Cleaner Prod. doi:10.1016/j.jclepro.2015.07.012.

CDER, 2011. Renewable energy and energy efficiency Algerian program: Centre de Développement des Energies Renouvelables (CDER).

EgyptERA, 2014. Energy pricing and renewable energy deployment: Egypt case study final report. Report prepared for the Regional Center for Renewable Energy and Energy Efficiency (RCREEE) and the International Renewable Energy Agency (IRENA), Egyptian Electricity Utility and Consumer Protection and Regulatory Agency (EgyptERA), Cairo, Egypt.

EIPR, 2015. The Egyptian economic development conference: is it a step forward or pursuit of policies of the past. Egyptian Initiative for Personal Rights (EIPR), Economic & Social Justice.

Fattouh, B., El-Katiri, L., 2012. Energy Subsidies in the Arab World. United Nations Development Programme, Regional Bureau of Arab States, New York.

GIZ, 2012. Renewable energy and energy efficiency in Tunisia – employment, qualification and economic effects. Deutsche Gesellschaft für Internationale Zusammenarbeit (GIZ).

IEA, 2012. World Energy Outlook 2012. Cedeux, International Energy Agency, Paris.

IMF, 2014. Energy subsidies in the Middle East and North Africa: lessons from reform. International Monetary Fund (IMF).

James, L.M., 2015. Recent Developments in Egypt's Fuel Subsidy Reform Process: Research Report. International Institute for Sustainable Development (IISD)/Global Subsidies Initiative (GSI), Geneva.

Johnson, O., Altenburg, T., Schmitz, H., 2014. Rent management capabilities for the green transformation. In: Pegels, A. (Ed.), Green Industrial Policy in Emerging Countries. Routledge, New York, USA and Abingdon, UK, pp. 9–37.

Lütkenhorst, W., Altenburg, T., Pegels, A., Vidican, G., 2014. Green industrial policy: managing transformation under uncertainty. Discussion Paper 28/2014, German Development Institute/Deutsches Institut für Entwicklungspolitik, Bonn.

Lütkenhorst, W., Pegels, A., 2014. Stable Policies, Turbulent Markets – Germany's Green Industrial Policy: The Costs and Benefits of Promoting Solar PV and Wind Energy. International Institute for Sustainable Development/Global Subsidies Initiative (IISD/GSI), Geneva.

MEMEE, 2011. Moroccan Energy Strategy, Overview – *Stratégie Énergétique, Bilan d'Étape*. Ministry of Energy, Mines, Water and Environment, Rabat.

MESP, 2015. Egypt: billions pledged for the energy sector at the Economic Development Conference.http://www.mesp.me/2015/03/23/egypt-billions-pledged-for-the-energy-sector-at-the-economic-development-conference/ (accessed 23.03.2015).

OECD, 2013. Renewable Energies in the Middle East and North Africa: Policies to Support Private Investment. Organisation for Economic Cooperation and Development (OECD), Paris.

REN21, 2013. MENA Renewables Status Report. Renewable Energy Policy Network for the 21st Century in collaboration with the UEA Directorate of Energy & Climate Change and International Renewable Energy Agency, Paris.

UfM, 2013. The Mediterranean Solar Plan: A Flagship Initiative for a Sustainable Mediterranean Region. Union for the Mediterranean (UfM), Barcelona.

UNEP, 2011. Towards a green economy: pathways to sustainable development and poverty eradication – A synthesis for policy makers. United National Environment Programme (UNEP).

UNEP, 2015. Solar energy in Tunisia (vol. green economy: success stories). United Nations Environment Programme (UNEP).

Vidican, G., 2012. Building Domestic Capabilities in Renewable Energy: A Case Study of Egypt. German Development Institute/Deutsches Institut für Entwicklungspolitik (DIE), Bonn.

Vidican, G., 2014. Reforming fossil-fuel subsidy regimes in the Middle East and North African countries. In: Pegels, A. (Ed.), Green Industrial Policy in Emerging Countries. Routledge, New York, USA and Abingdon, UK, pp. 148–178.

Vidican, G., 2015a. 2014 Article IV Consultations for Egypt: Input Paper for the BMZ 210 Division. German Development Institute/Deutsches Institut für Entwicklungspolitik, Bonn.

Vidican, G., 2015b. The emergence of a solar energy innovation system in Morocco: a governance perspective. Innovation and Development. 5(2), 225–240.

Vidican, G., Böhning, M., Burger, G., Regueira, E. de Siqueira, Müller, S., Wendt, S., 2013. Achieving inclusive competitiveness in the emerging solar energy sector in Morocco. Studies 79, Deutsches Institut für Entwicklungspolitik/German Development Institute (DIE).

Williamson, R., 2015. Sun shines on Egypt's solar startups. Wamda. http://www.wamda.com/memakersge/2015/04/sun-shines-on-egypt-solar-startups (accessed 30.04.2015).

World Bank, 2012. Inclusive Growth.http://web.worldbank.org/WBSITE/EXTERNAL/TOPICS/EXTDEBTDEPT/0,contentMDK:21870580~pagePK:64166689~piPK:64166646~theSitePK:469043,00.html (accessed 06.06.2013).

World Bank, 2014. Jobs or privileges: unleashing the employment potential of the Middle East and North Africa. Report No. 88879-MNA,The World Bank Group.

Chapter 5

The Renewable Energy Targets of the MENA Countries: Objectives, Achievability, and Relevance for the Mediterranean Energy Collaboration

Bernhard Brand
Research Group 1 "Future Energy and Mobility Structures" Wuppertal Institute for Climate, Environment and Energy, Wuppertal, Germany

1 INTRODUCTION

Establishing renewable energy targets has become an integral part of national energy policies in many countries throughout the world. What started as an initiative by some Organisation for Economic Co-operation and Development (OECD) countries in the early 2000s, has now reached a global dimension, with 164 countries having at least one type of renewable energy target on their policy agenda (IRENA, 2015). The Middle East and North African (MENA) countries are among those that are following this trend. Today, almost all countries in the region,[1] have announced targets for the deployment of renewable technologies in their future energy systems. Putting clarity into energy visions is a first, important step toward the development of concrete policies, roadmaps, and support schemes for renewable energy. Very often, renewable energy targets are intended to attract the financial support of investors, funding organizations, or other international stakeholders. Many MENA governments also regard the promulgation of renewable energy targets as a matter of national prestige due to the positive and "modern" image that is usually attributed to solar and wind technologies. But announcing renewable energy targets based solely on such aspirations can also create some risks. If the targets turn out to be too ambitious, too incoherent, or are not backed by appropriate policy actions, there is a risk that potential stakeholders will ignore the renewable energy policies, and they can lose part of their credibility or even completely fail. Therefore, a critical appraisal of the

1. Only Oman has not yet announced renewable energy targets.

Regulation and Investments in Energy Markets. http://dx.doi.org/10.1016/B978-0-12-804436-0.00005-9

targets is highly desirable, especially in the context of globalized energy markets, in which strategic decisions made by one country can influence the supply/demand situation in other countries. The Mediterranean basin is a region where traditionally there have been strong energy ties between EU and MENA countries. Currently, these ties are concerned mostly with the oil and natural gas supply, but in the future an increased collaboration is also foreseen for the electricity sector in the Mediterranean region (Medreg, 2015). How do we have to interpret the recent renewable energy targets of the MENA countries, in light of the aspirations of an overall greater convergence of energy policies, and potentially also more interconnected energy infrastructures in the Mediterranean region? This chapter scrutinizes the renewable energy goals of four Southern Mediterranean MENA countries: Morocco, Algeria, Tunisia, and Egypt. These countries are currently deemed to be the best choices for such an assessment because due to the political and security uncertainties in the region, they are among the few states on the southern shore of the Mediterranean, with realistic perspectives to pursue sovereign and long-range energy policies. The assessment begins with a brief overview of the key features of the renewable targets, followed by a qualitative analysis. The subsequent sections provide a discussion of the transnational, Mediterranean context of the renewable targets and a final conclusion.

2 BACKGROUND

The most widespread types of renewable energy targets worldwide are electricity targets. Also, the four MENA countries chosen for this analysis have focused primarily on the electricity sector[2] (Table 5.1). Algeria, Tunisia, and Egypt have formulated their overarching goals as a percentage of the total electricity generation (GWh of total generation), while Morocco has expressed its targets in terms of installed capacity (in MW). Some countries have chosen catchy formulations for their targets, that is, "20% by 2020" (Egypt), "30% by 2030" (Tunisia), and "2 GW – 2 GW – 2 GW by 2020" for installed wind, solar, and hydroelectric capacities (Morocco). Of course, announcing targets in such a manner is also practiced in other regions of the world, for instance in the EU's 2020 climate package known as the "20–20–20" targets.[3] In the case of the four MENA countries, the targets are also broken down into more precise and specific targets for the different renewable technologies, that is, solar, wind, and hydroelectricity, with Tunisia and Algeria also including biomass (Table 5.1).

Renewable energy targets are provided in the official master plan documents or roadmaps that are usually published on the Internet sites of energy ministries or government agencies. The publication of these targets often follows a

2. Some indicative solar water-heating goals are in place in Morocco (1.7 million m^2 of collector area by 2020), Algeria (490,000 m^2 by 2020), and Tunisia (1.0 million m^2 by 2016). Source: REN21 (2013).

3. 20% climate gas reductions, 20% increase in energy efficiency, and 20% from renewable energy by 2020 (EU, 2009).

TABLE 5.1 Key Features of the Renewable Energy Targets of Algeria, Egypt, Morocco, and Tunisia

Countries	Targets
Algeria	Overall target: 2030, 27% of electricity generation, 22 GW of installed capacity. Technology-specific targets (installed capacity): Solar CSP, 2000 MW by 2020, 2000 MW by 2030; Solar PV, 3000 MW by 2020, 13,575 MW by 2030; Wind, 1010 MW by 2020, 5010 MW by 2030; and Biomass, 360 MW by 2020, 2000 MW by 2030.
Egypt	Overall target: 2020, 20% of electricity generation; 2030, N/A. Technology-specific target (installed capacity): Solar CSP, 1100 MW by 2020; 2800 MW by 2027; Solar PV, 220 MW by 2020, 700 MW by 2027; Wind, 7200 MW by 2020; and Hydro, 2800 MW by 2020.
Morocco	Overall target: 2020, 42% of electricity installed capacity; 2030, N/A. Technology-specific target (installed capacity): Solar (CSP + PV), 2000 MW by 2020; Wind, 2000 MW by 2020; and Hydro, 2000 MW by 2020.
Tunisia	Overall target: 2020, N/A; 2030, 30% of electricity generation. Technology-specific target (installed capacity): Solar CSP, 500 MW by 2030; Solar PV, 140 MW by 2016, 1500 MW by 2030; Wind, 430 MW by 2016, 1700 MW by 2030; and Biomass, 40 MW by 2016, 300 MW by 2030.

Source: IRENA/LAS, 2014. CDER, 2015.

prominent/official announcement of the targets in a minister's speech or in a press release. Morocco and Tunisia began their renewable programs, called "Solar Plans," in 2009; Algeria followed in 2011 with a "Renewable Energy Development Plan,"[4] and the "Egyptian Solar Energy Plan" was promulgated in 2012 (IEA/IRENA, 2015). Following the widespread fashion of centralized top-down energy planning in most MENA countries, many national targets have essentially the character of project-based, renewable capacity expansion programs, that is, predefined renewable power plants at chosen sites with their own, specific capacities. Sometimes these projects even embrace large, national prestige projects, such as the Moroccan concentrated solar power (CSP) plant Noor 1 in Ouarzazate, which is slated to become Africa's largest power plant of this type, or Algeria's attempt to set up a national photovoltaic (PV) manufacturing

4. The initial plan (CDER, 2011) was amended in 2014 (CDER, 2015).

industry to supply PV modules for the projects. Little is known however, about the actual planning rationale behind the renewable targets. It appears that the targets were mainly decided by insiders from ministries, planning authorities, and electricity utilities, perhaps with some support from external consultants. Hardly any scientific publications or studies have been made available regarding the rationale for choosing the renewable energy targets, for example, whether they were chosen on the basis of resource assessments, electricity system modeling, or cost–benefit analyses. Moreover, none of the countries conducted a real public consultation process about the renewable roadmaps, and there was no noteworthy debate in the public media prior to the announcement of the goals.

3 ANALYSIS

It would have been highly desirable to conduct thorough public and scientific deliberations on the targets. The announced roadmaps are multibillion-euro proposals with significant macroeconomic impacts as well as environmental and social impacts in the affected MENA countries. In many cases, the proposed goals also require changes in legal and regulatory frameworks, and they involve issues related to grid integration and the security of energy (electricity) supply. Assessing these impacts should usually be based on modeling studies,[5] and the assessments require the input of independent experts, research institutes, and a transparent public debate, for instance, in the form of a "green paper/white paper" process to support the development of public policies. With the exception of Tunisia, where some sparse discussions on renewable targets took place (ANME, 2013a,b; WB-ESMAP, 2014), a stringent science-backed debate on (renewable) energy strategies has not been encouraged or performed in most MENA countries. Thus, there should be a robust and critical appraisal of the renewable energy targets, which have been established.

The analysis in this section is primarily qualitative, in which the following dimensions of the renewable energy targets were explored, that is, the underlying motivation behind the targets, achievability of these targets within specified timeframes, the extent to which these individual countries are committed to or bound by the goals, and the relevance of goals in the overall context of the countries' energy systems.

3.1 Motivation

Certainly, there is no single objective that has motivated MENA countries to announce renewable energy targets; rather there is a multitude of objectives. In addition to the previously mentioned strategic goals, that is, national prestige and raising the awareness of institutional investors, practical considerations also have had an important role. For countries that depend on importing fossil

5. The decision-making process of the European Commission concerning the development and implemenatation of future energy strategies was supported by the PRIMES energy model of the National Technical University of Athens, Greece.

energy, such as Morocco, or countries with dwindling fossil resources, such as Tunisia and Egypt, one key motivation for turning to renewable energy sources is their desire to decrease energy dependency by diversifying the energy mix and to substitute imported fossil fuels. Algeria, as the only fossil fuel exporter among the four countries, has a strategic incentive of saving its domestic fossil fuel reserves for future exports.

Irrespective of the motivations of individual MENA countries, one must ask whether their specific plans for expanding the production of renewable energy, as outlined in Table 5.1, are really the best way to develop their renewable energy capabilities, or whether there are other pathways that would be more economically efficient. Brand and Zingerle (2011) explored the national renewable energy targets of the Maghreb countries Morocco, Algeria, and Tunisia, with an electricity model, and they determined that the proposed pathways for expanding production capacity were not cost optimized. Theoretically, the targets could be improved by adapting them better to the available solar and wind resources, and by selecting a more cost-efficient technology mix, notably with regards to the chosen proportions of CSP, wind, and PV technologies. Morocco, for instance has become inclined to prioritize CSP technology (in spite of its higher investment costs) because it promises significant potential for local manufacturing and the creation of jobs (Fraunhofer and Ernst & Young, 2011). Generally many MENA countries intend to implement local content clauses in public procurement procedures to support national manufacturers of solar and wind components. As mentioned previously, even Algeria supports an industrial policy in order to promote the manufacturing of PV components in the country. Therefore, the generation of domestic value and promotion of local industries can be regarded as significant motivators in setting targets for the production of energy from renewable sources. How about environmental issues and climate change, which are an intrinsic driver for renewable energy policies in Europe? In the MENA region, there is very little evidence to suggest that these factors were seriously considered in setting up the existing renewable energy targets. A study of stakeholders' preferences in energy decision-making in Tunisia (Brand and Missaoui, 2014) indicated that environmental/climate considerations received relatively little attention in comparison to economic considerations, energy security, and generation of industrial value. Therefore, at this time, it is doubtful that climate protection is actually a true motivation for the MENA decision makers to support renewable energy goals.

3.2 Commitment

Even though Algeria, Egypt, Morocco, and Tunisia are signatories of the Kyoto Protocol, their international responsibility to reduce the emissions of greenhouse gases is not bound to any mandatory target. Hence, in the absence of external pressure, their renewable energy goals must be considered to be voluntary from a practical perspective. It also has to be noted that on an internal and national level, the commitment to establish goals is rather weak, given that MENA governments can hardly be held accountable if they do not achieve their own targets. Thus, in practice the targets are often revised or even withdrawn. For example, Algeria published a national renewable energy program in 2011, but only 3 years later, the program underwent an amendment that introduced significant changes in the overall targets and in proportions of different renewable technologies in the energy mix (CDER, 2015).[6] In Tunisia, deliberations are still going on, concerning the definition of the capacity targets to support 30% renewable penetration goal by 2030 (WB-ESMAP, 2014). There is no structured process of regular review and monitoring of the renewable energy achievements in any of the MENA countries, and no enforcement or compliance mechanisms are in place, such as sanctions if the targets are not met.

3.3 Achievability

Figure 5.1 shows that the achievements of these four countries are currently lagging far behind what was specified in the roadmaps. The voluntary and aspirational nature of renewable energy targets and their arbitrary amendments, as

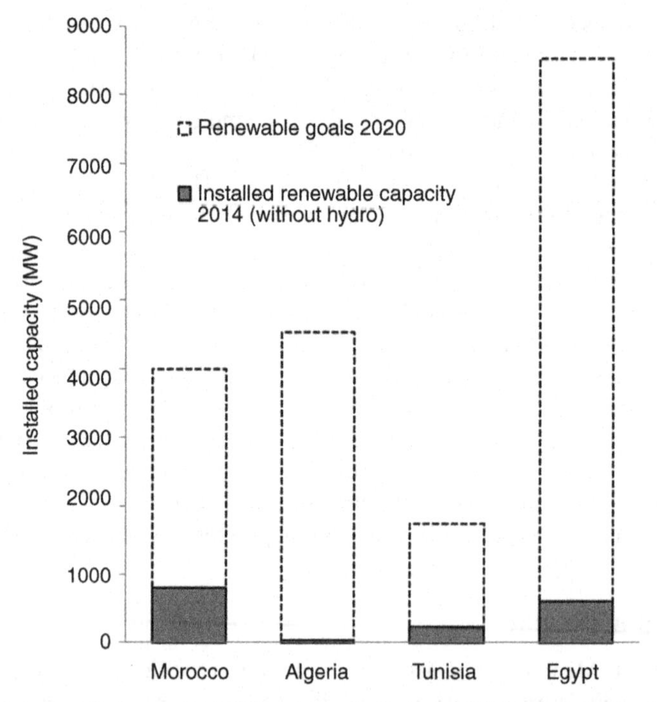

FIGURE 5.1 Installed capacity (2014) and targeted capacity (2020) of wind and solar power plants (in MW). *(Source: Target announcements, annual reports of national utilities.)*

6. The most noticeable change is a shift in the focus from CSP toward PV technology, probably due to the dramatic price drops of PV technology in recent years.

mentioned in the previous section, are certainly not conducive for an accelerated expansion of solar and wind technologies in the region. In addition, there are other reasons for delays in implementing the renewable energy programs in the MENA region. One reason is that most of the renewable energy projects[7] have so far been carried out by national agencies or state-owned electricity utilities,[8] which have received the support of international donors, development agencies, and finance institutions, such as the World Bank or the African Development Bank. This project-by-project approach, which usually involves an extensive phase that includes acquiring the funds, conducting negotiations with international partners, and, finally, a competitive procurement, was a good start for the initial renewable power projects, but it has also proven to be quite cumbersome.

It remains unlikely that with an approach, in which state entities act as project developers, the entire rollout of the renewable expansion programs can be realized in a timely manner. To achieve the targets more quickly and efficiently, other project development mechanisms must be applied, including the involvement of the private sector. Currently, Egypt is exploring a power purchase agreement that would allow private companies to apply for generation licenses under fixed feed-in tariffs (FiTs) for wind and solar power plants (Egyptera, 2014). Algeria also intends to attract private investments and has recently enacted an FiT scheme to boost solar and wind power projects (CDER, 2014a). However, the inclusion of private sector in the rollout of renewable energy programs also has limitations. Most MENA electricity sectors are controlled by very dominant national electricity utilities that often have vested interests in the oil and gas sectors. Thus, the implementation of market-based instruments, especially for renewables often faces resistance from monopolists. In Tunisia, for example, the proposal of a renewable energy law – foreseeing an increased participation of the private sector – has met strong opposition from the national electricity utility (L'Economiste Maghrebin, 2014). Monopolized electricity market environments certainly are one key impediment to achieving renewable energy targets, and the extent to which that impediment can be overcome depends on how the MENA countries proceed in liberalizing their electricity sectors and set up new regulatory models for renewable energies.

3.4 Relevance

A key question concerns the relevancy of renewable energy targets to overall energy system. Are the targets really a "system changer" that points to a broader transition from a fossil fuel dominated system to a renewable-based energy system in the MENA countries? With regards to Algeria, Egypt,

---

7. A bulk of so far installed capacities are wind power projects.

8. State utilities and national development agencies: Morocco, ONE (Office National d'Electricité) and MASEN (Moroccan Solar Energy Agency); Algeria, Sonelgaz (Sociéténationale de l'électricité et du gaz), NEAL (New Energy Algeria); Tunisia, STEG (Sociététunisienne de l'électricité et du gaz); Egypt, NREA (New and Renewable Energy Agency).

Morocco, and Tunisia, it is unlikely that the renewable targets are regarded in that way. All four countries still pursue very vigorous expansion strategies for conventional technologies, irrespective of their assertions that they support the renewable goals. In addition to its 22-GW renewable goal, Algeria also has announced a conventional capacity target, with almost 28 GW of new gas-fired power plants to be installed between now and 2025 (Sonelgaz, 2015). Morocco is about to significantly expand its coal power share in the electricity mix, with three new coal-fired power plants under construction, with a total capacity of 2.4 GW to be commissioned by 2018 (ONE, 2015). Likewise, Tunisia is considering coal power as a diversification option, while Egypt, which also has embarked on an ambitious expansion of its natural gas power plant program, is even considering nuclear energy for its future electricity system (Jewell, 2011). It must be understood that energy demand is increasing tremendously in all MENA countries, and many decision makers still take a very conservative stance toward energy planning. The national electric utilities appear to have little trust in sources of renewable energy, and their representatives often argue that intermittent wind and solar technologies cannot provide a sufficiently secure energy supply to safeguard the electricity system against power shortages or blackouts. In Europe, where electricity demand is stagnating, it is commonly accepted that renewables – in conjunction with intelligent grid integration measures – can gradually be substituted for conventional power plants. However, the North African planning paradigm perceives wind and solar technologies primarily as "add-ons" to the conventional power system rather than serious alternatives for permanent changes in the energy system. The widespread subsidies for fossil fuels that still prevail in the MENA region's energy sectors (IMF, 2014) underscore the low relevance that decision makers and planners attribute to renewable energy sources. Although renewable generation is becoming cost competitive in many electricity markets worldwide, the MENA countries still continue to privilege conventional power generation by enabling national utilities to procure fossil fuels at subsidized prices. According to El-Katiri (2014), this policy strongly "harms the cause of renewables," because it reinforces "the message that renewables are an expensive technology built only due to political will." For a serious and consistent commitment to renewable energies, MENA policy makers would have to support their renewable target aspirations with visible actions for the removal of subsidies for fossil fuels.

4 TRANSNATIONAL PERSPECTIVE

A closer look at the renewable targets of the four countries revealed another shortcoming: each national roadmap primarily is an isolated initiative that neither adequately considers the national roadmaps of the other (neighboring) MENA countries, nor the possibility of joint coordination for better use of renewable resources in the wider trans-Mediterranean context, that is, with

Europe. Generally, it is well known that interconnected electricity systems can have an enormously positive effect on facilitating the integration of renewable energy sources because they bridge the regional heterogeneities of resource availability and allow better balancing of intermittent power generation across larger geographical areas. Brand (2013) pointed the high economic benefits of the joint use of renewable resources through electricity interconnectors among North African countries. Other studies, such as Dii (2013), have shown similar advantages on an even larger scale, considering interconnected electricity systems and integration of electricity from renewable sources across the entire Euro-Mediterranean region. Why have such opportunities largely been ignored[9] in the target announcements of the four MENA countries? Perhaps the answer is that multilateral activities in general are difficult to initiate and sustain when it comes to electricity system planning in the MENA region. One major reason for this is that energy is viewed by many MENA governments, as a sensitive area that touches on national sovereignty. An increased interdependency with other countries' power systems or even a shared utilization scheme of (renewable) energy resources, is often perceived as a threat to the security of energy supply. Second, to date Euro-Mediterranean collaboration schemes on renewable energy have failed to produce positive results. None of the initiatives and programs announced over the past few years has succeeded in producing positive outcomes. One of the earliest attempts to implement such a renewable power collaboration scheme was the European Directive 2009/28/EC (Article 9), which was meant to support joint renewable electricity export projects between MENA countries and Europe by allowing EU member states to include these imports in their national renewable energy accounting. But this scheme was never implemented by any EU country.[10] Also the process of the Mediterranean Solar Plan (MSP), an initiative by the Union for the Mediterranean (UfM) to support a common framework of pan-Mediterranean power markets, has been lying idle since late 2013, after it met with strong opposition from southern European UfM member countries (Ansamed, 2013). In addition, the idea of renewable energy collaboration across the Mediterranean received a setback when the voice of its most prominent private supporter, the Desertec Industrial Initiative (Dii), fell silent.[11]

Given the circumstances described previously, it is no surprise that the renewable energy targets of the MENA countries have remained isolated initiatives that, for the most part, do not consider opportunities for transnational collaboration. But, are there ways to overcome this fragmentation? A fresh impetus might come from a new initiative of the League of Arab States (LAS) to set up

9. In its initial Renewable Energy Program of 2011, Algeria had foreseen 10 GW of renewable energy for exports to Europe (CDER, 2011). However, this option was not included in the revised renewable energy targets of 2015 (CDER, 2015).

10. The key barrier to the application of Article 9 of the EU Directive was its requirement of actual and physical electricity transfers to be carried out between third countries and EU member states.

11. After its key shareholders left, the Dii consortium dissolved in 2014.

a "Pan-Arab Renewable Energy Strategy" (IRENA/LAS, 2014). The related strategy paper, published with support of the International Renewable Energy Agency (IRENA) and the Cairo-based Regional Center for Renewable Energy and Energy Efficiency (RCREEE), advocates a common framework for the use of renewable energy among the 22 members of the LAS. Although the document is not yet a fully fledged roadmap for renewable energies in the region, it provides a common guideline for member states to formulate clear and structured National Renewable Energy Action Plans (NREAPs). The first NREAPS were intended to be published in 2016, after which they will be subjected to a regular monitoring process. It remains to be seen whether the authority of transnational organizations (like the LAS) can bring more coherency in the formulation of renewable energy targets in the MENA region. A good example of a similar process, in which several national energy policies have been put under the umbrella of a transnational organization, is the Economic Union of West African States (ECOWAS). In 2012, 15 ECOWAS member states published a Renewable Energy Policy document (ECREEE, 2012) that aggregated different national policies and targets into one jointly shared strategy for renewable energy in the West African region until 2030. Although it is too early to judge the outcome of the ECOWAS initiative, it must be acknowledged that at least it has been successful in terms of establishing a consensus of opinions. Certainly, the key catalyst that led to an agreement concerning a common renewable energy framework was that the West African countries were able to identify a set of shared objectives and visions for their energy future, that is, the reduction of energy poverty, universal access to electricity, and development of sources of clean energy for rural areas.

Clearly, the fate of intended Pan-Arab Renewable Energy Strategy will depend on whether such shared objectives can be formulated for the members of the LAS. Finding such consensus for the entire Mediterranean region – a mosaic of culturally and economically very disparate countries – undoubtedly will be an even more ambitious task. If we think in terms of greater convergence of (renewable) energy policies across the Mediterranean, the different motivations and expectations in these countries, as well as the different styles of political decision-making must be understood. For instance, the European narrative for renewable energies is dominated by the will to combat climate change, while southern Mediterranean countries – as the analysis of the targets in the previous section showed – are more concerned by the questions whether renewable energies can provide a secure supply at reasonable costs, and possibly even positive effects for socioeconomic welfare. Another structural divide is the development of energy demand. In the MENA region, the demand for electricity is increasing at very rapid rates with supply shortages in the horizon, while many EU countries have power generation overcapacities. In this situation, proposing exports of renewable electricity from the MENA countries to the EU appears to be paradoxical. The aforementioned examples should show that it is of paramount importance that before embarking on ambitious trans-Mediterranean renewable energy targets and master plans, a better mutual understanding must be developed concerning market structures, stakeholders, and even societal aspirations toward renewable energy.

5 CONCLUSIONS

Renewable energy targets have become a common feature of energy strategies in many countries in the MENA region. The present analysis, which focused on Algeria, Egypt, Morocco, and Tunisia, indicated that the targets have a rather normative, aspirational character, and it can be assumed that their key rationale was primarily to increase public attention, supposedly by sending signals to investors or development agencies to support the rollout of renewable capacity expansion plans. Since the targets are not binding, their nonachievement (which is very likely, at least for the short run until 2020) cannot be sanctioned. An additional problem is that the targets are not backed by strong energy sector reform measures. For instance, subsidies for fossil fuels are still in place, and most countries continue their conventional power plant expansion policies, even at accelerated paces. It is well known that increasing investments into fossil fuel technologies could, due to the long lifecycles of power plants, lock the MENA energy systems into a fossil fuel path, and thereby impose further barriers to renewable energy expansion. With regards to their international dimension, it must be mentioned that the MENA targets largely disregard the possible synergies of a joint, coordinated, renewable strategy – neither within the MENA region, nor in the context of Trans-Mediterranean collaboration with Europe. It can be argued that the design of targets mirrors, to a certain extent, the fragmented status of renewable energy policies in the Mediterranean region, where most initiatives, such as the MSP or Desertec, are lying idle. In order to overcome this impasse, and also for the sake of improved, more economically efficient national targets for renewable energy, a new dialogue about a viable renewable energy model in the Mediterranean region is a pressing need.

REFERENCES

ANME, 2013a. Nouvelle version du Plan Solaire Tunisien. Programmation, conditions et moyens de la mise en oeuvre. Agence Nationale pour la Maîtrise de l'Energie (ANME), Tunis.

ANME, 2013b. Stratégie nationale du mix énergétique pour la production électrique aux horizons 2020 et 2030. Agence Nationale pour la Maîtrise de l'Energie (ANME), Tunis.

Ansamed, 2013. Embattled UfM ministerial meeting focuses on energy. Rome, www.ansa.it/ansamed/en/news/sections/energy/2013/12/11/Embattled-UfM-ministerial-meeting-focuses-energy_9765069.html (accessed 11.12.2013.).

Brand, B., 2013. Transmission topologies for the integration of renewable power into the electricity systems of North Africa. Energy Policy 60, 155–166.

Brand, B., Missaoui, R., 2014. Multi-criteria analysis of electricity generation mix scenarios in Tunisia. Renew. Sust. Energ. Rev. 39, 251–261.

Brand, B., Zingerle, J., 2011. The renewable energy targets of the Maghreb countries: impact on electricity supply and conventional power markets. Energy Policy 39 (8), 4411–4419.

CDER, 2011. Renewable Energy and Energy Efficiency Program. Centre de développement des Energies Renouvelables, CDER, Alger. Available from: http://portail.cder.dz/IMG/pdf/Renewable_Energy_and_Energy_Efficiency_Algerian_Program_EN.pdf

CDER, 2014. Photovoltaïque: Les tarifs d'achat garantis en Algérie. Centre de développement des Energies Renouvelables, CDER, Alger. Available from: http://portail.cder.dz/spip.php?article3990

CDER, 2015. Objectifs nouveau programme des Energies Renouvelables en Algérie (2015–2020–2030). Centre de développement des Energies Renouvelables, CDER, Alger. Available from: http://portail.cder.dz/spip.php?article4565

Dii, 2013. Desert Power: Getting Started. Dii GmbH, Munich. Available from: http://www.desertenergy.org/publications/getting-started.html

Directive 2009/28/EC of the European Parliament and of the Council of 23 April 2009 on the promotion of the use of energy from renewable sources and amending and subsequently repealing Directives 2001/77/EC and 2003/30/EC. Official Journal of the European Union.

ECREEE, 2012. ECOWAS Renewable Energy Policy (EREP). ECOWAS Centre for Renewable Energy and Energy Efficiency (ECREEE), Praia, Cabo Verde. Available from: http://www.ecreee.org/sites/default/files/documents/basic_page/151012_ecowas_renewable_energy_policy_final.pdf

Egyptera, 2014. Renewable Energy – Feed-in Tariff Projects' Regulations. Egyptian Electric Utility and Consumer Protection Regulatory Agency (EgyptERA). October 2014. Available from: http://egyptera.org/Downloads/taka%20gdida/Download%20Renewable%20Energy%20Feed-in%20Tariff%20Regulations.pdf

El-Katiri, L., 2014. A Roadmap for Renewable Energy in the Middle East and North Africa. The Oxford Institute for Energy Studies, Oxford.

IEA/IRENA, 2015. Global Renewable Energy. IEA/IRENA Joint Policy Measures Database, Paris. Available from: http://www.iea.org/policiesandmeasures/renewableenergy/

IMF, 2014. Energy Subsidies in the Middle East and North Africa: Lessons for Reform. International Monetary Fund, Washington. Available from: https://www.imf.org/external/np/fad/subsidies/pdf/menanote.pdf

IRENA, 2015. Renewable Energy Target Setting. International Renewable Energy Agency, Abu Dhabi. Available from: http://www.irena.org/DocumentDownloads/Publications/IRENA_RE_Target_Setting_2015.pdf

IRENA/LAS, 2014. Pan-Arab Renewable Energy Strategy 2030: Roadmaps of Actions for Implementation. International Renewable Energy Agency (IRENA), Abu Dhabi. Available from: http://www.irena.org/DocumentDownloads/Publications/IRENA_Pan-Arab_Strategy_June%202014.pdf

Jewell, J., 2011. A nuclear-powered North Africa: just a desert mirage or is there something on the horizon? Energy Policy 39 (8), 4445–4457, doi:10.1016/j.enpol.2010.09.042.

L'Economiste Maghrebin, 2014. Tunis – Des cadres de la STEG dénoncent le projet de loi des énergies renouvelables. L'Economiste Maghrebin, Tunis. Available from: http://www.leconomistemaghrebin.com/2014/03/08/tunis-cadres-steg-denoncent-loi-energies/

Medreg, 2015. Interconnection Infrastructures in the Mediterranean. A challenging environment for investments. Report N MED15-19GA-4.6B, Mediterranean Energy Regulators, Milan. Available from: http://www.medreg-regulators.org/Portals/45/questionnaire/INV_report_final.pdf

ONE, 2015. L'Office renforce son parc de production. Office National d'Electricité, Casablanca. Available from: http://www.one.org.ma/FR/pages/interne.asp?esp=2&id1=5&t1=1

REN21, 2013. MENA Renewable Status Report 2013. REN21 Secretariat, Paris, France.

Sonelgaz, 2015. Programme de developpement – production d'électricité. Société nationale de l'électricité et du gaz, Alger. Available from: http://www.sonelgaz.dz/?page=article&id=13

WB-ESMAP, 2014. Une vision stratégique pour le secteur tunisien de l'énergie. Réflexion sur des thèmes prioritaires. World Bank Energy Management Assistance Program, Washington.

Chapter 6

Toward a New Euro-Mediterranean Energy Roadmap: Setting the Key Milestones

Manfred Hafner, Simone Tagliapietra
Fondazione Eni Enrico Mattei, Milan, Italy

1 INTRODUCTION: ENERGY AS A KEY PREREQUISITE FOR SUSTAINABLE REGIONAL DEVELOPMENT

When it erupted in 2011, the so-called "Arab Spring" unveiled the unsustainable economic and social situation of most southern and eastern Mediterranean countries (SEMCs).[1] Since the very beginning of the uprisings it was clear that only a new model of economic growth and social development would have represented a solid and sustainable solution to the various problems underpinning the revolts.

However, years after the explosion of uprisings those problems continue to remain largely unresolved. In fact, SEMCs continue to face a range of pressing socioeconomic challenges, including solving problems of poverty and high levels of structural unemployment in the context of fast demographic growth. Considering its macroeconomic and energy fundamentals, the region has considerable potential for triggering a new development process, and the persistent need to find sustainable solutions for regional, economic, and social stability could enhance such a new dynamic.

Energy is an essential commodity enabling socioeconomic development. The energy situation in SEMCs continues to be characterized by a rapid increase in energy demand, low energy efficiency, and low domestic energy prices due to the extensive deployment of universal consumption subsidies.

The patterns of energy supply and consumption in SEMCs greatly affect the main macroeconomic parameters of the countries, including fiscal balances and poverty trends. The volatility of global prices for energy commodities and

1. More specifically, SEMCs in this context refers to 11 countries namely: Algeria, Egypt, Israel, Jordan, Lebanon, Libya, Morocco, the Palestinian territories, Syria, Tunisia, and Turkey.

Regulation and Investments in Energy Markets. http://dx.doi.org/10.1016/B978-0-12-804436-0.00006-0

their relatively high levels in recent years constitutes a burden on the finances of many net-importing countries (NICs) and their customers. The volatility also negatively affects energy-exporting countries. Furthermore, rapid population growth, urbanization, and a low degree of energy efficiency add to rising energy demands, putting pressure on existing infrastructure and necessitating new large investments in electricity, oil, and gas.

In short, the current energy situation of SEMCs does not appear sustainable and poses several risks to the prospects of socioeconomic development in the region. This chapter intends to shed light on these crucial issues, first of all by providing an overview of the regional energy landscape, and then – on this basis – by setting the six fundamental milestones of a new Euro-Mediterranean Energy Roadmap aimed at enhancing regional cooperation in the field to achieve a truly sustainable development path.

2 THE EURO-MED ENERGY LANDSCAPE: AN OVERVIEW

2.1 Energy Consumption and Efficiency

The primary energy demand in SEMCs – calculated by the International Energy Agency (IEA) as the total primary energy supply (TPES)[2] – has progressively increased over the last few decades, most notably after the early 2000s. In absolute numbers, the primary energy demand in SEMCs grew from 89 million tons oil equivalent (Mtoe) in 1980, 157 Mtoe in 1990, 221 Mtoe in 2000, and 322 Mtoe in 2010.

Parallel to the growing primary energy demand, over the last few decades the total final energy consumption (TFC) of SEMCs also experienced a considerable growth. In fact, in absolute terms TFC in SEMCs grew from 66 Mtoe in 1980, 107 Mtoe in 1990, 156 Mtoe in 2000, and 222 Mtoe in 2010. Transport remained a predominant energy-consuming sector throughout the period of reference, followed by the residential sector, industry, and other consumption sectors (a category that combines segments of the economy such as agriculture, tourism, and commercial and public sectors) (Fig. 6.1).

In relative terms, from 1980 to 2010, the transport sector maintained its share of 27% in the region's TFC, while the residential sector declined from 30% to 25% and the industry also fell from 29% to 24%. Only the "category of other consumption" sectors actually increased their shares from 14% to 24%. This significant expansion was due to an increased use of energy in agriculture (+5%) and the commercial and public sectors (+5%) between 1980 and 2010 (Fig. 6.2).

When discussing energy consumption trends it is necessary to take into consideration how efficiently energy is actually consumed. It is thus necessary to analyze energy efficiency, which is – by the way – the most cost-effective

2. TPES is calculated by the IEA as production of fuels + inputs from other sources + imports – exports – international marine bunkers + stock changes.

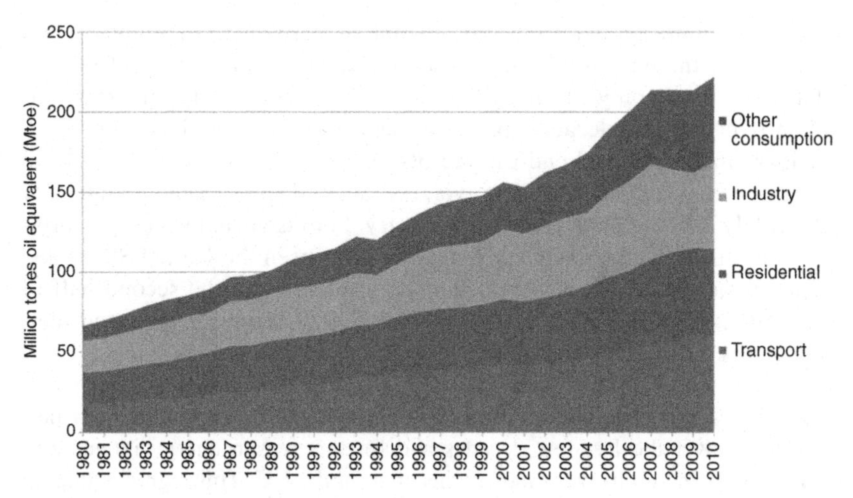

FIGURE 6.1 TFC by sector in SEMCs (1980–2010). *(Source: author elaboration on IEA, Extended World Energy Balances Database.)*

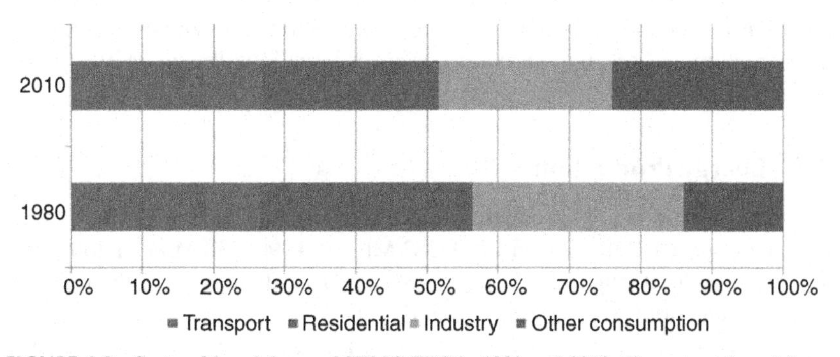

FIGURE 6.2 Sectoral breakdown of SEMC TFC in 1980 and 2010. *(Source: author elaboration on IEA, Extended World Energy Balances Database.)*

way to enhance the security of energy supply, to reduce emissions of greenhouse gases, and to enhance economic competitiveness in one fell swoop. For decades, the significance of energy efficiency was undervalued and its potential role in the energy mix discussed, sometimes dismissively so, all over the world. This was due to the fact that, in contrast to supply-side options, energy efficiency options are often obscured as efficiency is rarely traded or priced. Furthermore, improving efficiency involves a wide range of actions affecting a variety of energy services across different sectors – including buildings, industry, and transport – so that its overall achievement is often difficult to quantify. However, today energy efficiency has moved from contention to consensus. Environmental groups and energy companies agree that it should be at the top of the agenda regarding what needs to be done. Indeed, efficiency could well be called the "first fuel."

Energy intensity, defined as the amount of energy used to produce a unit of GDP,[3] is the indicator generally used to measure the energy efficiency of a nation's economy. As an overall trend, the world's energy intensity has fallen over the last decades, primarily as a result of efficiency improvements in the power and end use sectors as well as a transition away from energy-intensive industries. However, the rate of decline in energy intensity has widely differed from country to country. For instance, the best performers in terms of energy intensity reduction have been the United States and Japan, which started to lower their energy intensity in the second half of the 1970s, when the oil crises of 1973 and 1979 seriously impacted their economies.

On the contrary, SEMCs (albeit starting from a structurally lower level of energy intensity) have not considerably improved their energy efficiency performances over the last few decades and for this reason energy intensity in SEMCs remains up to two times higher than in the EU. This signifies that up to two times more energy is consumed in some SEMCs per unit of GDP PPP, relative to the best performers in the EU (Table 6.1).

Without additional efforts toward energy efficiency, the TFC in SEMCs will thus continue to grow substantially in the future, putting pressure on the existing infrastructure and creating an urgent need for large investments in electricity, oil, and natural gas.

2.2 Energy Production

The total primary energy production in the southern and eastern Mediterranean region has grown: 230 Mtoe in 1980, 283 Mtoe in 1990, 338 Mtoe in 2000, and 407 Mtoe in 2010 (Fig. 6.3). Three countries cover about 80% of the region's energy production: Algeria, Libya, and Egypt.

Oil has always been – and continues to remain – the first fuel produced in SEMCs. SEMC oil production has remained practically constant over time, ranging from 194 Mtoe in 1980 to 200 Mtoe in 2010. Over the last few decades natural gas has been the real game changer in SEMC primary energy production (Fig. 6.4). In fact, natural gas production boomed from 17 Mtoe in 1980 to

3. As Suehiro (2007) points out, it is impossible to accurately evaluate how advanced a country's energy conservation is and measure it against that of other countries, which are different not only in terms of their economies and welfare, but also in natural and social conditions. However, energy intensity of GDP is often used to see a country's energy conservation level as the approximate index. The problem is that, this index largely differs depending on the currency conversion rate. Conversion based on market exchange rates (MERs) tends to overestimate the GDP of countries that have higher prices, while conversion based on purchasing power parity (PPP) tends to overestimate the GDP of countries with lower prices. This means energy intensity based on MERs is advantageous for advanced countries with higher prices and that based on PPP is advantageous for developing countries with lower prices.

TABLE 6.1 Final Energy Intensity 1980–2008 in Thousands of Toe per US$ GDP (PPP) 2005

Countries	1980	1990	2000	2008	2000–2008 change (%)
Algeria	0.055	0.080	0.081	0.100	2.7
Egypt	0.112	0.115	0.105	0.112	0.8
Israel	0.080	0.082	0.078	0.069	−1.5
Jordan	0.121	0.207	0.195	0.154	−2.9
Lebanon	0.094	0.090	0.110	0.060	−7.3
Libya	0.134	0.274	0.121	0.106	−1.6
Morocco	0.083	0.072	0.085	0.086	0.2
Syria	0.134	0.215	0.165	0.121	−3.8
Tunisia	0.101	0.113	0.104	0.085	−2.5
Turkey	0.132	0.118	0.119	0.109	−1.1
EU-27	0.135	0.109	0.091	0.080	−1.6

Source: Blanc (2012) based on WEC global energy and CO_2 data prepared by Enerdata.

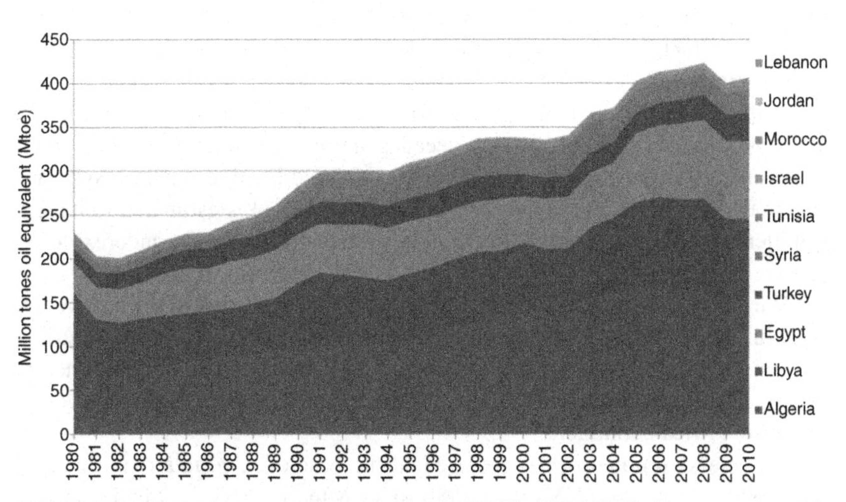

FIGURE 6.3 Total primary energy production in SEMCs (1980–2010). *(Source: author elaboration on IEA, Extended World Energy Balances Database.)*

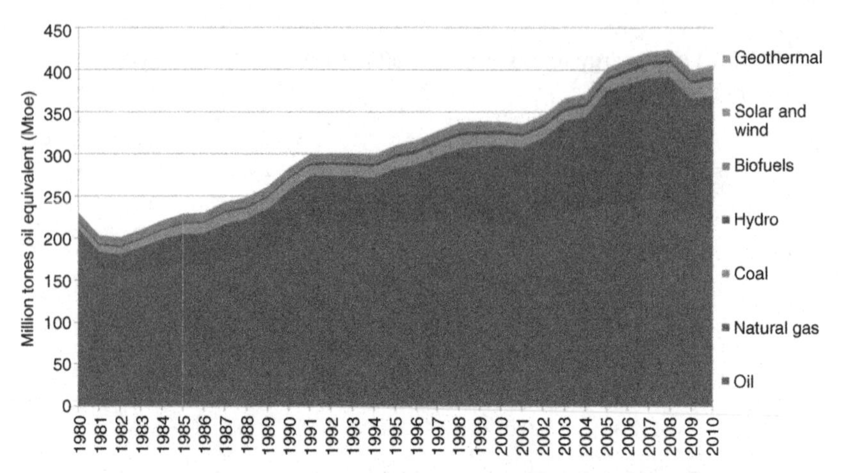

FIGURE 6.4 Total primary energy production by fuel in SEMCs (1980–2010). *(Source: author elaboration on IEA, Extended World Energy Balances Database.)*

149 Mtoe in 2010. The production trend of other fuels followed a flat dynamic over this period of reference.[4]

The oil production of SEMCs is mainly concentrated in Libya and Algeria. Libya owns 48.4 billion barrels (bbl) of proven oil reserves (BP, 2015) the largest in the entire African continent. This level of reserve might even grow in the future, considering that the US Geological Survey (USGS) estimated in 2011 that there might be almost 4 bbl of undiscovered oil across the Libyan Sirte Basin, the Tunisian Pelagian Basin, and western Libya (US Geological Survey, 2011). Despite such a huge bonanza, Libya's oil production has not increased substantially over the past few decades, remaining at a level far below its potential.[5]

With 12 bbl of proven oil reserves (BP, 2015) Algeria significantly increased its oil production over the last few decades and in particular after 2003, when substantial oil discoveries took place in the Ourhoud Field. Egypt and Syria are also major oil-producing countries. In particular, Egypt is estimated to have a significant untapped potential, as the country remains relatively underexplored (US Geological Survey, 2010).

The natural gas production of SEMCs has traditionally been concentrated in Algeria, the frontrunner of the regional natural gas market. In 2014 Algeria's proven natural gas reserves were estimated at about 4.5 trillion cubic meters (TCM; BP, 2015), around 30% of the estimated proven natural gas reserves of the entire African continent.

This massive bonanza has represented the base of Algeria's rising natural gas production (from 11 Mtoe in 1980 to 72 Mtoe in 2010), a trend likely to

4. For a detailed discussion of SEMCs' oil and gas markets, please refer Hafner et al. (2012a,b).
5. For an updated analysis of Libya's oil production trend please refer El-Katiri et al. (2014).

grow in the future as the country better assesses its shale gas reserves, considered the most extensive in North Africa.

But Algeria is not the only important natural gas-producing country in the region. In fact, since the early 2000s the region's traditional natural gas production landscape has rapidly evolved, with Egypt and Libya in the scene.

Egypt's proven gas reserves were estimated to be 1.8 TCM in 2014 (BP, 2015) representing the third largest reserves in Africa (after Nigeria and Algeria). Moreover, in 2010 the USGS released an assessment of the Nile Delta Basin Province, estimating that 6 TCM (mean estimate) of undiscovered, technically recoverable natural gas would be located in the Basin (BP, 2015). If confirmed, these figures would provide Egypt with a major natural gas production potential, to be added to its already distinguished production trend (from 1 Mtoe in 1980 to 50 Mtoe in 2010).

Libya's proven natural gas reserves were estimated to be 1.5 TCM in 2014 (BP, 2015) but recent new discoveries are expected to raise these estimates in the near term. The country's natural gas production has progressively grown over the last few decades (from a level of 4 Mtoe in 1980 to 13 Mtoe in 2010) and is expected to expand in the future. Natural gas production represents a high priority for Libya, in order to increase its exports – mainly to Europe – and to use natural gas instead of oil for domestic power generation (freeing up more crude oil for export).

Among SEMCs, coal is basically produced only in Turkey. The country's coal production is low (2 Mt in 2010), but the production of lignite rose from 45 Mt in 1990 to 65 Mt in 1999. After a sharp drop between 1999 and 2004 (46 Mt in 2004), it has recovered since then (77 Mt in 2010). In Turkey about 70% of the coal is used in electricity production, 15% is consumed by industry, and the remaining by residential and tertiary sectors (Blanc, 2012). Coal production in Turkey is likely to further expand in the future, as it is often considered to be a key tool to lower the country's huge energy dependency on imported fuels.

Turkey also plays a major role in the region's hydropower production. In fact, hydropower production in the country increased from 1 Mtoe in 1980 to 4 Mtoe in 2010. As in the case of coal, this increase is mainly due to the country's willingness to maximize its own energy resources in the face of a heavy dependence on imported oil and natural gas. Egypt also has a well-developed hydropower sector, which has tapped most of the Nile River's hydroelectric potential.

As far as the production of other renewable energy sources is concerned, Turkey is leading the trend mainly because of the substantial use of biomass and geothermal. In 2010, about 60% of the region's biomass production took place in Turkey, while in the same year the country also accounted for 100% of the region's geothermal production. Solar and wind thus account for a very minor share of renewable energy production, totalling only 2 Mtoe in 2010: 0.5% of SEMCs' total primary energy production.

3 THE FIRST EURO-MED ENERGY MILESTONE: ENHANCING HYDROCARBON COOPERATION IN THE REGION

Considering the historical evidence, it could be expected that Mediterranean oil and gas export sector will continue to develop without particular difficulties, as it is based on strong complementarities between the northern and southern Mediterranean: a capital – and technology-rich but energy-hungry north and a resource-rich but capital and technology-poor south.

However, it is primarily important to reinforce the regional hydrocarbon cooperation with a new win–win Euro-Mediterranean oil and gas strategy, which is able to tackle the barriers that are currently limiting SEMCs' oil and gas sectors, with the target of promoting local energy access and further exports to Europe in one fell swoop.

Such an exercise seems to be particularly important in the framework of the recently launched EU Energy Union. In fact, the new EU energy strategy calls for enhanced diversification of European energy supplies and it clearly targets the Mediterranean region as a crucial contributor in this strategy.

As previously illustrated, Algeria, Libya, and Egypt have a large amount of oil and gas resources. However, they are not able to fully exploit them due to governance and institutional constraints. For instance, in 2013 Algeria exported a total of 40 billion cubic metres (BCM) of natural gas, while its well-established natural gas export infrastructure has a capacity of 94 BCM. In the same year Algeria exported 12 BCM of natural gas to Italy, leaving 22 BCM of unutilized excess capacity in its existing infrastructure. These figures provide an initial idea of the untapped potential existing in the Euro-Mediterranean hydrocarbon cooperation.

Furthermore, it should be underlined that the Mediterranean region is not only important in terms of production but also in terms of transit of hydrocarbons. In fact, the Mediterranean is strategically located at the crossroads of hydrocarbon flows between the Middle East, Africa, and Europe, making it an important transit corridor for global energy markets. A new EU strategy toward the region would also be beneficial in terms of securing this transit, for the benefit of both shores of the Mediterranean. As far as this issue is concerned, a particularly important case is represented by the EU–Turkey natural gas partnership. In fact, Turkey represents the epicenter of various plans to deliver new natural gas supplies to Europe from the Caspian, the Middle East, the eastern Mediterranean and – last but not the least – from Russia (Hafner and Tagliapietra, 2013, 2015).

4 THE SECOND EURO-MED ENERGY MILESTONE: CHALLENGING THE PERSISTENCE OF ENERGY SUBSIDIES

In SEMCs, energy prices are held artificially low for all customers for social (but mostly political) reasons. These universal energy-consumption subsidies act as strong disincentives to a more rational and efficient use of energy and investment in the energy sector, including renewable energy. Moreover, energy subsidies pose heavy burdens for SEMC state budgets, especially in light of the oil price surge experienced over the last decade. As an example, in 2010 Egyptian energy subsidies accounted for 12% of GDP.

Such universal subsidies, which mostly benefit wealthy customers, generate imbalances, especially huge state deficits and debt. The reform of energy subsidies is not easy, but is possible. In fact, several examples exist for best practice (e.g., Jordan gradually phased out universal price subsidies, replaced by individualized/targeted subsidies and Turkey's fuel prices are among the highest in Europe).

The elimination of universal energy subsidies would provide a number of economic and environmental gains such as energy savings, lower levels of emissions, and reduced fiscal burdens. Furthermore, as the IEA emphasizes, "curbing the growth in energy demand through subsidy reform also has several important energy security implications. In net importing countries, lower energy demand would reduce import dependence and thereby spending on imports. For net exporting countries, removing subsidies would boost export availability and earnings. For all countries, it would also improve the competitiveness of renewable energy in relation to conventional fuels and technologies, further diversifying the energy mix. Lower energy demand would also alleviate upward pressure on international energy prices, yet the elimination of subsidies would make consumers more responsive to price changes, which should contribute to less volatility in international energy markets (IEA, 2011)."

However, economic and commercial rationalities do not always match with political requirements. Governments face difficult challenges in reforming energy subsidies also because of this scenario, as the IEA points out, "subsidies create entrenched interests among domestic industries advantaged by cheap energy inputs and those income groups that are accustomed to receiving this form of economic support. Such stakeholders can be expected to resist subsidy phase-out, particularly in the absence of clear plans to compensate losers or make the transition gradual. Resistance to fossil-fuel subsidy reform can be particularly strong in major fossil-fuel-exporting countries, where people may feel entitled to benefit directly from their nation's resource wealth (IEA, 2011)."

A reform of energy subsidies, for instance aimed at phasing out universal fossil fuel consumption subsidies and replacing them by individualized/targeted subsidies) would be essential to improve the energy, economic, and social sustainability of the overall southern and eastern Mediterranean region. Such a move would also favor the deployment of solar and wind energy by removing a key market distortion.

In the current regional geopolitical landscape such a reform process seems to be difficult, even if in the case of Egypt a recent reform has effectively reduced for the first time the country's unsustainable energy subsidy system.[6] In the medium term all SEMCs should advance an energy subsidy

6. For a comprehensive analysis of the implication of Arab uprisings on MENA oil and gas markets see El-Katiri et al. (2014).

reform process, phasing out universal fossil fuel consumption subsidies in favor of targeted subsidies aimed at effectively addressing the problem of energy poverty.

5 THE THIRD EURO-MED ENERGY MILESTONE: PROMOTING ENERGY EFFICIENCY

As previously illustrated, the level of energy efficiency in SEMCs remains very low. Indeed, the energy intensity in SEMCs is several times higher than in the EU. Without additional efforts for energy efficiency, the final energy consumption in SEMCs will supposedly be more than double by 2030.

Energy efficiency measures are cost effective, if they do not need to compete with subsidized energy prices, but even when they are cost competitive, the investment – though minor – often needs to be carried out at a household level. Yet households often lack the necessary information and advice as well as the initial capital needed for the investment. An effective policy agenda on energy efficiency and demand side management (DSM) needs to take into account the following issues:

1. *Public finance*. Energy efficiency and DSM measures enable significant improvements in public finances in SEMCs.
2. *Institutions*. Energy efficiency and DSM concern several sectors that depend on a number of institutions. At the national level, creating synergies among these various institutions will thus be particularly important. Specifically, dedicated agencies should be in place to implement the right technical choices – and the best financing tools – to enforce mandatory regulations. Such agencies should work at an interministerial level, under the guidance of the Energy Ministry and the auspices of the Finance Ministry or the Prime Minister's office.
3. *Hierarchy*. Considering the financial constraints of energy efficiency and DSM, the timing and hierarchy of the implementation measures matters greatly and thus should be developed within action plans that set implementation schedules and responsibilities. Policy priorities should first be placed on measures with high visibility, lower costs, and high rates of return.
4. *Domestic pricing*. A serious energy-efficiency and DSM agenda will need to deal with domestic pricing, including the elimination of universal subsidies. Indeed, energy pricing remains a key tool for modifying consumer behavior toward more efficient usage of energy.
5. *Sectors*. Households, SMEs, and the building sector should be the priority targets of an effective energy-efficiency and DSM policy. They represent a major share of energy consumption and they have substantial potential for energy-efficiency gains at low cost. In particular, the introduction of eco-labeling and technical, mandatory, standard regulations on consumption for equipment and appliances concerning cooling, heating, lighting, and industrial machinery have proven to be more effective and durable at low

(or even negative) costs. Support for the purchase/installation of proven small equipment based on renewable energy sources (solar water heaters and PV) by these sectors should also be at the top of the agenda.

6. *Support from the EU.* The EU has great potential to support the development of energy-efficiency and DSM measures in SEMCs. Technical assistance programs could easily transfer best practices on energy efficiency, especially concerning standards and labeling. The financing of small-scale projects and implementation of domestic funds dedicated to support investment by households and SMEs in energy efficiency would also help SEMCs to improve their energy efficiency. An example of this is already provided through the facilities of the European Bank for Reconstruction and Development in Bulgaria and Turkey (the Turkish Sustainable Energy Finance Facility and the Mid-Size Sustainable Energy Financing Facility), now under study for several SEMCs.

Taking into consideration all these elements, SEMCs should consider energy efficiency and DSM as an opportunity for economic development rather than a specific burden.

6 THE FOURTH EURO-MED ENERGY MILESTONE: UNLOCKING THE RENEWABLE ENERGY POTENTIAL

SEMCs are endowed with a huge potential for renewable energy as well as for significant energy efficiency and DSM. Thanks to the ongoing technological and institutional changes, all SEMCs could make use of this huge potential. Although in the past only hydropower potential was exploited (mainly in Turkey, Egypt, and Morocco), presently all countries are developing plans to enable them to also rely on other renewable sources, such as solar, wind, and biomass. The development of renewable energy projects in SEMCs offers a wide variety of advantages:

1. Renewable energy projects could initially be devoted to diversifying the energy mix, still largely dominated by fossil fuels and thus contributing to satisfy the rapidly increasing domestic demand for energy. This would free natural gas alternatively used in the domestic power generation sector for additional exports to Europe. Considering that an important but partly underutilized gas infrastructure, connecting North Africa with Europe is already in place, this choice would involve an immediate and significant economic return for SEMCs, just because of the growth in the export value of gas stocks.
2. Renewable energy projects, due to their intermittency (now well forecast days ahead), require the reinforcement of grids (especially with the use of software for grid management and weather forecasts) to enable their integration into larger, interconnected electricity networks and markets, therefore further fostering the integration of SEMCs.

3. Part of the renewable electricity could also be exported to Europe *via* HVDC (high-voltage direct current) electricity interconnections. This could allow SEMCs' renewable electricity to take advantage of European feed-in tariffs (FiTs). Such a scheme implies allowing SEMCs to be eligible for EU FiTs, with the advantage of EU helping it meet its decarbonization targets at a lower overall cost.
4. Renewable energy projects could develop significant new industry and service sectors (e.g., installers), leading to local job creation and manufacturing developments. By sharing manufacturing facilities and therefore exploiting larger economies of scale, south–south cooperation could be promoted. This is particularly important in a region that, according to the World Bank, presently has a low level of intraregional trade.
5. The economic and industrial development consequent to the large-scale implementation of renewable energy projects in SEMCs could have several positive spillovers for the EU, such as creating new markets and securing the existing energy infrastructure in the Mediterranean.

It is important to avoid focusing solely on large-scale renewable energy projects, but also to firmly develop decentralized systems, such as solar water heaters and rural PV systems. These systems are cost efficient, but nevertheless need to be promoted. Best practices already exist in some SEMCs, such as Israel, Palestinian Territories, Tunisia, and Morocco.

7 THE FIFTH EURO-MED ENERGY MILESTONE: PROMOTING A NEW INTERCONNECTED MARKET

The core challenge to the production and trade of renewable energy in SEMCs is that the development of the electricity supply system is limited by the lack of a regional market, largely due to energy price gaps and subsidies. The rigidities that this imposes mean that existing infrastructure is not used optimally, investment in new infrastructure is distorted and probably hindered, and the development of renewable energy is delayed.

For renewable energy to contribute most effectively to the development of SEMCs, it must be embedded in a functioning, regional electricity market that permits the exchange of power in substantial volumes, has no barriers to trade, and is friendly to private investment. The exchange of energy is to the benefit of both buyer and seller: it enables both parties to balance portfolios of generating assets, it can alleviate some of the disadvantages of nondispatchable and intermittent supplies, and it can permit joint ventures to share risks. Such a market does not yet exist across SEMCs. There is neither the infrastructure nor the regulatory and legislative framework that would be necessary for a regional market to function correctly.

Indeed, electricity interconnection remains a key issue for energy cooperation in the region. It is of crucial importance to reinforce the national transmission lines in SEMCs, which are often weak, as well as interconnections between these countries. Since the late 1990s, the two shores of the Mediterranean have been connected through a line across the Strait of Gibraltar; however, the electricity interconnection between the two shores needs to be further reinforced. Moreover, nontechnical (commercial) distribution losses remain at very high levels (up to 40% in Lebanon and 20% in Algeria) at the expense of paying customers and distributors.

Furthermore, another key problem of renewable energy development in SEMCs is to ensure their financing. For this reason innovative financing methods are needed. At the same time, SEMCs seem to be endowed with significant carbon market opportunities for investments in both energy efficiency and renewable energy.

8 THE SIXTH EURO-MED ENERGY MILESTONE: FINANCING THE SUSTAINABLE ENERGY TRANSITION

The Clean Development Mechanism (CDM) is a potential source of additional revenue streams for investments in energy efficiency. In fact, the CDM is designed to assist developing countries achieve sustainable development by allowing the entities among the Annex I Parties under the UN Framework Convention on Climate Change (UNFCCC) to participate in low-carbon projects and obtain certified emission reductions in return. Nonetheless, carbon financing through the CDM can only cover a small share of total investment (around 10%) and require specific expertise.

Till date, SEMCs have not fully tapped into the vast potential for CDM projects, as several barriers to the development of CDM projects in the region persist: (1) the lack of capacity for operation and management, (2) the lack of regional coordination, and (3) the lack of engagement of the private sector. Possible factors leading to a successful entry into this mechanism would be the capacity for data collection and management as well as experience in baseline setting and crediting. In this respect, the electricity generation sector is considered suitable for testing a nationally appropriate mitigation action (NAMA) or a sectoral mechanism.

The EU could negotiate bilateral sectoral agreements with its neighboring countries as part of the Union for the Mediterranean (UfM). Both EU and the non-EU countries would benefit from engaging the private sector in a consultation process, as the latter have the data, technology, and know-how that are essential for the implementation of a sectoral mechanism. Actually, the Mediterranean region offers an interesting test case for an integrated approach to carbon markets: there is an institutional setup (the UfM), a financial facility (the Mediterranean Carbon Fund), and a region-wide initiative with substantial potential for energy related emission reductions (the Mediterranean Solar Plan (MSP)) that could fit into a new market-based mechanism. The outcomes of the MSP could feed into not only the mid-term scenario-building process for SEMCs across policy areas, but also the ongoing process of elaborating new market-based mechanisms at UNFCCC and EU levels.

With regard to the need for new financing sources and instruments for the development of renewable energy projects in SEMCs, the EU could also play an important role in facilitating investments by the Gulf Cooperation Council (GCC) in SEMCs. Notably, a strong complementarity exists between these regions in the field of renewable energy. The wide availability of capital in the GCC, the great renewable energy potential of SEMCs, with the possibility (considering their geographical proximity) to export some of it to Europe, and the institutional support of the EU, could represent the three main pillars of a new "triangle of growth". Private and public investors (such as sovereign wealth funds) from the GCC are increasingly focusing their investments on the renewable energy sector, with the aim of transforming oil wealth into technological leadership in renewable energy. Some of the investment could be directed toward SEMCs, whose potential for solar energy is among the highest in the world and who are already promoting several large-scale renewable energy projects. The EU should facilitate the implementation of this process by providing institutional support (in terms of both regulation and public finance) and technological know-how.

9 CONCLUSIONS: THE NEED FOR A NEW EURO-MEDITERRANEAN ENERGY ROADMAP

Efforts for more sustainable energy development in SEMCs could represent the key element in EU foreign energy policy toward the region. This policy could provide important dividends to both the EU and SEMCs, as far as energy security, sustainable development, economic growth, and job creation are concerned. If the EU is committed to improve cooperation with SEMCs, it is important not to be solely perceived as a hydrocarbon buyer, but also as a fully fledged partner, notably to foster regional cooperation. In the framework of the European Neighborhood Policy (ENP), a broad set of capacity-building and technical assistance measures for regional energy projects in the southern Mediterranean (MEDSTAT, MED-ENEC, MED-EMIP, and PWMSP) have already contributed extensively to regional energy cooperation and provided support to national and regional efforts.

The current transition phase in SEMCs could provide EU with the opportunity to play a more meaningful role in the region in the future. Efforts to assist SEMCs in facilitating energy efficiency, DSM, renewable energy, and energy interconnections could represent the main components of EU foreign energy policy toward the region.

To conclude, an integrated scheme of energy cooperation designed to function as a catalyst for reinforcing Euro-Mediterranean economic, political, and social integration should rely on three main pillars:

1. *A long-term strategy for domestic socioeconomic development*, based on a robust institutional setup and enhanced public governance, and including an oil revenue management and poverty reduction strategy with targeted support instead of universal consumption–price subsidies.

2. *An integrated energy and climate policy*, elaborated as part of a global energy strategy covering energy security, energy access, regulatory reforms toward energy prices that are fully cost reflective, energy efficiency, and renewable energy action plans in synergy with climate policies (carbon financing).
3. *Regional energy cooperation* (intra-SEMCs and EU–SEMCs) to focus on sustainable policy development with the Regional Center for Renewable Energy and Energy Efficiency (RCREEE), which is the regional focal reference for both SEMCs and GCC countries on energy efficiency and renewable energy deployment. This regional energy cooperation should also focus on regulatory (tariff) and social reforms (targeted subsidies), infrastructure (e.g., power and gas interconnections), and markets (e.g., the renewable electricity market for the EU/SEMCs), fostered by the MSP and integrated interregional financing.

In short, energy is a topic of crucial importance for both the EU and SEMCs. In view of their geographical proximity, further market integration would be in the interest of both sides. SEMCs and the EU furthermore have many challenges in common in their endeavors to secure a sustainable energy transition for the overall Euro-Mediterranean region.

Taking into consideration the past evolutions in energy relations between the EU and Russian Federation, and also between the EU and GCC, it is evident that EU–SEMCs' energy cooperation could be enhanced and better developed by creating a new Euro-Mediterranean Energy Roadmap.

Such a roadmap should aim at designing a sustainable energy transition for the overall Euro-Mediterranean region, with the underlying idea that a wider framework of energy cooperation could greatly contribute not only to the economic development and environmental performance of the entire Euro-Mediterranean region, but also to its social and political stability.

REFERENCES

Blanc, F., 2012. Evolution tendancielle et prospective de l'efficacité énergétique dans les pays méditerranéens, MEDPRO Technical Report, MEDPRO Deliverable D. 4b. 3. Forum Euro-Mediterranéen des Instituts de Sciences Economiques (FEMISE), Marseille.

BP, 2015, BP Statistical Review of World Energy, London.

El-Katiri, L., Fattouh, B., Mallinson, R., 2014. The Arab Uprisings and MENA Political Instability: Implications for the Oil & Gas Markets, OIES Paper MEP 8. Oxford Institute for Energy Studies, Oxford.

Hafner, M., Tagliapietra, S., El Andaloussi, E.H., 2012. Outlook for Electricity and Renewable Energy in Southern and Eastern Mediterranean Countries, MEDPRO Technical Report No. 16, Centre for European Policy Studies (CEPS), Brussels.

Hafner, M., Tagliapietra, S., El Andaloussi, E.H., 2012. Outlook for Oil and Gas in Southern and Eastern Mediterranean Countries. MEDPRO Technical Report No. 18, Center for European Policy Studies (CEPS), Brussels.

Hafner, M., Tagliapietra, S., 2013. A New Euro-Mediterranean Energy Roadmap for a Sustainable Energy Transition in the Region. MEDPRO Policy Paper No. 3, Center for European Policy Studies (CEPS), Brussels.

Hafner, M., Tagliapietra, S., 2015. Turkish Stream: What Strategy for Europe? Fondazione Eni Enrico Mattei, Milan, Nota di Lavoro 50.

IEA, 2011. World Energy Outlook 2011. OECD/IEA, Paris.

Suehiro, S., 2007. Energy intensity of GDP as an index of energy conservation. Problems in international comparison of energy intensity of GDP and estimate using sector-based approach. Institute of Energy Economics, Tokyo.

US Geological Survey, 2010. Assessment of Undiscovered Oil and Gas Resources of the Nile Delta Basin Province, Eastern Mediterranean, Fact Sheet 2010-3205. USGS, Reston, VA.

US Geological Survey, 2011. Assessment of Undiscovered Oil and Gas Resources of Libya and Tunisia, Fact Sheet 2011-3105. USGS, Reston, VA.

Chapter 7

Toward a Mediterranean Energy Community: No Roadmap Without a Narrative

Gonzalo Escribano
Energy Programme, the Elcano Royal Institute; Applied Economics, Spanish Open University (UNED), Spain

1 INTRODUCTION

The Mediterranean region has experienced significant transformations over recent years. The economic and financial crisis in its northern shore was mirrored by political turmoil and instability in the south. As a result, the Mediterranean energy picture has completely changed in several respects. The European energy demand has stagnated, and in some southern member states is now lower than 5 years ago. Idle capacity and renewables' support fatigue have plagued European energy markets. The US shale revolution has further eroded European competitiveness, and simultaneous geopolitical instability in its southern and eastern vicinities seriously challenged the EU's energy security.

This chapter approaches the discussion on a Mediterranean Energy Community in three steps. First, it starts by describing the different pathways that could lead toward it, highlighting the need for differentiation of both at the geographical (or energy corridor) and energy source levels. Second, it tries to contextualize such an institutional construct within the harsh and volatile southern Mediterranean geopolitical realities. Third, it presents some elements required for a functioning Mediterranean Energy Community, starting by being supported by a truly single European energy model. It concludes by calling for a renewal in the EU's energy narrative toward its Mediterranean neighborhood to increase its attractiveness to the region.

2 PATHWAYS TOWARD A MEDITERRANEAN ENERGY COMMUNITY

The EU first talked about a Mediterranean Energy Community in the aftermath of the Arab Spring. The March 2011 Communication "A partnership for democracy and shared prosperity" included the possibility to extend the Energy

Regulation and Investments in Energy Markets. http://dx.doi.org/10.1016/B978-0-12-804436-0.00007-2

Community Treaty to Mediterranean Southern Partners, or at least build on its experience. The idea was to advance Euro-Mediterranean energy integration as a driver for Mediterranean Southern Partners' development, with the final goal to contribute to shared prosperity, the hallmark of Euro-Mediterranean relations since the 1995 Barcelona Conference.

In its original formulation it was presented as another example of outward Europeanization in the field of energy, with Mediterranean partners having to almost passively adopt the energy related *acquis communautaire*. Energy convergence was simply understood as downloading EU energy software (norms) without fully considering the divergence in energy preferences among and between EU member states and Mediterranean Partner Countries. Recent EU documents point to either an extension of the Energy Community Treaty toward those few Mediterranean Southern Partners ready to adopt a significant part of the EU *acquis*; or to the creation of something like a EU–North Africa Energy Community, based on a differentiated and gradual approach (Glachant and Ahner, 2013, p. 9).

Pathways toward a Mediterranean Energy Community ask for differentiation at two levels: from other EU external energy initiatives (Energy Community Treaty, Energy Charter, Mediterranean Solar Plan (MSP), or whatever remains of it); and by country and sector. Any Mediterranean Energy Community should consider country specificities and asymmetries: asking for full reciprocity in countries heavily dependent on hydrocarbon exports is extremely asymmetrical because for southern Mediterranean oil and gas producers the hydrocarbon sector is key to its economy and its social fabric. Nonproducing countries can advance faster in this domain. A gradual and differentiated sector-by-sector approach can also entail starting with the renewable sector, less institutionally entrenched and economically vital, and therefore more prone to institutional innovation (Escribano, 2010).

Additionally, any roadmap for a Mediterranean Energy Community should address Mediterranean Partner Countries' preferences such as fighting energy poverty and promoting energy development, optimizing the contribution of local energy resources to economic (and industrial) development, technical cooperation and technological transfers, as well as access to EU markets. Shared prosperity is a political final goal that should inspire every step and prevail over exclusively technical approaches. More normative approaches regarding other new issues like sustainability and better energy resource governance in the neighborhood are also vital to long-term energy security.

As a normative actor, the EU has developed a soft power narrative regarding global energy. Its aim is to offer the EU energy model (renewable, decarbonized, efficient, and secure) as an inspiration to other countries in order for them to follow the same path. If sustainability and transparency start at home, the next logical step is to promote such value-laden drivers in the immediate vicinity. Promoting sustainability abroad, especially to its neighborhood, is a key feature to legitimize the EU's aspirations toward the design and implementation of external energy governance. Decarbonization further projects a responsible EU, pioneering the fight against climate change.

Improving transparency in the management of energy resources and a fair redistribution of its income are not only needed to prevent resource conflicts like the one that is going on in Libya, but also to provide populations in the southern shore of the Mediterranean with the opportunity to profit from its energy resources. While energy resource governance has focused on oil and gas, it should also be applied to renewable energy in the Mediterranean. To date, the EU approach has not sufficiently included the latter's socioeconomic and developmental impacts. For instance, at least part of the failure of the MSP has been due to its Eurocentric emphasis on promoting the EU's renewable industry as well as their engineering firms: this led to some doubts about who the MSP was being a driver for development – Mediterranean populations or European companies (Escribano and San Martín, 2011).

During recent decades, Euro-Mediterranean energy relations have developed to include several issues from physical interconnections (the Euro-Mediterranean electricity and gas rings) to normative issues included in both Barcelona's Association Agreements and European Neighbourhood Policy (ENP) Action Plans. The Union for the Mediterranean (UfM) added the now almost defunct MSP, and EU renewable regulation explicitly contemplates renewable energy source (RES) flows from the region.[1] Since 2005, under the dialog promoted by the Italian Regulatory Authority for Energy, a voluntary platform was established to harmonize technical standards and regulatory frameworks. MEDREG (the association of Mediterranean Energy Regulators) and MED-TSO (the Association of the Mediterranean Transmission System Operators) were created and are now part of the EU Commission Mediterranean platforms for gas, electricity, and renewables, together with the Observatoire Méditerranéen de l'Énergie (OME) (Rubino, 2015).

A detailed account of these initiatives is beyond the scope of this chapter, which on this point only tries to (1) highlight the need to consolidate such processes and to further explore their opportunities and (2) more importantly for the purposes of this chapter, to distinguish between three complementary pathways to achieve a Mediterranean Energy Community: physical integration, regulatory convergence, and normative deployment.

Both elements include reviewing its successes and failures, and redesigning its foundations and narratives when needed (Escribano and San Martín, 2014). The need to improve and enlarge physical Euro-Mediterranean corridors and harmonizing regulatory frameworks is the subject of several chapters in this

---

1. The failure of the DESERTEC initiative and Spanish opposition to implementation of the MSP brought it to the brink of waning. Spain blocked the MSP roadmap arguing that before building new interconnections with North Africa (Spain has two electricity interconnectors with Morocco and two gas pipelines with Algeria) the EU should work on attaining interconnections between France and Spain, the small scale of which acts as a protectionist entry barrier for Spanish RES into the rest of Europe.

book, and will not be dealt with here. Instead, this chapter concentrates on the EU's soft power, the normative attractiveness of its energy narratives to Mediterranean neighbors as the third element required to succeed in integrating Mediterranean energy markets, and to achieve a functioning and inclusive Mediterranean Energy Community that really deserves that name. As with physical integration and regulatory convergence, such a normative narrative may be differentiated on individual issues, the energy sector, and country level. However, it also requires taking into account the geopolitical and geoeconomic energy challenges the Mediterranean faces. The next section addresses these realities before elaborating on some elements that could form a consistent but flexible EU energy narrative toward its Mediterranean neighborhood.

3 HIGH EXPECTATIONS, HARSH REALITIES

The recent literature is full of analyses predicting the emergence of a new geopolitical balance in the international energy system due to the global impact of the shale gas and tight oil revolution. This section tries first to nuance the direct and indirect impacts of such developments on the Euro-Mediterranean energy space before addressing the Mediterranean dimension itself.

At a global level, the first interesting feature in the emerging energy landscape is its bias toward the unconventional energy power of the United States, while the alternative European vision of a renewable, low-carbon future seems increasingly unappealing to strategic thinking. A renewables-based energy mix seems too much of a soft-energy power, whose appeal is fading in an energy world dominated by the hard narrative of hydrocarbon resources, whether conventional or unconventional. When dealing with alternative energy futures, unconventional oil and gas are from Mars, while renewables are from Venus, to put it in Robert Kagan's (2003) terms.

Along these lines, it has been argued that US geopolitical interests in the Middle East and North Africa will almost disappear. "Saudi" America would lead world energy markets, pushing conventional producers to the fringe of international geopolitics, while Europe would stay alone in providing for its energy security, increasing European energy vulnerability. Although unconventional resources have changed the balance of global energy geopolitics, at least in the short and medium term, its development does not push conventional hydrocarbon suppliers to the fringe of energy geopolitics. The recent price war, started and sustained by Saudi Arabia and its Gulf Cooperation Council (GCC) partners, has shown to what extent conventional producers are still key actors in the global energy scene.

This is why Russia has received most of the attention from European policy makers, who are now trying to balance the policies of engaging the country and diversifying away from it. Although Russia is the EU's main supplier, its energy policies had always raised concerns due to the use of petropolitics to attain its geopolitical objectives. The Mediterranean is the closest alternative energy–exporting region to Russia. Therefore, a deepening of Euro-Mediterranean energy relations advocated by this chapter is the best and more geographically immediate option to erode Russian geopolitical hard-energy power.

In this regard, the unconventional revolution does not directly affect the energy security prospects of Euro-Mediterranean countries, which under normal circumstances would keep Middle East and North African producers as their main energy suppliers. Nor does it affect EU central European countries, which in spite of the current strategic competition between the EU and Russia and claims for diversification, will remain highly dependent on Russian supplies.

Indeed, in the Mediterranean, Algeria and Libya are thought to hold significant unconventional reserves. If we compare two EIA (2011, 2013) world estimations of shale gas recoverable resources, there are significant differences. Algerian unconventional recoverable resources were multiplied by three in the last report due to the inclusion of six additional basins, ranking third in the world's unconventional resources after China and Argentina. Libyan resources were downgraded due to the application of a more severe organic threshold in shale gas estimation. Finally, Egypt was not assessed in the former report, but it is supposed to have resources of similar magnitude to Libya.

The impact, on Mediterranean Europe's energy security, of an eventual US strategic neglect of the Middle East is also difficult to envision. First, while unconventional oil and gas is making the United States an unconventional (energy) power, no country is immune to the evolution of global markets as has been seen in the recent tumbling of oil prices or before in its sustained high levels. Second, the 2014 Ukrainian crisis could have caused a new geopolitical shift, as the United States has realized that Europe cannot help to contain Russia if this country supplies approximately one third of European oil and gas imports and one quarter of the European consumption of oil and gas. Therefore, no sudden US strategic neglect of the Gulf region and the Middle East (or Europe) seems to be expected.

Can the same be said about the Mediterranean region? It is doubtful for Southern Europe in the short term, or for the whole of the EU in the medium to long term. In fact, over recent years, events in the region have seen the geopolitical deepening of a wider Mediterranean that now extends well beyond its shores to reach the Middle East, with the Syrian conflict having spilled over to Iraq and threatened GCC countries; and to the Gulf of Guinea through the linkages between the turmoil in Libya, the Sahel, and countries like Nigeria that face the Boko Haram threat, who recently expressed allegiance to Islamic State.

European member states, especially in Mediterranean Europe, were repeatedly affected by energy disruptions coming from the apparently old-fashioned energy pivot of the Middle East and North Africa. The sequence of events that followed is well known. First, the 2011 Tunisian revolution affected the maintenance of the trans-Mediterranean gas pipeline exporting Algerian gas to Italy. Then, the Egyptian revolution threatened Egyptian supplies and transit through the Suez Canal. Neither of these caused serious disruptions, but reminded the

EU of the strategic importance of their oil and gas imports from the Gulf and North Africa and through the Mediterranean. Since 2012, the Iranian embargo has forced Southern Europe, which was importing almost 5% of its crude oil from Iran, to rely on new oil imports from the Persian Gulf and Nigeria.

The war in Libya has also caused serious disruption, halting the country's oil and gas exports. Since then Libyan hydrocarbon production has behaved analogously as a roller coaster; a former stable provider has turned into a failed state destabilizing its neighbors. An already deteriorating situation in the Sahel has worsened due to the security vacuum left in Southern Tunisia (as recently seen in the tragic attacks in the Bardo Museum and at the Sousse beach resort) and Libya, and the mounting pressure on Mali's territorial integrity, forcing French military intervention. In January 2013, a jihadist attack against the In Amenas gas plant, producing about 10% of all Algerian gas, significantly reduced Algerian gas supplies to Italy through the trans-Med and affected Algerian gas production for more than a year.

Renewable energy deployment in the region could bring a more positive outlook and a better-compensated energy interdependency pattern between Europe and its Mediterranean partners. Mediterranean geopolitical and geoeconomic dynamics are rapidly evolving and the energy sector needs to adapt to it. Nevertheless, while Arab revolts and regime change in Tunisia, Egypt, and Libya disrupted supplies from Tripoli to the Sinai Peninsula, the North African energy pivot that is Algeria remained untouched, although at a considerable fiscal cost.

With this background, the extension of the Sahel crisis to Southern Mali and the terrorist attack on the In Amenas gas facility threatened to develop into a long-time feared nightmare scenario. To put it in a classical geopolitical manner, energy "rimlands" from the Mediterranean to the Gulf of Guinea would have been threatened simultaneously by a diffuse power emanating from the Sahelo-Saharan "heartland," whose borders suddenly and directly reached European interest.

There is for sure a lot to nuance in this interpretation, especially in three respects. First, the generalization of terrorist attacks to energy facilities is difficult to foresee, especially in Algeria, which has extensive experience in dealing with terrorism and protecting energy infrastructures. Second, it remains to be seen if the Libyan capacity to recover production levels and to exploit its unexplored potential under more transparent rules and institutions concerning domestic oil wealth redistribution and contract terms with international majors. Third, the influence of Boko Haram on Nigerian oil and gas supplies also remains to be seen, as does the extent to which a Somalia-like piracy situation could affect shipping in the Gulf of Guinea.

In any case, and on a more positive stance, the energy preferences of Euro-Mediterranean countries call for a deepening and widening of their energy interdependency with North Africa, in the sense of both further developing existing initiatives and projects, and diversifying energy partnership to new areas.

In security affairs this may call for an integral strategy toward North Africa, the Sahel, and even western Africa, also entailing further widening of Europe's geopolitical borders. If Mediterranean Europe's shared strategic challenge regarding energy security is to be the geopolitical deepening of a wider Mediterranean, both deepening and widening call for more resources devoted to cover more issues over a larger geographical area in a more integrated manner. Contingent to the evolution of EU–Russian rivalry, this strategic shift may well extend to the whole of Europe.

4 MANAGING INTERDEPENDENCY: ELEMENTS FOR A MEDITERRANEAN ENERGY COMMUNITY[2]

These recent Mediterranean energy developments entail a different pattern of interdependency between Europe and the region. This section addresses policy implications for the EU, which comprise at least the following actions: the EU doing its homework first, consolidating current energy relations, launching new initiatives in new domains, improving energy governance, and addressing energy-related hard security threats, before concluding on the need to jointly develop a credible Euro-Mediterranean energy narrative.

4.1 Homework First

New Mediterranean geopolitical and political economy balances imply deeper consequences for European energy supplies, either current hydrocarbon resources or future renewable electricity flows. In the short term such a drastic change in the wider Mediterranean's energy scenario implies an asymmetric energy shock for Euro-Mediterranean member states. This is also true for other security dimensions such as terrorism, trafficking, irregular migration, or money laundering. However, sooner or later, and depending on the evolution of Russian–EU energy relations, it may also affect most European countries.

The convergence of European energy preferences in the Mediterranean is so evident that the lack of at least a shared and more coherent reflection strikes most observers. True, the energy dimension is not an exception regarding the difficulties that Southern European officials have found in projecting their shared preferences toward the EU or in dealing with them bilaterally in a more fruitful manner, especially in the Mediterranean. The geopolitical deepening of the southern Mediterranean is perhaps the clearest example of a shared strategic challenge that demands a much more proactive approach.

The Euro-Mediterranean energy dimension is to be promoted in a more consistent manner at the EU level, despite the bilateral track being explored more fully, from policy coordination to information exchanges. Other schemes, like the 5 + 5 or new initiatives between Euro-Mediterranean members and

---

2. This final section follows Escribano and San Martín (2014).

Mediterranean Partner Countries, can also offer channels of cooperation and merit exploration. Coordination in the security dimension may involve not only the 5 + 5 initiative and NATO's Mediterranean Dialogue, but also emerging schemes to deal with the Sahel crisis and its North African spillovers. A similar situation happened in the Levant regarding its links to other Middle Eastern countries (especially those in the Gulf).

Three factors need to be considered here: first, EU officials and policy makers need to be aware of the existence of common preferences and shared challenges in the region. Second, there needs to be a conviction that cooperation between EU member states is itself crucial because Europe is perhaps the region most affected by these Mediterranean developments. This is especially so in the energy domain, where EU countries share the eventual consequences of an asymmetrical energy shock coming from the region, albeit to different extents. Third, there needs to be the will to invest political capital to change the prevailing discourse on Euro-Mediterranean energy cooperation.

This includes acting by example rather than by default and practicing what is being preached. For instance, promoting a Euro-Mediterranean energy ring when the EU has proved unable in the past to resolve the isolation of the Iberian Peninsula or the Baltic States from European energy markets. The EU needs to take the initiative in Euro-Mediterranean matters. Energy is no exception here; on the contrary, it offers a clear opportunity for credible cooperation schemes.

4.2 Consolidating Energy Relations and Launching New Initiatives in New Domains

Over the years, Euro-Mediterranean energy relations have developed to include several issues from physical interconnections (the already mentioned Euro-Mediterranean electricity and gas rings) to regulatory issues included in both Barcelona's Association Agreements and ENP Action Plans. The UfM added the MSP and the EU renewables regulation that explicitly contemplates the flow of renewables from the region. There has been an increasing energy dialog in different areas, from energy regulators (MEDREG) to companies (Observatoire Méditerranéen de l'Énergie, OME). As a result, a good relationship has been built between regulators, grid operators, and utilities across the Mediterranean.

A detailed account of the various initiatives and their evolution is beyond the scope of this chapter, and is addressed by Vantaggiato (chapter 2) in this book. This chapter's purpose is to consolidate energy relations and to explore the opportunities they already offer. This includes reviewing successes and failures, and redesigning foundations wherever needed. The existence of such foundations and bridging gaps identified between expectations and achievements should be highlighted in any narrative aimed at renewing the European energy discourse for the region. Joint reflection on existing initiatives should be conducted at various EU, multilateral, and bilateral levels.

Wherever significant gaps in the existing energy framework are identified, the EU should immediately step in and launch new initiatives. This can entail deepening the interdependency in some fields where cooperation already exists – such as physical interconnections or technical and regulatory harmonization – or widening its scope to cover new issues – such as extending interdependency to the deployment of renewables or securing energy infrastructures. It would be interesting for both shores of the Mediterranean to identify new domains on which they could have convergent energy preferences and the ways to advance them. This includes diversifying economic relations within the energy sector (deployment of renewables, cross-investments, security of facilities, etc.) and outside it, to build a bilateral relationship less polarized on oil and gas imports.

Broadening the agenda and revitalizing it should however be balanced with the need to provide real progress and not just rhetoric. The problems of both deepening and widening are well known for observers of EU integration, who tend to think that one happens at the expense of the other. This tends to be true under the assumption of fixed resources, but this chapter argues that the new geopolitical weight of a wider Mediterranean needs more resources to conduct both a deepening and widening of EU policies toward the region. Regarding the renewal of the EU's energy discourse, the idea that energy cooperation has "solid" foundations needs to be complemented with the idea that there is dynamism and a broad potential to do more things together in the future, as well as the operational and financial capacity to attain more ambitious goals.

4.3 Focusing on the Good Governance of Energy Resources

Good energy governance is a key driver of the future evolution of energy security in the Mediterranean, and should be included in any credible narrative on Euro-Mediterranean energy interdependence. The time consistency of southern Mediterranean energy policies will critically depend on whether energy governance is improved to provide a more inclusive growth and investment pattern, or remain anchored to internal and external power politics and their related economic inefficiencies. For this chapter's purposes, energy governance has two distinct aspects: the domestic dimension of escaping the resource curse, and the international imperative of avoiding resource competition between international companies.

Contrary to the political science concept of the resource curse, economists like to approach energy governance problems as a (Dutch) disease.[3] There are different reasons and implications behind each figure, but a fundamental one

3. Dutch macroeconomic mismanagement of newly found North Sea gas revenues became the well-known Dutch disease economics textbook case (Corden and Neary, 1982). The usual prescription is monetary sterilization and prudent fiscal and exchange rate policies, together with microeconomic measures to prevent deindustrialization. However, while Dutch disease has an orthodox economic cure (admittedly a macroeconomic one), for political scientists resource wealth tends to be considered a curse with few policy solutions other than increasing revenue transparency.

is that "curses" require some kind of witchcraft while "diseases" can have a scientific cure. Oil and gas producers like Algeria have learned a lot about the macroeconomic management of their resource wealth over recent decades. Macroprudency, monetary sterilization, oil wealth funds, and (admittedly modest) microeconomic reforms have allowed them to obtain better economic results than in the past.

However, institutional reforms making the management of these resources more transparent and inclusive were slow in coming, and were not able to prevent rent-seeking behavior at both the economic and political levels. Resource allocation has been led by clientelism, rather than efficiency. Redistribution through subsidies and public employment distorted markets and ultimately became a political trap. Political distortions also affected international company ecosystems and citizen's perceptions of the role they perform in their countries. Greater transparency adds legitimacy to foreign energy companies that behave according to rules, such as those provided by the Natural Resource Charter or Extractive Industries Transparency Initiative. Furthermore, the recent enactment by the EU of new disclosure requirements could help to offer a more attractive interdependence framework for southern Mediterranean populations who often do not fully benefit from the energy resources of their own country.

In this regard, invocating good energy governance (including the current regressive structure of energy subsidies) in Mediterranean producers represents a course of action that would take into account the preferences of Mediterranean citizens on transparency and rule of law as well as a fair redistribution of resource wealth and its impact on their economic development. Only then can the need of governments to redesign energy policies and international company claims on access to resources be addressed in a politically sustainable way. The latter can expect to obtain better results in a rule-of-law scenario than in an energy sector dominated by political influences.

4.4 Addressing Energy-Related Hard Security Threats

One of the most pressing Mediterranean energy security challenges for the EU is to address the issue in a way that helps secure supplies without oversecuritizing energy relations. One way to approach this has been called "outward securitization," which means both projecting security risks toward third countries and confining security cooperation to borders. To date, the logic of bilateral dependencies has overplayed the opposing preferences of producers and consumers.

The geopolitical transformations in the region offer an opportunity to overcome such a reductionist approach by making explicit how interdependency spills over from Mediterranean Partner Countries to southern neighbors. It is not producer countries that threaten security of supply, nor consumer countries that threaten security of demand. Instead, both are threatened by third parties, reinforcing the logic of interdependency vis-à-vis shared risks and making the need for cooperative security self-evident and more palatable from a political perspective.

However, given that security cooperation is a highly sensitive issue, pragmatism should prioritize measures at the border, not inside the countries themselves. Measures can combine different aspects, from training to information-sharing initiatives. For instance, training formats have been tested with some success within the framework of the Africa Partnership Station and could be pursued at a Euro-Mediterranean level in the domain of oil platform and maritime security. In both cases the training would constitute a capacity-building opportunity regarding operational skills – such as force protection equipment, platform defence equipment, and facility protection – and information management – such as Maritime Domain Awareness (MDA) technology and information-sharing systems.

As a result of events in the Sahel, it makes sense to consider even more ambitious security cooperations, something akin to a SIVE-like system for North African Saharan and Sahelian borders.[4] Drone sharing in the region is another option. The US African Command (USAFRICOM) has established a drone base in Niger to monitor the activities of Boko Haram and their links with Al Qaeda in the Maghreb (AQIM) and now with Islamic State. However, despite the United States already having other capacities in the region, this did not impede Algeria from denouncing the lack of information coming from them. Cooperative security does not only require agreeing on deploying technical capacities, they also have to be shared to be fruitful, which in turn implies setting up procedures and priorities in a mutually cooperative manner.

5 CONCLUDING REMARKS: DEVELOPING A CREDIBLE EURO-MEDITERRANEAN ENERGY NARRATIVE

This chapter has tried to show why North Africa remains a key energy partner for the EU despite a rapidly changing energy global landscape. Mediterranean energy geopolitics are becoming more important and demanding for Europe. New issues should be included in a comprehensive strategy aimed at a geopolitically expanded Mediterranean region. EU countries (Mediterranean member states in the short term, but also the rest of Europe in the medium to long term) do share energy preferences in the Mediterranean that are barely reflected in common positions or even common strategic thinking toward the region. This situation overlaps with the seemingly everlasting European economic crisis that makes it difficult for policy makers to devote resources and public opinion to pay attention to what is going on in their southern neighborhood.

Things are definitely changing in the Mediterranean. New social, political, and economic balances are in the making. Europe should be prepared to offer a credible vision of how energy interdependency can be managed to be mutually beneficial for both shores of the Mediterranean. From a political economy

4. SIVE is the Spanish acronym for *Sistema Integrado de Vigilancia Exterior*, or Integrated External Surveillance System, used to monitor and watch Spain's southern maritime borders.

perspective, the Arab Spring has made it imperative that these benefits reach all citizens – not only their elites or European firms. Turning to geopolitics, the situation in Libya, Syria, the Sahel, or the Gulf of Guinea clearly needs a European security response. Collective energy security is made of constituent parts combined in an appropriate manner. The EU should think about promoting a new approach to energy cooperation in the region, including shared preferences toward Mediterranean outer borders. To be convincing on both shores of the Mediterranean, a credible narrative for energy cooperation in the region should be developed, a narrative able to match their respective preferences. This chapter concludes by pointing up admittedly highly speculative elements for inclusion in such a renewed Euro-Mediterranean energy narrative.

The first element is the development of regional and subregional narratives. EU countries should work together to promote and support energy initiatives in the Mediterranean along the lines presented earlier, either under Euro-Mediterranean, 5 + 5, NATO Mediterranean Dialogue, bilateral, or multilateral formats. Regional visions are always more appealing and can have greater political traction, but it is the subregional and bilateral dimensions that are key to fine-tuning cooperative schemes and reducing transaction costs. The main elements of any such vision should be consistent in order to project a coherent global strategy. Polycentric governance requires strong links among the nodes (Goldthau, 2014; Ostrom, 2010).

Second, the EU discourse should focus on managing energy interdependence rather than merely reducing dependence. Interdependence is a more propitious conceptual environment for cooperation than dependence, and the EU should base its regional energy cooperation on providing such an alternative narrative of cooperative interdependency rather than continue to insist on the competitive dependency approach.

Third, the EU should foster a shift in the reasoning of southern Mediterranean producers for them to focus on development instead of rent. European preferences are consistent with promoting energy as a driver for development and cooperation in the region. EU countries are interested in preserving socioeconomic stability in the Mediterranean and accessing its energy resources under transparent rules and procedures.

Fourth, fostering a shift in the discourse toward democratic energy governance. After the Arab Spring, energy resource management in Mediterranean producing countries is being increasingly scrutinized by their populations. Energy policies will be subjected to more open scrutiny, which is essentially a good thing, but risks leading to a more populist stance. Emphasizing good governance allows for a compromise among different institutional elements that could serve as a focal point for energy policy reform.

Fifth, the regional energy narrative should be refreshed by promoting sustainability. This can be achieved by deploying renewable forms of energy, energy efficiency, technological transfers, and training, technical, and industrial cooperation. Renewables are more security neutral than conventional forms of energy, provide innovation opportunities, and offer a softer approach to energy cooperation. Current difficulties in the MSP illustrate the lack of a common narrative in such a symbolic area as renewable energy and sustainability.

Sixth, cooperative energy security is the best way to confront shared risks in the Euro-Mediterranean vicinity. Projecting security threats beyond North African geographical borders allows space for more cooperation on the southern shore of the Mediterranean itself. True, hard-energy security threats should be addressed, but solutions should seek consensus and results. Solutions should not only be embedded in both a broader and well-diversified strategy, but also in operational measures to improve security in the region. It would be advisable, however, to confine hard-energy security issues to specialized forums like NATO's Mediterranean Dialogue, 5 + 5 initiatives, and bilateral relations.

Finally, any narrative supporting the creation of a Mediterranean Energy Community should be based on the Energy Community Treaty becoming pan-Mediterranean. Most of the soft elements mentioned earlier could be addressed under the framework of an extended Energy Community Treaty, which could be opened up selectively to North African countries. Morocco and Tunisia could be the first countries to benefit from it. This would entail interesting long-term prospects such as participating in the European energy single market or even exporting virtual "green electricity" through green certificates. Opening up the EU's energy market to oil and gas producers in exchange for good energy governance and better sustainable economic development performance could bring about a warmer climate for pan-Mediterranean energy cooperation. It is just as important for political agreements to include technical cooperation (Tholens, 2014).

However, for a Mediterranean Energy Community to be credible it should be coupled (even preceded) with clear signals from the EU demonstrating it is practicing what it is preaching. To effectively promote energy integration abroad requires first achieving it at home. To be credible in pursuing a Mediterranean energy ring the EU must show it can fulfill its commitments to build interconnections across the Pyrenees or with the Baltic States. In a similar vein, fostering investments into exporting renewables toward the EU from Mediterranean Partner Countries would first require operationalizing flexibility measures inside what we now call the Energy Union. Finally, the Energy Union itself should fully include, in a well-specified manner, the linkages between currently confusing and diffuse initiatives: the Energy Community proposal, the MSP, and the new Mediterranean energy platforms. Building a coherent Euro-Mediterranean energy narrative requires consistency between not only the internal and external dimensions of the Energy Union, but also among the different European energy strategies (be it communities, plans, or platforms) toward the Mediterranean region.

REFERENCES

Corden, W.M., Neary, P.J., 1982. Booming sector and de-industrialization in a small open economy. Econ. J. 92, 825–848.

EIA – Energy Information Administration, 2011. World Shale Gas Resources: An Initial Assessment of 14 Regions Outside the United States. EIA, Washington, DC.

EIA – Energy Information Administration, 2013. World Shale Gas Resources: An Initial Assessment of 14 Regions Outside the United States. EIA, Washington, DC.

Escribano, G., 2010. Convergence towards differentiation: the case of Mediterranean Energy Corridors. Mediterr. Polit. 15 (2), 211–230.

Escribano, G., San Martín, E., 2011. Morocco and the Mediterranean Solar Plan: a driver for the development of whom? In: Morata, F., Solorio, I. (Eds.), European Energy Policy: The Environmental Dimension. Instituto Universitario de Estudios Europeos, Barcelona, pp. 209–230.

Escribano, G., San Martín, E., 2014. Managing energy interdependency in the Western Mediterranean. Paix et Sécurité Internationales 2, 81–102.

Glachant, J.M., Ahner, N., 2013. In Search of an EU Energy Policy for Mediterranean Renewables Exchange: EU-Wide System vs. 'Corridor by Corridor' Approach. Florence School of Regulation, EUI, Florence, Policy Brief, Issue 2013/06, October.

Goldthau, A., 2014. Rethinking the governance of energy infrastructure: scale, decentralization and polycentrism. Energy Res. Social Sci. 1, 134–140.

Kagan, R., 2003. Of Paradise and Power: America and Europe in the New World Order. Alfred A. Knopf, New York.

Ostrom, E., 2010. Beyond markets and states: polycentric governance of complex economic systems. Am. Econ. Rev. 100 (3), 641–672.

Rubino, A., 2015. Three platforms for no Mediterranean (energy) policy. ISPI Energy Watch Comment, May 4. Available from: http://www.ispionline.it/it/energy-watch/three-platforms-no-mediterranean-energy-policy-13227

Tholens, S., 2014. An EU–South Mediterranean Energy Community: the right policy for the right region? Int. Spect.: Ital. J. Int. Affairs 49 (2), 34–49.

Part II

Challenge of Market-Based Regulation

Chapter 8

EU Pressures and Institutions for Future Mediterranean Energy Markets: Evidence from a Perception Survey

Carlo Cambini*, Alessandro Rubino**
*Politecnico di Torino, DIGEP Department of Management, Torino; IEFE, Bocconi University, Milan, Italy; **DISAG Department of Business and Law Studies, University of Bari Aldo Moro, Bari, Italy

1 INTRODUCTION

The Mediterranean basin is a densely populated area that spans three continents and includes 24 countries.[1] This region has traditionally been the center of intense trade that has brought together different cultures and traditions offering the opportunity for mutual growth but also occasions for strong disagreement and conflict. Energy has added to this complex picture an essential geopolitical dimension that contributed to defining the relationship between the shores of the Mediterranean Basin for several decades. The presence of energy-exporting countries (mainly Algeria, Libya, and Egypt) has meant large quantities of oil and gas have been traded from southern Mediterranean countries to the industrialized demand hubs in the north creating a south–north flow of energy.

Notwithstanding these intense trading activities and vast economic interdependencies, countries in the area fail to show convergence in terms of macroeconomic and demographic fundamentals, and a significant disparity is notable between the two shores of the Mediterranean Basin. Following OME (2013) analysis, simply by looking at the gross domestic product (GDP) and the population distribution of countries belonging to the region, we discover some interesting diverging patterns that can be highlighted.

1. Albania, Algeria, Bosnia and Herzegovina, Croatia, Cyprus, Northern Cyprus, Egypt, France, Greece, Israel, Italy, Jordan, Lybia, Lebanon, Malta, Montenegro, Morocco, Palestine Authority, Portugal, Syria, Slovenia, Spain, Tunisia, and Turkey.

Regulation and Investments in Energy Markets. http://dx.doi.org/10.1016/B978-0-12-804436-0.00008-4

Southeast Mediterranean countries[2] (SEMCs), although currently accounting for only 25% of total GDP of the region, are expected to grow at twice the rate of north Mediterranean countries (NMCs) until 2030, when they will account altogether for around one third of total GDP of the region. In terms of population, we observe a similar trend: the population will grow in SEMCs at a faster rate than the north and this will imply that, by 2030, 60% of the population will be based in the countries situated along the south shore of the basin.

These patterns are also mirrored in the energy sector by the most accredited forecast (Obervatoire Mediterranéèn de l'Energie and Medgrid, 2013) and describe an interesting evolution of the demand/supply landscape in the years up to 2030. Depending on the scenarios considered,[3] SEMCs will consume between 43% and 46% of total electricity demand, growing at an annual rate in the range of 2–2.7%, against 1.4% in the conservative scenario (the most encouraging) for northern countries.

The bulk of the growth in demand will take place in SEMCs and will be mostly concentrated in Egypt and Turkey (accounting for over 60% of expected demand in 2030) in stark contrast with the situation in 2009 when NMCs[4] accounted for over 70% of total demand, implying a rapid and radical transformation of the energy pattern in the region. In this landscape it is notable how renewable energy sources (RES) generation will play an increasing role as part of the energy mix up to 2030 in the region, independently of the scenario considered. Nonprogrammable renewable generation will represent the majority of newly installed capacity up to 2030 (accounting for 50% in the conservative and 80% in the proactive scenario, respectively) meaning that, according to the conservative scenario, in excess of 230 or even 300 GW of RES will be generated in the coming 15 years in the Euro-Mediterranean (Euro-Med) area.

These projections describe a radical change in electricity patterns that started to occur in 2008, when the financial and economic crises and the ongoing demographic decline radically reduced electricity demand outlook in most countries in Europe. The situation since then has not changed and most NMCs have now to cope with considerable installed capacity that is underutilized or mothballed. This panorama has also required a sharp revision of most major industrial initiatives in the Mediterranean Basin (such as DESERTEC and Medgrid) which envisaged the possibility of installing new RES generation in SEMCs, mostly to provide considerable additional capacity to be exported to NMCs, and a significant increase in electricity consumption (particularly in the demand for clean electricity supply from NMCs). RES flow, according to these industrial projects,

2. Morocco, Algeria, Libya, Egypt, Jordan, Palestine Authority, Syria, Lebanon, Israel, and Turkey.

3. The *Conservative Scenario* (CS) takes into account past trends, current policies, and ongoing projects but with a slow rate of adoption of new policy measures and planned projects. The *Proactive Scenario* (PS) assumes the implementation of strong energy efficiency programs and great diversification of the existing energy supply mix (Obervatoire Mediterranéèn de l'Energie and Medgrid, 2013, p. 25).

4. Portugal, Spain, France, Italy, Greece, Slovenia, Croatia, Bosnia and Herzegovina, Montenegro, and Albania.

would have mirrored primary energy fuel flows, and should have continued to move northward (Zickfeld and Wieland, 2012). This paradigm, which has shaped major industrial initiatives in the electricity sector, is partly under question. The majority of new electricity demand will take place in SEMCs in the coming years. Considering the abundant underutilized capacity installed, the region should also experience significant north–south flows (at least up to 2030), which have an inverse direction compared with those initially projected.

A new electricity paradigm is therefore taking shape, characterized by two main factors. First, significant additional generation capacity as a result of the increase of demand in southern Mediterranean countries will need to be installed, which will mean reorganizing electricity flows in the region. This, in turn, will have major implications for the structure and direction of financial flows within the region. Second, RES technologies are quickly developing (IRENA, 2013). This, together with increasing rates of RES penetration, is likely to alter the electricity supply industry (ESI).

These two factors entail redefining and recalibrating industrial initiatives in the region. In addition, significant financial support in excess of €700 billion will be needed in the coming 16 years to meet the investment needs of the region (Obervatoire Mediterranéèn de l'Energie and Medgrid, 2013, p. 30). Moreover, this will call for a new and different understanding of energy policies and what this means in terms of the revised regional energy initiatives in the area as well as their wider social impact (Coady et al., 2010; Cottarelli et al. 2013; Fattouh, El-Katiri, 2012). These dynamics will also be strongly influenced by emerging technology that will become pivotal in the future energy mix in the region: RES generation will significantly affect the future market structure, the regulatory approach, and definition of the energy paradigm in the Mediterranean region. In this chapter, we investigate EU action as rules promoter in the energy sector (Lieberman and Doherty, 2008). We also identify the capacity of the emerging regulatory framework to attract the level of investment needed, taking into consideration three rules diffusion patterns: bottom-up, top-down, and network approaches. We do so by investigating the pressure applied by the EU in shaping future energy markets in the Mediterranean. The research method adopted consists of a perception survey directed at 20 energy experts coming from 11 non-EU Mediterranean countries.[5]

The chapter is organized as follows: in Section 2, we introduce the EU external energy policy and briefly illustrate the literature on norm diffusion and regulatory change in the energy sector. Section 3 outlines the methodology proposed and the results of the perception survey administered to selected energy experts.[6] Section 4 offers up some final considerations and highlights the policy implications of the results.

5. Albania, Algeria, Bosnia and Herzegovina, Croatia, Egypt, Israel, Jordan, Libya, Montenegro, Palestine Authority, and Turkey.

6. Participants of the training course "New Challenges for Energy System in the Mediterranean Region" held in Venice, May 24, 2013.

2 NORMATIVE DIFFUSION IN THE ENERGY SECTOR

The EU external energy policy toward its Mediterranean neighbors has been constant in recent decades and has gained significant momentum with the treaty of Lisbon, which entered into force in 2009. The treaty endowed the community with formal competences in the energy field. However, EU energy policy has a longer history and is part of its international dimension characterized by normative power (Manners 2002). Normative power refers to the way the EU reinforces and maintains its international legitimacy by defending (and exporting) its norms, regulations, and institutions. According to this approach, the EU community is a social construct, which unlike traditional nation-states, has to build its identity by shaping the legal, moral, and regulatory dimensions of its member states. This also represents EU's main value when dealing with external actors and when operating in the international domain. EU policy in international relations is predominantly characterized by the intention to export its internal values and norms outside EU borders (Del Sarto, 2010). Extending the concept of "normative power" to the energy area is a complex task and represents an unresolved jigsaw, when considering the overlap with member state external energy policies and the countless bilateral relationships with non-EU countries (at the national level) existing in the energy field. It is worth underlining that national executives still hold a strong grip on energy matters both in the EU and in non-EU countries in the region. The promotion of energy norms and regulations, in particular, can be described by means of three main routes (Cambini and Franzi, 2014): bottom-up pressures, hierarchical top-down approaches, and network pressures for rules change and adoption.

EU "normative power" in the energy sector has been reaffirmed through different policy instruments that utilize a range of these three approaches and implement them with differing degree of success. We will now briefly provide an overview of these instruments and programs in an effort to analyze how normative diffusion has been translated in EU external energy policy.

The first route stems from the Euro-Mediterranean Partnership[7] (EMP). The EMP aims at enhancing integration both at the regional and bilateral level, by establishing an area of peace, security, and economic development. However, social and political unrest in the area as well as unsatisfactory regional integration in most of these vital economic and social dimensions called for the design and implementation of a broader all-encompassing program that could foster deeper integration and harmonization among the EU and its closest neighbors.[8] In 2004, the European Neighbourhood Policy (ENP)[9] was launched toward 16 countries that represented the external borders of the EU. The ENP offered financial support and a stake in the internal market in exchange for

---

7. Also known as the "Barcelona Process."

8. Albania, Armenia, Azerbaijan, Belarus, Egypt, Georgia, Israel, Jordan, Lebanon, Libya, Moldova, Morocco, Occupied Palestinian Territory, Syria, Tunisia, and Ukraine.

9. COM(2004) 373 final, Brussels, 12.05.2004.

consolidation of the harmonization process. "Everything but membership" was therefore obtainable to promote transformations in accordance with the *acquis communautaire.* Progressive consolidation of the internal EU energy market called for a durable external energy policy, mainly addressed at reinforcing energy security (Tholens, 2014). The ENP consisted in action plans established bilaterally between the EU and each partner country, complemented by three additional multilateral policies (Vantaggiato, 2014). These actions were directed at promoting cooperation platforms, such as the Association of Mediterranean Energy Regulators (MedReg) and the Association of the Mediterranean TSOs (Med-TSO), which seek to encourage cooperation and convergence in energy regulation and electricity transmission network management and interconnection expansion across the Mediterranean region.

The second route envisaged the creation of a more institutionalized instrument with the remaining neighboring countries; namely, those in the southeast and west Balkan region. This initiative led to the creation of a full-fledged top-down initiative, the Energy Community[10] (EnC), which aimed at consolidating the South Eastern European (SEE) energy market and connecting it with the EU internal market via a legal tool (the Energy Community Treaty) which could bring together preaccession countries and countries with membership perspectives. The EnC is the institutional model that has also informed the current commitment to the creation of a south Mediterranean Energy Community, based on the transposition of EU law. This proposal was recalled in the May 2011 Communication on "Democracy and Shared Prosperity" (European Commission, 2011, p. 9) and more recently reintroduced in the December 2013 Union for the Mediterranean Ministerial Meeting on Energy.[11] According to Tholens (2014), this confirms the existence of this ambition and interest in "high-level policy circles."

The third traditional route to advancing institutional change and rule adoption is via convergence toward a target model represented by international standards (possibly the EU model) proposed by domestic actors (Escribano, 2010). This envisages voluntary convergence promoted by the executive powers responsible for energy policies in Mediterranean countries. This process required the contemporaneous verification of two sets of conditions (Berbè et al., 2009): the existence of a credible model able to attract convergence and the desire (sometimes the necessity) to demonstrate a convincing commitment toward system reforms for reasons of system stability (e.g., to attract foreign direct investment or to promote restructuring of the sector for social and economic reasons).

In the remainder of the chapter we will explore how these factors and the dynamics at play are perceived among energy experts and whether they could be

10. The treaty establishing the Energy Community was signed on October 25, 2005 in Athens by the European Community and then nine contracting parties from South East Europe. Following ratification, the treaty entered into force on July 1, 2006 (EC, 2006).

11. A topic on the agenda was "Political discussion on a Mediterranean Energy Community."

conducive to a unifying conceptualization of the functional and political rationales that underpin the formation and design of regulatory networks and could explain the move (or lack thereof) toward a harmonized energy Mediterranean framework.

3 PERCEPTION OF RULES PROMOTION: RESULTS FROM A SEMISTRUCTURED SURVEY

In addition to official declarations, policy decisions, and formal positions that have been extensively analyzed in the literature,[12] we consider it relevant to analyze how stakeholders, involved in the final application of rules discussed and promoted within the energy context, perceive the entire process and the distinct steps that have been taken by Mediterranean countries from the perspective of the dynamics mentioned in the previous section.

Perception surveys are increasingly utilized to understand the impact of regulatory intervention. They are commonly used to acquire information from an end user point of view. The information collected from perception surveys might be used by policy makers for (1) regulatory policy evaluation and design and (2) communication purposes. Perception surveys therefore represent useful diagnostic tools to identify areas of concern to business and citizens and to inform future regulatory reforms.

Our survey is based on a tested questionnaire that was utilized in an earlier exercise (Cambini and Franzi, 2014) carried out in May 2011 during the MedReg training seminar on "Mediterranean Energy Regulation" held in Florence. The study was based on analysis of just 5 countries whereas the scope of the current study is to extend the survey to 11 countries. The questionnaire investigates the perception of energy experts and regulators of the role that both the EU and energy networks (such as MedReg) have in terms of rule adoption.

The survey is structured according to OECD recommendations (OECD, 2012) and aims particularly at evaluating the regulatory reform process. It has been tested and reviewed, in its original version, by experts from the Oxford Institute for Energy Studies, the Italian International Affairs Institute, and from RES4MED. The questionnaire asked respondents to evaluate various elements considered on a scale from 0 to 5. The results collected in this and in the previous exercise pass the median answer test (Kwon and Shin, 2010) and therefore can be considered relevant for the level of information reported. The survey was submitted to 20 energy experts, working in ministries, regulatory authorities, and energy companies from 11 non-EU Mediterranean countries: namely, Albania, Algeria, Bosnia and Herzegovina, Croatia, Egypt, Israel, Jordan, Libya, Montenegro, the Palestinian Territories, and Turkey. These energy experts participated in the training course for energy experts entitled "New Challenges for Energy System in the Mediterranean Region" held in Venice, May 20–24, 2013. Respondents

12. For a thorough revision, see Cambini and Rubino (2014).

were well informed about the network role and competences and all possessed at least 5 years of seniority in the energy field. The participants therefore possessed in-depth knowledge of many relevant aspects of the energy supply industry as well as of its institutional framework.

In addition, we coupled the survey with a questionnaire requesting an opinion around the most urgent challenges that the Euro-Mediterranean energy sector was facing, the priority action that should be put in place to properly address these challenges, and the impact of energy matters on other dimensions such as social, environmental, and macroeconomic ones.

The results of the survey are analyzed in depth in the remainder of this section, starting with an analysis of the perception of top-down approaches and network diffusion processes. This is then followed by an evaluation of the role and relevance of local and domestic actors.

3.1 Top-Down and Network Pressures for Rules Promotion

Our survey initially investigated the role that the EU could play as energy rule promoter. We asked respondents to express their evaluation on a scale from 0 to 5 to assess participant perception of direct (top-down) pressure compared with EU indirect pressure. The question was: *On a scale from 0 to 5, how do you perceive the EU (i.e., European Commission Directorate for Energy, EU Development Cooperation Office mainly) methods for the energy rules promotion?* Respondents had to choose from three different options: (1) direct pressure (top-down); (2) indirect (horizontal, participatory model); and (3) absence of either direct or indirect pressure.

Figure 8.1 shows the results. It is interesting to notice the significant variability of the perception, which was not evident in Cambini and Franzi (2014). Countries belonging to the EnC seem to perceive a stronger EU role, both via direct and indirect pressure. Not surprisingly, energy experts from Egypt and Algeria reported the lowest perceived level of EU pressure. The result also shows a generalized declining impact of the EU role in rule promotion compared with the results of the previous perception survey.

We then went on to explore the issue of direct and indirect pressure applied by the EU, with a second question: *In your opinion, the EU action in the energy sector is more effective when: acting directly in bilateral relations, or when promoting energy rules indirectly through regulatory networks, such as MedReg?* Ten respondents expressed a preference for direct bilateral action, emphasizing that the EU should also promote direct investment in the energy sector in the region. However, nine respondents considered indirect, networking initiatives more effective. The remaining respondent did not have a clear opinion. This response confirms that both methods are appreciated by energy experts and operators, suggesting a slight preference for direct EU engagement. However, these results mark a significant rebalance of perception compared with the response obtained in the first edition, which was overwhelmingly in

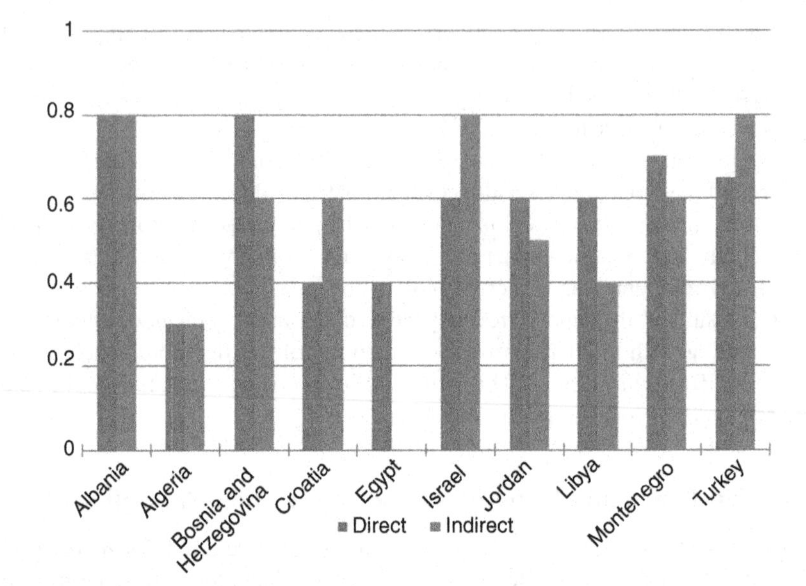

FIGURE 8.1 EU pressure for rule change.

favor of "direct EU action." These responses are in line with reduced confidence in EU efficacy in promoting economic policies outside its borders, as recalled by Tholens (2014).

We then paid special attention to the EU's role in direct rule promotion, with special reference to energy regulation. In our third question we asked: *On a scale from 0 to 5, how would you say that the cooperation with the EU impacts in terms of rules adoption or rules change in the following areas?* The question referred to: (1) the setting of tariffs, (2) retail market competition, (3) unbundling, (4) third party access (TPA) regime, (5) energy efficiency programs, (6) incentive for renewables, (7) import risk analysis (IRA) political independence, (8) attention to vulnerable customers, and (9) IRA stakeholder independence.

Figure 8.2 reports mean respondent perception for each of the areas considered. The strongest impact was experienced in those aspects related to the creation of the conditions for market opening such as TPA and national regulatory authority (NRA) independence. Moreover, climate-related energy policy was significantly impacted by EU actions. It is remarkable that EU pressures resulted in a greater score for countries participating in the EnC (Albania, Croatia, Bosnia and Herzegovina, Montenegro).[13] Unbundling, retail, and customer protection remain areas where rules adoption has exerted the smallest impact and therefore need to be considered areas of concern where specific direct action should be identified.

13. See Table 8.1 for a detailed report of the results.

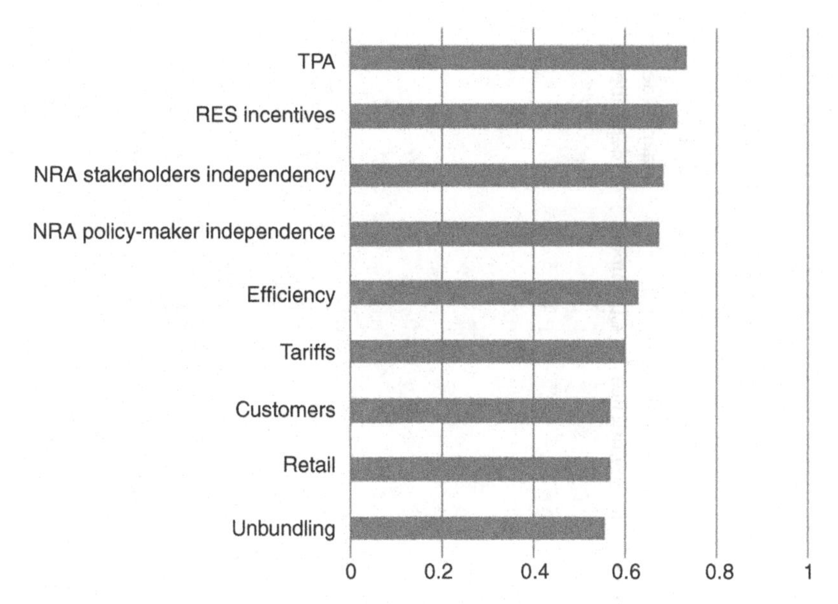

FIGURE 8.2 **EU role in rules adoption.**

To conclude the section dedicated to the role the EU plays in rule promotion, we asked respondents to evaluate on a scale from 0 to 5 their agreement with the following statement:

1. *Rules for energy sector organization are based on the EU model.*
2. *Energy sector organization is based on Internationally Recognized Standards (IS).*
3. *EU energy rules are very often mentioned in the public discourse of domestic decision makers.*
4. *Taking into consideration energy, domestic decision makers tend to conform their behavior, to EU recommendation.*

These questions intend to assess the level of adherence, in the Mediterranean region, to standards and recommendations proposed by the EU. Answers to this question provide information on the influence of the EU as a normative power, not only in the formal adoption of norms but also in its behavioral dimension (Schimmelfennig and Sedelmeier, 2005), where the former refers to the inclusion of EU rules in national domestic legal systems and the latter refers to consideration of the EU as a model and reference for domestic actors.

Figure 8.3 reports the average impact of participant responses. The question assessed the degree of adherence to the model recommended by the EU. The results show that international standards, rather than the EU model, have the largest impact on the domestic rule system. The aim is to understand the degree of Europeanization that Middle East and North African (MENA) countries are experiencing.

TABLE 8.1 EU Role in Rules Adoption

	Tariffs	Retail	Unbundling	TPA	Efficiency	RES incentives	NRA policy-maker independence	Customers	NRA stakeholder independence
Albania	0.8	0.8	1	1	0.8	0.8	0.6	0.6	0.8
Algeria	0.7	0.6	0.8	0.5	0.9	0.8	0.8	0.3	0.8
Bosnia and Herzegovina	0.8	0.6	0.6	1	0.6	0.8	0.8	0.8	1
Croatia	0.8	0.6	0.8	0.9	0.8	0.9	0.8	0.5	0.7
Egypt	0.3	0.7	0		0.4	0.5	0.8	0	0.6
Israel	0.4	0.6	0.6	0.6	0.8	0.6	0.4	0.6	0.8
Jordan	0.6	0.6	0.5	0.6	0.5	0.6	0.6	0.6	0.6
Libya	0.6	0.4	0.4	0.6	0.6	0.6	0.6	0.6	0.4
Montenegro	0.7	0.6	0.6	0.7	0.7	0.7	0.7	0.8	1
Palestinian Authority	0.2	0	0	0.6	0	0.8	0.4	0.6	0
Turkey	0.7	0.75	0.8	0.85	0.8	0.75	0.9	0.85	0.8
Average	0.60	0.57	0.55	0.74	0.63	0.71	0.67	0.57	0.68

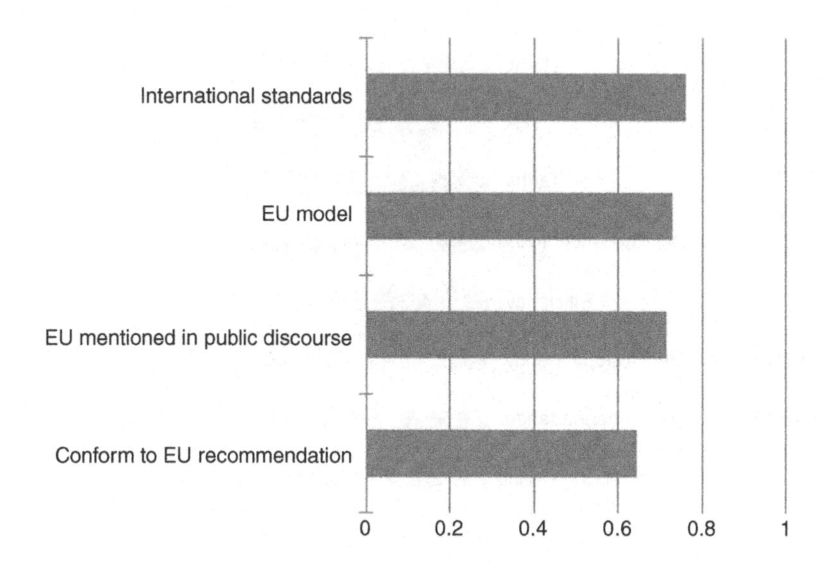

FIGURE 8.3 Conformity with EU rules system.

It is evident that the presence (and increasing direct activity) of other external actors is able to shape the energy regulatory framework but, although the EU is sufficiently present in the public debate, the domestic rule system does not entirely conform to EU recommendations.

The survey then moved on to explore the rule of networks in rule diffusion. In order to compare the relative impact of the energy network, mainly MedReg, we asked respondents to evaluate the same regulatory aspects we investigated earlier for the EU. The question was: *On a scale from 0 to 5, how would you say that the cooperation within energy networks impacts in terms of rules adoption or rules change in the following areas?* Like the earlier question, this one also referred to: (1) the setting of tariffs, (2) retail market competition, (3) unbundling, (4) TPA regime, (5) energy efficiency programs, (6) incentive for renewables, (7) IRA political independence, (8) attention to vulnerable customers, and (9) IRA stakeholder independence (Fig. 8.4).

In general, the impact of energy network is perceived significantly lower than the impact of direct or indirect EU action.

It is interesting to see that the largest impact is typically reached in areas where the EU has been less effective (see Fig. 8.3) showing interesting complementarities. TPA remains an area where regulatory networks have also exerted a significant impact highlighting its relevance in market opening and that TPA is a prerequisite to initiating liberalization policies. A common feature of the EU and energy network role is the scant impact on customer protection and unbundling. Both areas remain unresolved issues in many Mediterranean countries and are perceived as themes where neither the EU nor regulatory networks

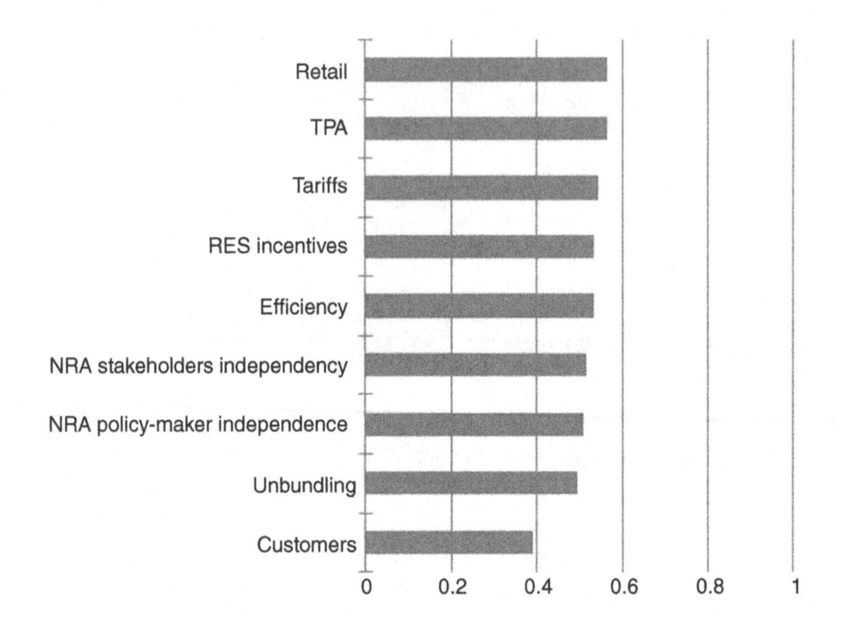

FIGURE 8.4 Network role in rules adoption.

play an effective role. Unbundling vertically integrated utilities and protection of vulnerable customers remain areas of activity under the direct influence of national policy makers. Careful analysis of the results shows the existence of a diverse range of perceptions among the experts interviewed. The low impact recorded for Turkey stands out (see Table 8.2). These findings are in line with those of Cambini and Franzi (2014).

The following question further explored the main features of network governance and investigated the type of relations that characterize MedReg activities. We asked: *On a scale from 0 to 5, how would you say that the following options characterize members' relations within MedReg?* The proposed options were: (1) a participatory model of decision-making; (2) codified procedural rules (regarding the definition of meeting agendas, voting systems, etc.); (3) monitoring and control procedures (regarding principles and rules to be implemented in the energy sector); and (4) resource sharing (primarily sharing expertise and know-how) (Fig. 8.5).

The answers to this question are summarized in Fig. 8.6, which reports the impact network governance has on rule promotion. As Cambini and Franzi (2014) illustrate, network rule promotion is typically characterized by a participatory model, in which lessons learned and acquired expertise are shared among members. The answers suggest that MedReg function is in line with findings in the literature. In fact, MedReg is characterized by significant structural flexibility; therefore, the relatively limited impact of monitoring procedures and codified procedural rules is not surprising. The voluntary nature of MedReg and its current

TABLE 8.2 Network Role in Rules Adoption

	Tariffs	Retail	Unbundling	TPA	Efficiency	RES incentives	NRA policy-maker independence	Customers	NRA stakeholder independence
Albania									
Algeria	0.7	0.6	0.7	0.7	0.8	0.8	0.8	0.5	0.7
Bosnia and Herzegovina	0.4	0.2	0.2	0.4	0.4	0.4	0.2	0.4	0.4
Croatia	0.7	0.6	0.7	0.6	0.7	0.6	0.5	0.4	0.6
Egypt	0.6	0.6	0.6	0.6	0.5	0.7	0.8	0	0.7
Israel	0.4	0.5	0.5	0.5	0.6	0.6	0.5	0.4	0.6
Jordan	0.6	0.7	0.6	0.6	0.7	0.6	0.5	0.6	0.6
Libya	0.8	0.8	0.6	0.6	0.6	0.6	0.8	0.6	0.6
Montenegro	0.8	0.8	0.6	0.6	0.8	0.8	0.6	0.6	0.6
Palestinian Authority	0.2	0.2	0	0.6	0.2	0.2	0.4	0.4	0.2
Turkey	0.25	0.65	0.45	0.45	0.05	0.05	0	0	0.15
Average	0.55	0.57	0.50	0.57	0.54	0.54	0.51	0.39	0.52

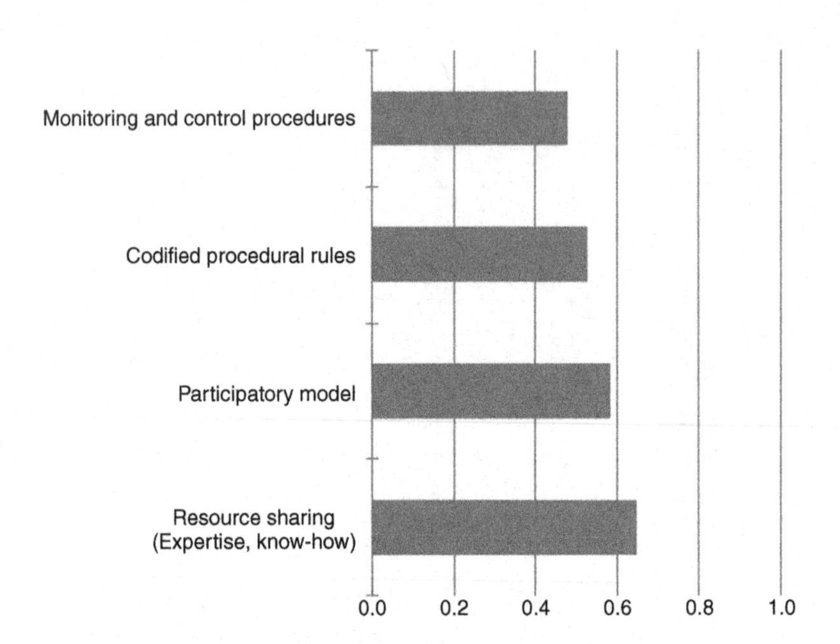

FIGURE 8.5 Energy networks – rules promotion.

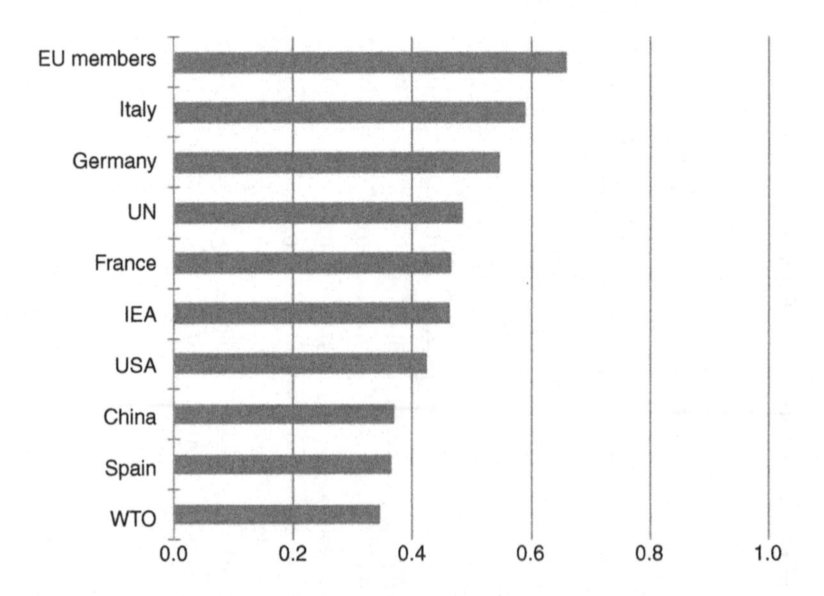

FIGURE 8.6 Perception of other international actors.

mandate confirm the perception of energy experts interviewed and are in line with Casey and Lawless (2011). In the absence of significant sanctioning ability and/or capacity to provide considerable incentives to cooperate, the impact of MedReg remains marginal in rule making, monitoring, and legal enforcement and can only rely on voluntary agreement. Significant steps need to be taken in this area in order to allow the association to play a significant role in shaping regulator actions. However, resource-sharing and capacity-building initiatives represent interesting areas to further MedReg activities and reinforce its role in the region.

3.2 Bottom-Up Pressures for Rules Promotion

The perception of respondents of the impact other international actors, regional networks, and domestic stakeholders have on energy rules adoption was also investigated. On a scale from 0 to 5, the respondents were asked to compare the role of the three types of actors: other international actors (with the exception of EU bodies previously investigated), networks (MedReg included), and domestic actors. The answers to these additional three questions are illustrated in this section.

The survey investigated the impact of other international actors and regional networks. The question was: *On a scale from 0 to 5, how do you perceive the pressure in favur of energy rules adoption and rules change of the following actors?*

We evaluated the role of other international actors, but took into consideration that the EU does not always speak with "one single voice." Therefore, we evaluated (see Fig. 8.7) whether other member states play a role in determining rule adoption in the energy sector. As a result the efficacy of other alternative cooperation and donor programs is also evaluated.

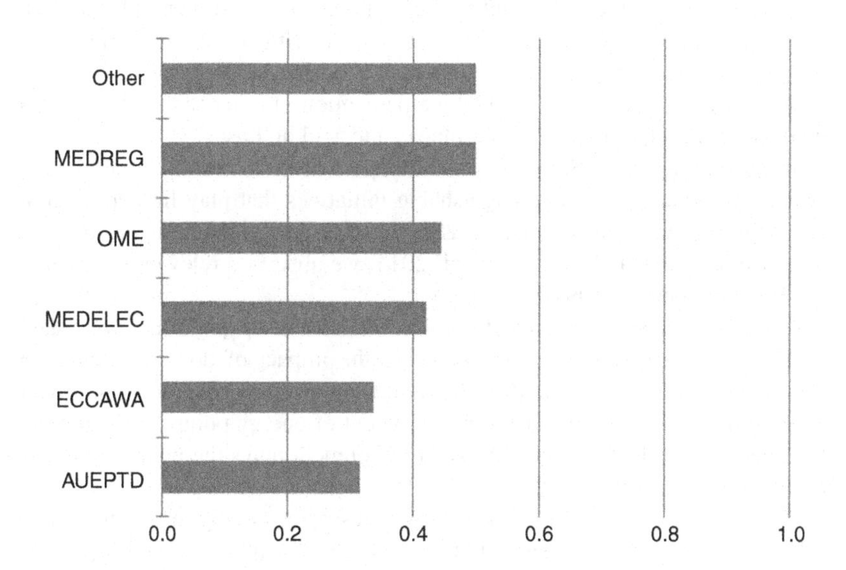

FIGURE 8.7 Role of regional networks.

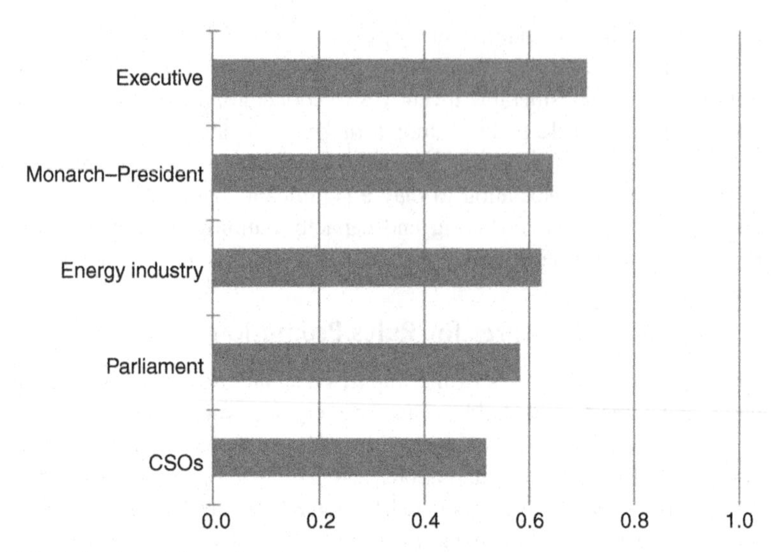

FIGURE 8.8 Role of domestic actors.

It comes as little surprise that member states have bilateral energy relations within the region, which overlap and/or replace EU external energy policy. The Italian and German influence is perceived as being particularly effective in the region. It is interesting to note that, although China is not perceived as being too influential globally, its pressure is strongly felt in certain countries (such as Algeria and Egypt). This possibly suggests that China has selected some target countries to most effectively implement its global energy strategy (Table 8.3). Germany's relevance comes as no surprise, considering its economic impact in Europe and in the region, as well as its direct and indirect actions in the MENA.

A subsequent question evaluated the perception of other regional networks of cooperation active in the Mediterranean and Middle East.

In general, as Fig. 8.8 points out, the impact of networks of cooperation is weak, with a number of indistinguishable initiatives that play limited or little role in the region. MedReg emerges as the most visible initiative. If we compare these results with Cambini and Franzi (2014) we observe a relevant increase in MedReg perception of about 10% points.

Finally, the survey evaluated the role of bottom-up pressure. We asked respondents to express how they perceive the impact of domestic actors in promoting rule changes and rules adoption. Bottom-up pressure was confirmed as being the most relevant for the development of energy policy in the region. As expected, the domestic institutional environment plays the main role in defining sectoral reform.

The relevant role played by the energy industry, especially in countries such as Egypt and Algeria, Eberhard and Gratwick (2007) explains the limited progress achieved in unbundling existing vertically integrated utilities (see Table 8.4).

TABLE 8.3 Perception of the Role of Other International Actors

	EU members	France	Germany	Italy	Spain	USA	China	UN	WTO	IEA
Albania	0.8		1	1		1				0.8
Algeria	0.5	0.5	0.5	0.6	0.6	0.6	0.8	1		0.6
Bosnia and Herzegovina	0.8	0.4	0.2	0.8	0.2	0.4	0.2	0.2	0.2	0.4
Croatia	0.9	0.8	0.8	0.7	0.5	0.1	0.2	0.4	0.3	0.3
Egypt	0.6	0.5	0.6	0.4	0.5	0.3	0.8	0.5	0.2	0.6
Israel	0.8	0.6	0.8	0.6	0.6	0.6	0.4			0.8
Jordan	0.2	0.2	0.2	0.2	0.2	0.2	0.2	0.2	0.2	0.2
Libya	0.4	0.4	0.4	0.4	0.4		0.4	0.4	0.4	0.4
Montenegro	0.8	0.5	0.5	0.7	0.5	0.5	0.5	0.5	0.6	0.6
Palestinian Authority	0.6	0.4	0.2	0.4	0	0	0	0.6	0.2	0
Turkey	0.85	0.35	0.8	0.7	0.15	0.55	0.2	0.55	0.65	0.4
Average	0.66	0.47	0.55	0.59	0.37	0.43	0.37	0.48	0.34	0.46

TABLE 8.4 Role of Domestic Actors

	Executive	Monarch/ President	Parliament	CSOs	Energy industry
Albania	1	1	1	1	0.8
Algeria	0.7	0.3	0.3	0.4	0.7
Bosnia and Herzegovina	0.8		0.6	0.4	0.6
Croatia	0.8	0.9	0.7	0.7	0.4
Egypt	0.7	0.6	0.6	0.5	0.8
Israel	0.7	0.7	0.6	0.5	0.6
Jordan	0.5	0.5	0.5	0.6	0.6
Libya	0.6	0.8	0.6	0.4	0.6
Montenegro	0.8		0.6	0.6	0.6
Palestinian Authority	0.4	0.4	0.2	0	0.6
Turkey	0.8	0.6	0.7	0.6	0.55
Average	0.71	0.64	0.58	0.52	0.62

This indication might also be considered in order to identify the obstacles hampering market harmonization. This is also a reminder that energy sector reforms have an important institutional dimension, in addition to the economic one (Glachant and Perez, 2007). At the same time, these results together with the indication concerning the role of international actors seem to confirm that globalization,

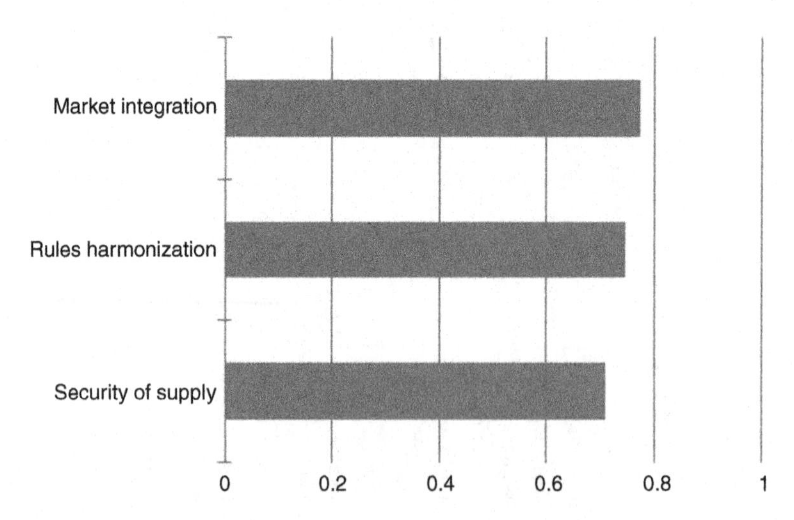

FIGURE 8.9 Drivers in Euro-Mediterranean energy cooperation.

described as the combined effect of international pressures and standards coupled with domestic instances, is currently combined to shape energy policies in the MENA.

Finally, we asked participants to evaluate the main driver able to shape Euro-Mediterranean energy cooperation. The results, summarized in Fig. 8.9, indicate that the three possible drivers are considered almost equally important in the region. However, the slightly larger impact that market integration registered allowed us to consider it as the main driver behind energy cooperation in the region, confirming that a technical process could promote convergence more effectively than a political process when it comes to targeting full harmonization (Tholens, 2014).

4 CONCLUSIONS

This chapter analyzed the main results of a perception survey addressed to energy experts from the Mediterranean region. Survey responses came from 20 energy experts from 11 countries in the region. The panel was therefore representative of countries involved in Euro-Mediterranean cooperation in the energy sector. The survey was aimed at assessing the way in which energy rule promotion is taking place in the Mediterranean region. Therefore, we explored three sources of pressures fostering rule change and transition: hierarchical (top-down), network, and bottom-up.

When we look at top-down pressure, the results confirm that respondents consider direct bilateral action from the EU as most effective. These results are significantly lower than those from the previous perception survey. They could partially be explained because of the limited progress harmonization has made so far. Comparing the EU and the network role, the survey shows a great degree of complementarity. Whether this is the effect of partitioning of responsibility and roles or the effect of EU supremacy rather than network actions which result in incompatibility between these two sources of pressure needs to be further explored.[14]

In contrast with the results of Cambini and Franzi (2014), our results, while confirming the remarkable role rule changes play in the region, register an even larger (and growing) impact of International Standards (other than EU standards). This is also in line with the significant role that emerging superpowers are now playing in larger markets in the region (Egypt and Algeria). MedReg, together with other energy networks, while increasing its impact in the region, still fails to play an effective role in terms of rules promotion. Nevertheless, its role and visibility has steadily increased since 2011. It is larger than other energy networks in the Mediterranean. MedReg is considered to be particularly effective at promoting knowledge sharing and as a platform to disseminate know-how.

14. For a revision of the conceptualization of the EU regulatory network see Blauberger and Rittberger (2014).

The results confirm that the Euro-Mediterranean region is still fragmented, where domestic actors play a leading role in promoting rule adoption and institutional change. In this framework, vertically integrated utilities and national champions exercise significant veto power and are able to halt or slow down creation of an integrated regional market. The trends and dynamics mentioned in this chapter represent features well known in the regulatory and governance literature and can be reconciled with the effect of so-called "European Regulatory Space" (Thatcher and Coen, 2008). In particular, the existence of regional regulatory networks, which represent a complex and multidimensional phenomenon, can be appropriately explained and described, according to Blauberger and Rittberger (2014), by the concept of orchestration. Orchestration refers to the soft and indirect mode of governance in which an "international organisation empowers intermediary agency to further its governance goals" (Abbott et al., 2015). According to this new approach, the entire EU external energy policy can be reassessed, taking into consideration that the orchestration focus is not necessarily finalized in formal structures (such as an energy community), but can also take place through soft and informal elements (such as MedReg). Future work will consist in better understanding the institutional dynamics and the relationship between the orchestrator (EC), its intermediaries (MedReg and EnC), and target actors (NRAs).

ACKNOWLEDGMENT

Alessandro Rubino gratefully acknowledges Enel Foundation for the support provided for the collection of the results during the training course "New challenges for energy system in the Mediterranean Region" held in Venice on May 24, 2013.

REFERENCES

Abbott, K.W., Genschel, P., Snidal, D., Zangl, B., 2015. Orchestration: global governance through intermediaries. In: Abbott, K.W., Genschel, P., Snidal, D., Zangl, B. (Eds.), International Organisation as Orchestrator. Cambridge University Press, Cambridge.

Berbè, E., Costa, O., Harranz Surallés, A., Johansson-Nogues, E., Natorski, M., Sabiote, M.A., 2009. Drawing the neighbours closer to what? Explaining emerging patterns of policy convergence between the EU and its neighbours. Coop. Confl. 44, 378–399.

Blauberger, M., Rittberger, B., (2014). Conceptualizing and theorizing EU regulatory networks. Regulation & Governance. doi: 10.1111/rego.12064.

Cambini, C., Franzi, D., 2014. Assessing the EU pressure for rules change: the perceptions of southern Mediterranean energy regulators. Mediterr. Polit. 1. 59–81.

Cambini, C., Rubino, A., 2014. Regional Energy Initiatives: Medreg and the Energy Community. Routledge, London.

Casey, D., Lawless, J.S., 2011. The Parable of the Poisoned Pork: Network Governance and the 2008 Irish dioxin contamination incident. Regulation & Governance 5(3), pp. 333–349.

Coady, D., Gillingham, R., Ossowski, R., Piotrowski, J., Tareq, S., Tyson, J., 2010. Petroleum Product Subsidies: Costly, Inequitable, and Rising. International Monetary Fund, Washington, DC.

Cottarelli, C., Sayeh, A.M., Ahmed, M., 2013. Energy Subsidy Reform: Lessons and Implications. IMF Policy Paper.

Del Sarto, R., 2010. Borderlands: the Middle East and North Africa as the EU's Southern Buffer Zone. In: Bechev, D., Nicolaidis, K. (Eds.), Mediterranean Frontiers: Borders, Conflicts and Memory in a Transnational World. I.B. Tauris, London, pp. 149–167.

Eberhard, A., Gratwick, K.N., 2007. From state to market and back again: Egypt's experiment with independent power projects. Management Programme in Infrastructure Reform & Regulation, University of Cape Town Graduate School of Business.

EC: Council Decision of 29 May 2006 on the conclusion by the European Community of the Energy Community Treaty. 2006/500/EC.

Escribano, G., 2010. Convergence toward differentiation, the case of Mediterranean energy corridor. Mediterr. Polit. 15, 211–229.

European Commission, 2011. A Partnership for Democracy and Shared Prosperity. COM(2011), Brussels.

Fattouh, B., El-Katiri, L., 2012. Energy Subsidies in the Arab World. United Nations Development Program, Regional Bureau for Arab Studies, Arab Human Development Report.

Glachant, J. M., & Perez, Y. (2007). Achieving electricity competitive reforms as a long term Governance Structure problem. Working paper GRJM, available on www. grjm. net.

IRENA, 2013. Renewable Power Generation Costs in 2012: An Overview. Irena Secretariat, Abu Dhabi.

Kwon, Y., Shin, K., 2010. Regulatory Reform Satisfaction Survey: Method and Result. OECD, Paris.

Lieberman, D., Doherty, S., 2008. Renewable Energy as a Hedge Against Fuel Price Fluctuation. Commission for Environmental Cooperation, Montreal.

Manners, I., 2002. Normative power Europe: a contradiction in terms? J. Common Market Stud. 40, 235–258.

Obervatoire Mediterranéèn de l'Energie and Medgrid, 2013. Towards an Interconnected Mediterranean Grid: Institutional Framework and Regulatory Pespectives. OME-MEDGRID, Paris.

OECD, 2012. Measuring Regulatory Performance, a Practitioner's Guide to Perception Surveys. Manual. OECD, Paris.

Rubino, A., 2014. A Mediterranean electricity co-operation strategy. Vision and rationale. In: Cambini, C., Rubino, A. (Eds.), Regional Energy Initiatives, MedReg, the Energy, Community. Routledge, London.

Schimmelfennig, F., Sedelmeier, U., 2005. The Europeanization of Central and Eastern Europe. Cornell University Press, New York.

Thatcher, M., Coen S D., 2008. Reshaping European regulatory space: an evolutionary analysis. W. Eur. Polit. 31, 806–836.

Tholens, S., 2014. An EU–South Mediterranean Energy Community: the right policy for the right region? Int. Spect. Ital. J. Int. Affairs 49 (2), 34–49.

Vantaggiato, F.P., 2014. Mechanism and outcomes of the EU external energy policy. An alternative approach. In: Cambini, C., Rubino, A. (Eds.), Regional Energy Initiatives. MedREG and the Energy Community. Taylor & Francis, Routledge, London, pp. 45–62.

Zickfeld, F., Wieland, A., 2012. 2050 Desertec Power: Perspectives on a Sustainable Power System for EUMENA. Dii GmbH.

Chapter 9

Analysis of Future Common Strategies Between the South and East Mediterranean Area and the EU in the Energy Sector

Pantelis Capros, Panagiotis Fragkos, Nikos Kouvaritakis
Department of Electrical and Computer Engineering, National Technical University of Athens, Athens, Greece

1 INTRODUCTION

It is widely accepted that the south and east Mediterranean (SEM) area has not yet exploited the vast opportunities for cooperation both within the region and with the European Union (EU). Energy constitutes an essential commodity enabling socioeconomic development especially for developing regions. At the same time, it can be a field of wide cooperation between the EU and the SEM. Mutual benefits can arise for both regions specifically due to the potential sharing of energy resources, both conventional and renewable. In particular, the context enabling such cooperation is twofold: supporting a sustainable energy transition in the SEM region through extensive renewable energy source (RES) deployment and accelerated energy efficiency improvements; and enhancing EU–SEM cooperation regarding the exploitation of hydrocarbon resources.

The south Mediterranean hydrocarbon sector has already developed large exchanges with the EU, which has strongly participated in the capital and technological infrastructure for hydrocarbons. An extensive oil and gas infrastructure system has been developed linking the two shores of the Mediterranean (gas pipelines, liquefied natural gas terminals, and oil refineries). The SEM region itself has asymmetric hydrocarbon resources, as only some of the countries are large hydrocarbon producers and exporters (Algeria, Egypt, and Libya). The producers have an interest to sustain hydrocarbon exports in the long term; they face the problem of a strongly rising domestic demand driven by artificially low hydrocarbon prices in the domestic market. The consuming countries are strongly depending on hydrocarbon imports, and have an interest in reducing consumption.

Regulation and Investments in Energy Markets. http://dx.doi.org/10.1016/B978-0-12-804436-0.00009-6

In fact, in both groups of countries, the historical trends show a rapid increase in domestic energy consumption. This is related to large and persistent energy inefficiencies and it is also due to low energy prices owing to extensive subsidization in hydrocarbon producing countries. Extrapolation of these trends into the future provides evidence that the economic benefits from fossil fuel exploitation are not sustainable in hydrocarbon-producing countries and also that the energy bill in hydrocarbon-importing countries will undermine their economies. Such trends threaten socioeconomic development in both groups of countries. Significant reorientation of energy strategies are required, in order to improve energy efficiency, remove distortions arising from subsidized energy prices, and introduce new efficient and clean technologies and infrastructure both for demand and supply of energy.

All south Mediterranean countries have great untapped potential in renewable resources, which are sufficient not only to cover local needs but also to increase exportable hydrocarbon surpluses. The potential is so large that it can help the EU in achieving the Energy Roadmap decarbonization targets at lower overall cost than otherwise. For example, large-scale centralized concentrated solar power (CSP) plants in the Sahara region and large-scale wind farms and photovoltaics combined with high-voltage direct current (HVDC) lines connecting the two shores of the Mediterranean can transform the region into a significant net exporter of green electricity to the EU.

The thesis of this chapter is that the SEM and the EU have a common interest to exploit the southern renewable resources cooperatively. To this end, the analysis investigates alternative strategies based on quantification using a detailed energy system model for the countries of the Mediterranean area. The cooperation provides benefits to the EU allowing a cost decrease in GHG emission reduction efforts by 2030 and 2050 and also benefits the SEM from exploitation of the renewables and exports of additional amounts of hydrocarbons. To implement such cooperation, market and system integration has to develop implying mutual confidence and long-term regulatory and market predictability. Such conditions do not exist today as instability prevails especially in the southern part of the Mediterranean region and the predictability for future policies and economic conditions is extremely poor. In addition, SEM–EU cooperation is currently hampered by a lack of a common energy and regulatory framework, the persistence of energy subsidies in most SEM countries, and the lack of an electricity interconnection infrastructure. These multiple nonmarket barriers and coordination failures can only be overcome through system-wide cooperation between the EU and the SEM region with common long-term aspirations.

There is growing awareness that SEM and EU regions could bring together significant benefits from closer cooperation in the energy sector, and in recent years many initiatives have been proposed in this direction (DLR, 2005; Plan Bleu, 2008). Several studies have evaluated the potential cooperation of Mediterranean countries with the EU in the fields of energy and sustainable development (Trieb and Müller-Steinhagen, 2007; Folkmanis, 2011). The evolution of the SEM energy system has been extensively investigated, especially with regard to the evolution of the oil and gas production sector (Hafner et al., 2012), the potential for massive RES deployment (Plan Bleu 2008; Supersberger and Führer, 2011), the role of nuclear power (Supersberger and Führer, 2011), and the costs and benefits of energy efficiency investments (Jablonski and Tarhini, 2013; Blanc, 2012). Another strand of the literature assesses the issue of fossil fuel subsidies (Coady et al., 2010; Schwanitz et al., 2014) and the market distortions and systemic misallocation of resources caused by energy subsidization (Fattouh and El-Katiri, 2013). Moreover, the International Energy Agency (IEA, 2012) and Mediterranean Observatory of Energy (OME, 2011) have quantified projections for the evolution of final energy demand, power generation mix, and hydrocarbon supply for the SEM countries.

The objective of the current analysis is to evaluate in quantitative terms the implications of alternative strategies regarding the configuration of the energy demand–supply system in the SEM–EU region. Toward this end, a technology-rich energy system model has been used, which provides a detailed numerical evaluation of the SEM–EU energy system with explicit coverage of sectoral final energy demand, fuel subsidies, hydrocarbon production and trade, power supply, and cost–supply curves for energy efficiency and renewable sources. Contrasting with other analyses which focus on specific aspects of the energy system, this chapter provides a comprehensive holistic model-based assessment of future alternative energy strategies, especially with regard to the degree of cooperation with the EU, and explores the potential synergies and trade-offs between multiple energy policy objectives (security of energy supply, sustainable development, reduction of import dependence, CO_2 emissions reduction, and RES deployment).

2 MODEL DESCRIPTION

In the current study, a large-scale energy demand and supply model (E3M-Lab, 2012), has been employed in order to evaluate the impacts of pursuing alternative strategies regarding the development of the SEM energy system, the associated carbon emissions, and to quantify the costs and benefits incurred for the SEM–EU region. The MENA–EDS model estimates the quantities demanded and supplied by the main energy system actors in a comprehensive manner, simulates the formation of prices in energy markets, and projects energy-related CO_2 emissions and RES deployment, driven by environmentally oriented policy instruments and emission abatement technologies. The model is designed for medium and long-term projections and produces analytical quantitative results for each country until 2030[1] (Fragkos et al., 2013); at this stage, the model is applied to the SEM (Algeria, Morocco, Tunisia, Egypt, Libya, Israel, Lebanon,

1. Model calibrated to historical energy, economic, and emissions data as derived from the IEA and ENERDATA databases until 2011–2012.

and Jordan[2]) and links to the PRIMES energy model[3] scenarios, which cover all individual member states of the EU. The model determines endogenously the power generation mix, investment, and flows over the interconnected grid and also calculates energy system costs (including power generation costs), so as to assess the impacts of alternative energy strategies and policy instruments such as taxes and subsidies, energy tariffs, carbon prices, and incentives promoting energy efficiency and renewable sources.

The model is dynamic with annual resolution and produces yearly projections until 2030 and beyond. The framework of the assessment is based on partial equilibrium energy system modeling in which the evolution of demographic and socioeconomic drivers (GDP and sectoral activity figures, covering the major energy-consuming sectors in industry, households, services, and transportation) is projected into the future based on another model (GEM-E3 model of E3MLab extensively used in the MEDPRO research program[4]). The macroeconomic and sectoral growth scenarios (Paroussos et al., 2015) retained for the analysis in this chapter, envisages the return of the south Mediterranean region in a context of stability and economic growth which enables a rise in energy demand domestically and an attraction of foreign investment. The energy model projects energy demand and supply in detail and also determines energy prices by type of commodity on a national scale taking into account country-specific characteristics, such as production, transport and distribution costs, taxes, and fuel subsidies. Economic activity forecasts together with national energy prices influence the evolution of the energy demand by sector and in this way the model closes the loop with energy supply and demand at a national level.

Final energy demand is projected for five main sectors: industry, households, services, agriculture, and transport. Depending on data availability, energy demand in specific subsectors is also represented (especially for industrial sectors and alternative transport modes). The MENA–EDS model formulates interfuel substitutions and the possibility of energy efficiency progress in each demand sector. Final energy demand uses natural gas, electricity, refined oil products, coal, traditional biomass, steam, and biofuels (in transport). The model incorporates explicitly several energy technologies, which are specific by sector (especially for passenger transport) and have capital and operation and maintenance (O&M) costs and energy efficiency evolving over time. The evolution of passenger car stocks in the SEM is projected to be driven by income growth and behavioral changes which influence car ownership, the rate of turnover of car fleets, and the purchasing of newer more efficient vehicles. The model represents possible saturation effects depending on income growth and the development of car stocks. For road transportation, which is expected to play an increasingly important role in the SEM region, specific technologies are modeled including

2. Syria is excluded from the current analysis due to the current political and social situation.

3. See www.e3mlab.eu for a description of the PRIMES model.

4. See http://www.medpro-foresight.eu/

conventional vehicles (using gasoline, diesel, or biofuels), hybrids (both standalone and plug-in), and electric vehicles. The new options, fuels (e.g., electricity), and technologies are developed smoothly over time and depend on supporting policies and relative costs compared with conventional options. In the absence of specific targeted policies, new fuels and powertrains for passenger cars, such as electric cars, hardly emerge in the future (up to 2030).

The energy model includes a detailed representation of the power generation sector, taking into account electricity load profile, own consumption of power plants, electricity trade between countries, and transmission and distribution losses in each country. Using a bottom-up method, load profiles of individual electricity uses, projected to the future, are aggregated into a system-wide power load curve (Fragkos et al., 2013) by taking into account the fact that demand in energy-intensive industrial sectors is mainly base load, while pronounced peaks characterize the demand from households. Unit commitment is derived from minimization of system marginal costs of generation under technical, operational, and resource availability constraints. Profitability of new power plants drives new investment dynamically taking into account growth of demand, change of relative total production costs (that include capital, O&M, and fuel costs), possible decommissioning of old and inefficient power plants, and electricity infrastructure. The model incorporates a wide variety of power generation technologies, including fossil fuel fired technologies (coal, oil, and gas combined with thermal, open cycle, integrated gasification, and combined cycle options), nuclear power, and a spectrum of renewable generation technologies (hydroelectricity, wind onshore and offshore, CSP, photovoltaics, and biomass) for which their limited potential is also taken into account. Investment decisions are driven by long-term marginal costs (that include the annualized and discounted capital costs, the fixed and variable operation and maintenance costs, and fuel costs) in combination with expectations about demand evolution and the formation of the yearly load curve. Variable operating and fuel costs drive capacity utilization of power technologies at each segment of the load curve and hence they determine fuel consumption and associated carbon emissions from the power generation sector. The price of electricity is determined as a function of long-term average marginal costs and taxes/subsidies. Electricity prices are differentiated between sectors (industry, households, commercial) taking into account differential distribution costs for each sector and the different load profiles of demand in each sector.

In order to quantify the impacts of alternative energy strategies on the hydrocarbon trade balance, primary production of fossil fuels is estimated as a function of reserves, investments in productive capacity, and demand. For crude oil it is assumed that the world market can absorb the amounts to be produced by SEM countries. Reserves are determined by a motion equation that calculates net additions in terms of new discoveries minus annual production. The rate of discovery depends on the evolution of fuel prices and the economically recoverable hydrocarbon resources as estimated by geological experts (BGR, 2009).

The difference between primary consumption and primary production determines net imports or exports. Natural gas trade between countries explicitly takes into account existing pipeline and liquefied natural gas (LNG) infrastructure and projects their future evolution largely based on the assessments carried out in Hafner et al. (2012). The comprehensive modeling of natural gas trade is an important component of the model-based analysis as it concerns a particularly important sector for SEM economies. Large hydrocarbon producers aiming at increasing profits from export surpluses have an interest in controlling domestic demand rises, while net importers struggle to reduce their dependency on imported fossil fuels. Furthermore, natural gas trade is a domain in which the EU already cooperates with North African countries, the latter being large exporters of gas based on LNG and gas pipelines (Algeria and Libya). Cooperation between the two shores of the Mediterranean in the field of gas trade can be further enhanced in the future, with North African exports substituting Russian gas.

3 SCENARIO DESCRIPTION

3.1 The Current Situation of the SEM Energy System

The SEM region consists mostly of developing economies that are characterized by different stages of development and different values of GDP per capita. Income differences influence the values of energy indicators shown in Table 9.1. With the exception of Israel, which is directly comparable with the EU-28 in terms of GDP per capita and energy demand indicators, all other SEM countries have particularly low values for energy per capita indicators especially regarding electricity consumption in 2013. Morocco is characterized by very low values for energy indicators, as it has not followed an energy-intensive mode of development and experiences high energy consumer prices due to limited energy subsidization and lack of hydrocarbon resources. On the other hand, Libya is characterized by a highly energy-intensive mode of development, as its consumption of primary energy per capita is comparable with the EU; therefore, the energy intensity of GDP is twice as high in Libya compared with the average European indicator. The average energy intensity indicator in SEM countries stands at 0.15 toe/$1000, and ranges between 0.10 and 0.12 in Morocco, Tunisia, Israel, and Lebanon and between 0.22 and 0.23 in Jordan and Libya, respectively.

Carbon intensity of primary energy in SEM countries in 2013 was over 2.5 tCO_2/toe, which is high in comparison with the EU average. Differences among countries mostly reflect the balance between the use of oil and natural gas, the fact that some countries use coal for power generation (Israel and Morocco), and the extent to which renewable energy sources have developed (Egypt, Tunisia). On the other hand, most countries in the region (with the exception of Israel and Libya) have much lower emissions per capita compared with the EU, and there is a clear risk that as economic development and living standards rise in these countries, CO_2 emissions will also rise in the absence of

TABLE 9.1 Summary of Energy Indicators for the SEM Region for 2013

	Algeria	Morocco	Tunisia	Egypt	Libya	Israel	Lebanon	Jordan	SEM	EU-28
GDP (US$1000, 2005 at constant PPP/person)	7.90	4.71	8.59	5.58	11.26	27.56	13.10	5.35	7.33	27.71
Primary energy demand per capita in 2010 (toe/person)	1.32	0.59	0.96	0.95	2.64	2.75	1.38	1.15	1.11	3.20
Electricity per capita in 2010 (kWh/person)	1713	898	1733	1952	4764	7046	3671	2647	2086	6367
Energy intensity of GDP in 2010 (toe/US$1000)	0.17	0.12	0.11	0.17	0.23	0.10	0.11	0.22	0.15	0.12
Carbon intensity of primary energy in 2010 (tCO_2/toe)	2.61	2.89	2.15	2.41	2.61	2.87	2.88	2.75	2.58	2.05
Carbon emissions per capita in 2010 (tCO_2/person)	3.45	1.70	2.07	2.29	6.91	7.91	3.97	3.17	2.86	6.55
Net exports as % of primary production if net exporter (2013)	67.0			6.1	77.4				32.6	
Net imports as % of primary demand if net importer (2013)		96.6	27.0			89.9	97.1	97.0		56.3

Source: Authors' own calculations using historical data from the IEA and Enerdata databases (the latest available data are used).

emission reduction policies. In terms of availability of hydrocarbon resources, the region can be subdivided into the following groups: major energy exporters (Algeria and Libya and to a lesser extent Egypt), a minor net importer (Tunisia), and predominantly net importers (Jordan, Lebanon, Morocco, and Israel).

A particularly important factor for the evolution of the SEM energy system is the subsidization of domestic energy prices. The common rationale behind the application of energy subsidies is alleviation of energy poverty, as energy subsidies are conceived as a means of improving the living conditions of poor citizens allowing access to modern energy sources (e.g., LPG and electricity instead of traditional biomass; Coady et al., 2010). Low domestic energy prices have also been used as a means of attracting foreign investment in energy-intensive industrial sectors. On the other hand, energy price subsidies imply an economically inefficient allocation of resources and market distortions, incentivize fuel smuggling, and weaken competitiveness of RES and energy-efficient technologies in demand sectors. The subsidies also imply a fiscal burden for the public budget in most SEM economies (Fattouh & El-Katiri, 2013). Although energy subsidies are usually intended to help lower income classes, the greatest benefit typically goes to those who consume high energy amounts per capita (i.e., those who can afford to own cars, electrical and heating/cooling appliances, etc.). The IMF has estimated that 80% of the total benefits from petroleum subsidies in 2009 accrued to the richest 40% of households, while in the SEM region the upper quintile of households receive 45% of total subsidies and the poorest 20% only 8% of subsidies (Coady et al., 2010). The IEA has evaluated fossil fuel subsidization rate in a large number of countries as a proportion of the full cost of fuel supply (IEA, 2013). Some of thc highest subsidization rates globally are found in Libya (80%), Algeria (57%), and Egypt (53%), where energy subsidies represent a significant fiscal drain on the economies. The situation is particularly critical for Egypt, as increasing domestic fuel consumption requires increasingly large amounts of hydrocarbon imports, and energy subsidies account for 12% of its GDP (in 2010). High subsidies directed to energy consumption exert crowding out effects on other public programs, which aim at more welfare-enhancing activities such as economy modernization, ICT, education, and infrastructure development (Fattouh and El-Katiri, 2013).

3.2 Alternative Energy Strategies for the SEM Region

The future energy system development of SEM countries is surrounded by a large number of uncertainties, among which are political and social conditions, the continuation of conflicts and Arab Spring uprisings, the degree of cooperation with the EU, geopolitical concerns, energy price reform, and opening up of the region's economies. Business-as-usual trends, reflected in the design of the Reference Scenario, involve continuation of economic growth in the future despite fluctuations experienced at present, limited cooperation with the EU, gradual but slow rationalization of energy-pricing regimes, and cautious implementation of energy policies that are currently in the political agenda of SEM countries. Alternative energy system strategies for the SEM region may increase cooperation with the EU and be used as leverage for enabling shifts and accelerating investment to exploit the vast domestic potential of forms of clean energy.

The Mediterranean region is of strategic importance to the EU both in economic and political terms. Europe committed itself to promoting Euro-Mediterranean cooperation within the Barcelona Process and adopted the ambitious European Neighbourhood Policy (ENP) that seeks to strengthen relations with its neighboring countries (High Representative of the Union for Foreign Affairs, 2011). SEM and EU countries also cooperate in the field of energy, environmental protection, and sustainable development (Plan Bleu, 2008). Increased political will for cooperation across the Mediterranean and the deepening of EU–SEM relationships are conceptualized in the "SEM–EU Cooperation Scenario," which assumes that large-scale energy projects like DESERTEC and the Mediterranean Solar Plan (MSP) will materialize by 2030 and that the European Emissions Trading System (EU-ETS) will expand to include SEM energy-intensive industries and power production plants under specific terms including free allocation of emissions permits to SEM countries. Therefore, SEM countries would not face a carbon cost, but may generate revenues by reducing their carbon emissions and by selling their allowances to EU countries.

In this context, the cooperative SEM–EU exploitation of emissions reduction opportunities would involve large-scale investment enabling RES exploitation in centralized applications (CSP technology) with mainly export orientation and concentration on specific geographical areas (mainly Algeria and Libya) which offer the best cost-effective prospects for electricity exports, as explicitly evaluated by the German Aerospace Center (DLR, 2005). Renewable electricity exports will require investments in HVDC interconnectors (with a capacity of 5 GW each) linking SEM countries with the southern countries of Europe. The scenario assumes that green electricity exports from the SEM to the EU start after 2020 with the reinforcement of the Spain–Morocco AC line, while by 2030 four HVDC lines will be constructed linking Algeria to the south of Europe, two linking Libya with Italy, one linking Morocco with Spain, and one linking Tunisia with Italy. This translates into 40 GW of CSP export capacity to the EU, while in 2030 electricity exports reach 235 TWh and cover 6.7% of the EU's electricity requirements. The exported renewable electricity is charged at predefined fixed tariffs which are set at a sufficient level to allow recovery of total capital and operating costs and to allow a reasonable rate of return on capital (at a 9% discount rate). Finally, the scenario assumes gradual strengthening of RES facilitation policies and acceleration of energy price reform especially for the industrial and the electricity sectors.

In the context of the 2030 energy and climate policy framework, the EU can take advantage of the large-scale development of centralized export-oriented RES in the SEM in order to reduce the requirements for domestic emissions

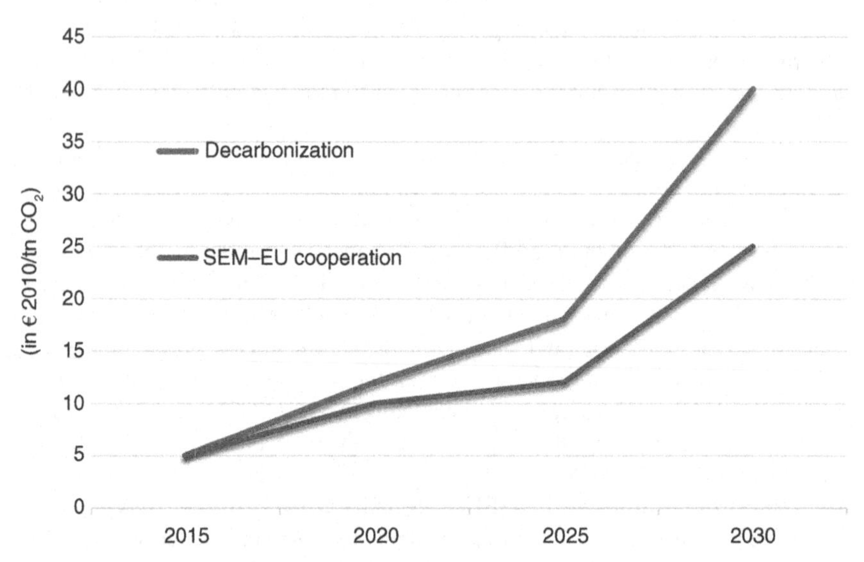

FIGURE 9.1 EU-ETS carbon price. *(Source: PRIMES and MENA–EDS models.)*

reduction and penetration of RES under conditions described in European Commission (2014); therefore under EU decarbonization conditions, the ETS carbon price is projected to decline from 40 €/tCO_2 to 26 €/tCO_2 in 2030 (Fig. 9.1). This means that the EU-ETS sectors will pay about €120 billion less for purchasing emissions allowances in the "SEM–EU Cooperation" Scenario in the period 2015–2030 in cumulative terms, while the cumulative value of electricity exports amounts to €78.2 billion by 2030.

According to DLR (2005) and Hafner and Tagliapietra (2013), there is a huge potential for improvements on several policy areas in the SEM region including economic and energy reforms, infrastructure upgrading, energy efficiency improvements, trade liberalization, and sustainable development. The "Decentralized Actions" Strategy assumes that SEM countries individually undertake vigorous policies and measures in order to promote energy efficiency and accelerate deployment of RES, thus leading to a reduction of import dependence in energy-importing countries and an increase in the export capability of hydrocarbon-exporting countries. Contrasting the export-oriented and highly centralized RES development assumed in the cooperation case, "the Decentralized Actions" Strategy assumes multiple decentralized actions regarding the exploitation of renewables, the complete removal of all current market and energy pricing distortions by 2020, improvements in domestic electricity grids allowing penetration of decentralized RES, and the establishment of incentives, standards, and measures for energy savings affecting all energy consumers. The materialization of these policies and market reforms can only be realistic in the context of a rapid political normalization and stabilization.

4 THE REFERENCE SCENARIO

This section summarizes the main results of the Reference Scenario in quantitative terms as simulated with the energy system model. The Reference Scenario takes a cautious view with regard to future energy and economic developments in the SEM region and especially with regard to the removal of energy subsidies, the exploitation of hydrocarbon resources, cooperation with the EU, and the deployment of carbon-free energy sources.

4.1 Exogenous Assumptions

Reference population projections are derived from the medium fertility variant of world population prospects of the United Nations (United Nations, 2011), according to which total population in the SEM is projected to increase from 183 million in 2010 to 230 million in 2050. The projection implies a marked slowdown in population growth for all countries in the region (from 1.7% per annum on average in the 1990–2010 period to 1.2% per annum in the 2010–2030 period). The Reference Scenario also assumes a continuation of the trends in terms of urbanization with marked consequences for changes in lifestyles and energy consumption patterns.

Economic activity growth in SEM countries over many years has been strongly hampered by political instability. Nonetheless, in the period from 1990 to 2010 growth has occurred at high rates. For the Reference Scenario political normalization is assumed leading to increased trade and cooperation with the EU, which combined with continuous productivity increases leads to a relative acceleration of GDP per capita growth (Paroussos et al., 2015) from 2.1% per annum in the 1990–2010 period to 2.7% in the 2010–2030 period.

In principle, international fuel prices, which are based on the latest PROMETHEUS projection (Fragkos et al., 2015), should be reflected in domestic consumer prices. However, the SEM region is characterized by a large variety of pricing policies. Looking at transportation fuels, Israel and Morocco have prices and taxation comparable with EU prices. In Tunisia and Lebanon transport fuel taxation is very low. The other countries in the region subsidize transport fuels and consumer prices are lower than tax-free international spot prices. Furthermore, in most SEM countries, the prices of refined oil products used in stationary energy applications and in electricity production do not reflect international oil prices, while natural gas prices are even lower than export netback prices. Clearly, this situation is not economically rational as the prices at which the fuels could be sold on international markets constitute essentially an opportunity cost for domestic production (Coady et al., 2010). Industrial and particularly residential electricity prices in most SEM countries are far lower than long-term marginal generation costs and therefore capital cost recovery is not possible. This is an indirect subsidization, which is commonly justified on the basis of public service arguments and also because expansion of electricity uses is among the main modernization measures aimed at improving living conditions and attracting new technologies in business.

Subsidization threatens public finance stability and the long-term viability of utility companies. Therefore, the SEM countries increasingly realize the need for price reform and removal of energy subsidies, as the present situation is clearly unsustainable. A program for the reduction of energy subsidies has already been attempted but stalled in Egypt, as this reform tends to be unpopular among energy consumers. It is recognized that the transition to a new pricing regime must be gradual so as to mitigate major adverse effects on poorer households and businesses. Along these lines, the Reference Scenario assumes gradual evolution toward more rational energy pricing in the period 2010–2030; price reform policies are assumed to be implemented at a different pace by country and in particular more slowly in Algeria, Libya, and Egypt. In the latter, pump prices are assumed to equal 80% of tax-free spot prices by 2030, while natural gas prices nearly double in the 2010–2030 period but they are assumed to remain below free on board (FOB) export prices by 2030. Subsidization of electricity and fossil fuels in stationary uses is assumed to gradually decline until 2030.

4.2 Energy Demand

The evolution of energy demand depends on a number of key factors among which are macroeconomic developments, urbanization prospects, evolution of energy prices, and the economic structure of each country. The SEM region in the recent past has registered very high growth rates of energy demand. This has been particularly true of Egypt, Morocco, and Algeria, in which primary energy consumption per unit of GDP strongly increased in the period 2000–2010. There is broad scope for SEM countries to increase their energy consumption per capita in line with income growth and increased standards of living. The region has mild climatic conditions and thus saturation levels of household energy demands will be lower than in Europe. The persistence of low energy prices in most SEM countries has attracted energy-intensive industries and as a consequence specific energy consumption tends to increase (e.g., Algeria).

The Reference Scenario assumes a reversal of this growing energy intensity trend as a result of gradual reforms, such as improved energy efficiency and abolishment of subsidization to some degree. Nonetheless, Reference Scenario assumptions do not fully remove the distortions and do not fully exploit the energy efficiency potential of the region. Therefore, most SEM countries are projected to experience reductions in energy intensity of GDP in the decade 2010–2020, with the exception of Algeria and Morocco. For the period 2020–2030, the Reference Scenario projects acceleration of efficiency improvements especially in such countries as Tunisia, Egypt, Libya, and Israel, as a result of gradual reduction of fossil fuel subsidies and restructuring of the economy toward a higher share of services and a slowdown in the growth of industrial sectors. Overall, the Reference Scenario shows a continuous decrease in energy intensity of GDP in the SEM region taken as a whole; the decline is on average 0.5% per annum in the period 2010–2030 (Table 9.2).

TABLE 9.2 Evolution of Primary Energy Intensity of GDP in the Reference Scenario % annual

	2000–2010 (%)	2010–2020 (%)	2020–2030 (%)
Algeria	0.1	0.3	0.2
Tunisia	−1.1	−0.5	−0.8
Morocco	−0.1	0.1	0.1
Egypt	1.0	−0.4	−0.5
Libya	−1.2	−1.4	−1.6
Israel	−0.7	−1.4	−2.1
Lebanon	−1.6	−1.2	−0.8
Jordan	−1.3	−0.5	−0.7
SEM	0.0	−0.4	−0.6
EU-28	−1.2	−2.0	−1.0

Source: MENA–EDS model.

Further penetration of electricity in demand sectors also has a great potential in most SEM countries. The Reference Scenario projects a strong increase in the share of electricity in final industrial demand (from 25% to 38% during 2010–2030) as a result of accelerated penetration of electrical industrial processes and increased demand for specific electricity needs, such as electric motors, machinery, and cooling. Another important feature of the projection is the increased penetration of natural gas for heat and steam uses mainly at the expense of oil (Fig. 9.2). This substitution occurs primarily for economic reasons: with the expansion of the natural gas grid there is greater potential for lower cost gas to substitute residual fuel oil; Libya, Algeria, and Egypt, the main exporters of natural gas, naturally register the highest gas shares (more than 60% in 2030). In the residential sector, natural gas is projected to increase its share in water heating and cooking applications at the expense of traditional biomass and LPG as facilitated by the expansion of the gas distribution grid. The use of traditional biomass is projected to decline driven by increased urbanization and rising standards of living. The latter also drives higher penetration of electricity in the residential sector and in services. Electricity demand in households is projected to increase by a factor of 2.9 by 2030 relative to 2010, mainly due to the rapid penetration of electrical appliances, such as refrigerators and deep freezers, air conditioners, television sets, washing machines, etc., as well as the expansion of cooling. Electricity will play an increasingly important role in the energy mix in the SEM and its share in the final energy demand is projected to increase from 22.5% in 2010 to 25.7% in 2020 and 29.1% in 2030 (Fig. 9.2).

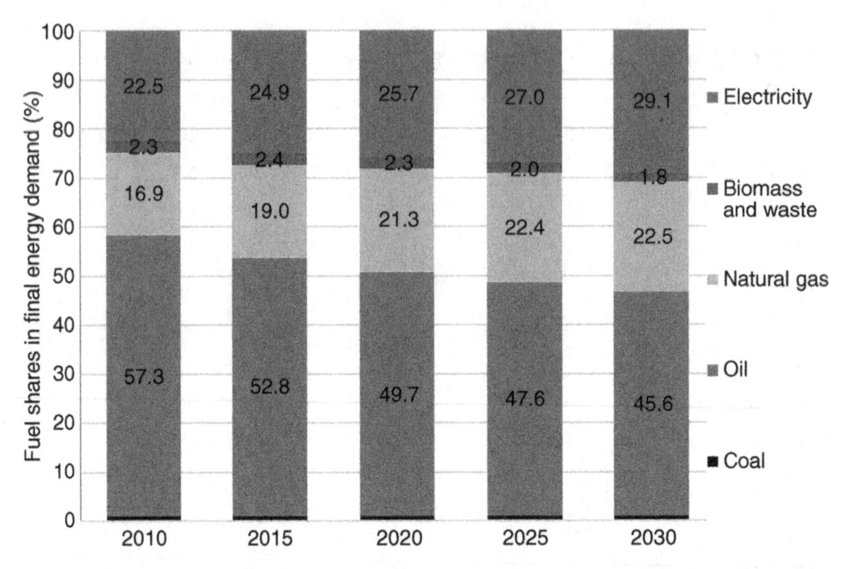

FIGURE 9.2 Fuel shares in final energy demand in the SEM region's Reference Scenario. *(Source: MENA–EDS model.)*

The evolution of car ownership rates in different countries is considered as the most important driver for the evolution of transportation energy demands. Nowadays, most countries have very low car ownership rates (with the exception of Lebanon), but rising standards of living and increased urbanization are expected to drive significant growth of ownership rates, which may more than double between 2010 and 2030 (from 78 cars/1000 inhabitants in 2010 to 156 in 2030), while the total stock of private vehicles in the SEM region is projected to increase from 14.3 million passenger cars in 2010 to 36 million by 2030. Despite factors which would drive a reduction of oil consumption by cars, such as the reduction in vehicle utilization rates as motorization increases, the increases in pump prices, and the constant improvements in vehicle efficiency, oil consumption for transport is projected to increase very significantly until 2030. The buoyant demand for oil for transportation purposes, combined with its diminishing role in the industrial, residential/commercial, and especially the power-generating sector, means that transport becomes the dominant market for oil products and accounts for 75% of primary oil consumption in the SEM region in 2030.

4.3 Power Generation

Over the period 1990–2010, the GDP elasticity of demand for electricity has been more than two times higher in the SEM region than in the EU (an elasticity value of 1.64 in the SEM compared with 0.71 in the EU). This implies that electricity growth in SEM demand sectors maintained a strong momentum, which

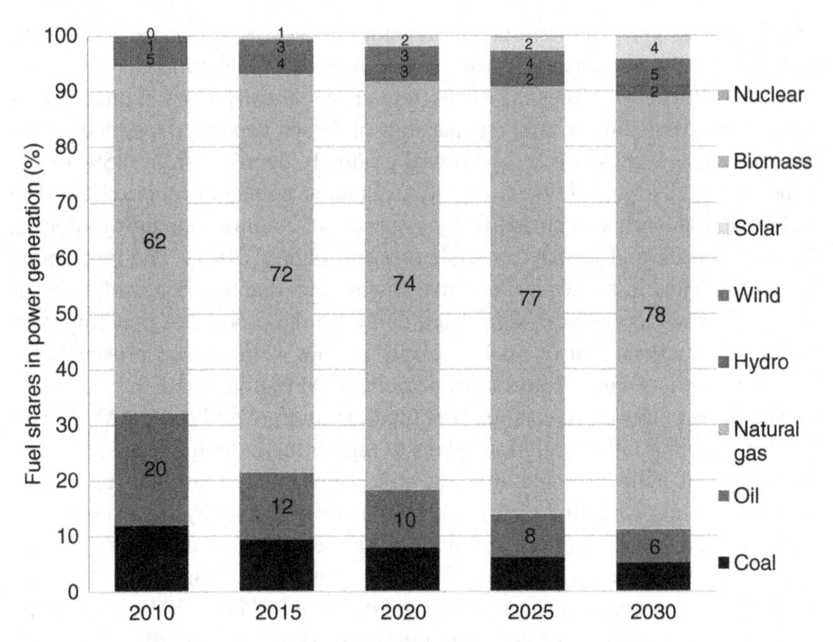

FIGURE 9.3 Fuel shares in power generation in the Reference Scenario. *(Source: MENA–EDS model.)*

is projected to persist in the medium term. The annual growth of electricity demand is projected to attain 5% on average in the period 2010–2030, a growth rate that is higher than that of GDP. Therefore, the SEM region will become a large electricity market by 2030 (920 TWh) requiring a significant expansion of productive capacity (from 83 GW in 2010 to 216 GW in 2030).

The Reference Scenario projects that natural gas will dominate the power-generating sector of the SEM region getting increasing shares in total generation, reaching 78% in 2030 up from 62% in 2010 (Fig. 9.3). In gas-producing countries the share of gas in total power generation is projected to reach 85% in 2030 (Algeria, Egypt, and Tunisia). The dominant option for the new gas-fired capacity is combined cycle gas turbine technology, which combines relatively low capital costs with high efficiency rates. Gas-fired power generation is convenient in complementing intermittent RES power production and facilitates load management especially when renewable shares are relatively high. Gas prices for power generation are projected to remain relatively low, despite a gradual decline of subsidies; market conditions do not favor the development of capital-intensive plants such as nuclear, and the local environmental burden does not favor development of coal plants, except in a few cases. The share of coal in power generation will substantially decrease from 12% in 2010 to 5% in 2030, as new investments in coal-fired power plants are only assumed in Morocco, while no new investments are projected for Israel since domestically produced gas is expected to fuel the major part of its additional power capacities. Gas penetration in power generation also substitutes oil-based generation, a trend that is visible already today in most SEM countries.

The hydroelectric potential of the region has already been exploited to a large extent, and only a minor expansion of hydro capacity is projected for Morocco; hence the share of hydro will gradually decrease from 4.5% to 2% in the 2010–2030 period due to rapidly expanding demand for electricity. The nuclear option has been considered by a number of countries including Algeria, Morocco, Israel, and Jordan (Supersberger and Führer, 2011). The Fukushima accident resulted in higher nuclear investment costs due to stricter safety regulations. This combined with significant delays in the process of planning, tendering, and construction of nuclear plants implies very limited prospects for implementation of nuclear investments in the SEM region by 2030.

The SEM countries have already started exploiting wind power. Moreover, they are pursuing rather ambitious plans to further increase this renewable. The availability of many suitable sites for wind power (allowing high utilization rates) and wind-promoting policies – such as investments by state-owned enterprises, feed-in tariffs (FiT), and quotas – are expected to drive a significant expansion of wind capacity until 2030. The Reference Scenario projects an increase of wind generation from 3 TWh in 2010 to 44 TWh in 2030, accounting for about 5% of total power generation in the SEM region by 2030.

Generally, the region as a whole, but especially its Saharan parts, is considered to offer highly suitable sites for CSP development. Despite a large sunshine potential, CSP with storage is expected to remain significantly more expensive than gas combined cycle technology in terms of levelized power generation costs. However, the decreasing trend of photovoltaic costs implies that photovoltaics will soon be economically very attractive provided that low-voltage grids have sufficient coverage and reliability, which currently is among the factors limiting the expansion of photovoltaics in the region. However, the main factor hampering investment in solar technologies is the lack of capital financing and the uncertainties surrounding foreign direct investment. In very recent years special support systems aiming at promoting CSP and photovoltaic investment have started to be put in place. There is considerable interest in the promotion of hybrid CSP projects (in combination with natural gas) in Algeria, Morocco, Tunisia, Egypt, Israel, and Jordan. The Reference Scenario assumes gradual establishment of a suitable framework promoting solar power in the SEM region addressing its domestic needs. No large-scale exports of renewable electricity to the EU are assumed under the conditions of the Reference Scenario. Therefore, solar power is projected to account for about 4% of electricity generation under reference conditions in 2030.

5 ALTERNATIVE EU–SEM STRATEGIES

This section presents model results for the two alternative strategies examined and especially their implications for energy demand, fuel consumption, power generation mix, carbon emissions, and trade of hydrocarbons for the period 2015–2030.

5.1 Energy Demand

The alternative scenarios assume more aggressive price reforms than in the Reference Scenario. Thus, energy prices will have to increase as subsidies are removed; this influences energy demand improving the rational use of energy. Phasing out subsidies on fossil fuels helps promote a sustainable transition for the domestic energy system via two basic causal chains leading to both lower energy consumption and substitutions of fossil fuels by low-carbon technologies (Schwanitz et al., 2014).

The "SEM–EU Cooperation" Scenario specifically targets power generation and energy-intensive industries (which participate in the common ETS system). Holding allowances given for free to SEM industries provides an incentive to improve efficiency and reduce emissions so as to gain from allowances seen as opportunity costs. Therefore, the scenario projects industrial final energy demand to decline by 13% in 2030 relative to the Reference Scenario. This reduction is particularly pronounced in countries with low industrial prices in the Reference Scenario, such as Algeria, Egypt, and Libya. In other SEM countries which experience higher energy prices, the impact of the cooperation strategy is relatively small. Other final energy demand sectors are affected to a much lower extent since they are not included in the ETS. Efficiency gains in non-ETS sectors are mainly driven by accelerated price reform.

The "Decentralized Actions" Scenario leads to larger energy savings in households compared with the "SEM–EU Cooperation" Scenario due to the development of renewables in a highly decentralized scale. Energy efficiency in industry progresses in the "Decentralized Actions" Scenario more than in the "SEM–EU Cooperation" Scenario despite the absence of ETS incentives. As carbon emissions are not priced in the SEM countries, the only driver of efficiency gains in industry comes from accelerated price reform and specific measures promoting energy efficiency assumed in the "Decentralized Actions" Scenario.

Despite the accelerated removal of energy subsidies, the Reference Scenario projections do not show full achievement of the efficiency gains targeted by current SEM country policy targets; for example, a 20% reduction in final energy demand in Tunisia in 2030 (World Energy Council, 2010), an Energy Efficiency Plan in Lebanon (NEEAP, 2010), and energy efficiency policies in Egypt and other SEM countries (World Energy Council, 2010). Nonprice barriers hamper full achievement of the target and therefore the price reform has to be accompanied by specific measures aimed at making efficiency investments more attractive by removing financial and technology access barriers. Command and control policies, such as the adoption of standards on energy performance of equipment and labeling practices are needed as a complement to the price reform. The evolution of energy efficiency progress takes into account that the energy efficiency cost curves are highly nonlinear, especially when energy savings approach the maximum technical potential. Following Blanc (2012), who developed a bottom-up methodology to evaluate the costs for energy savings for

SEM countries, the scenarios calculate that cumulative investments required to meet the target of a 20% reduction in energy demand compared with reference projections are estimated at €23 billion by 2030. On the other hand, Jablonski and Tarhini (2013) estimated that the required cumulative demand-side investment amounts may need to be substantially higher, reaching €90 billion by 2030. Under favorable financial conditions, such costs can be effectively recovered from a reduction of energy-purchasing costs. Similarly, setting standards and specific actions in industry are assumed in the "Decentralized Actions" Scenario, allowing reduction in final industrial energy demand above performance in the Cooperation Scenario.

The impacts of the Cooperation Scenario are very limited for energy demand in the transport sector (Table 9.3). By contrast, the "Decentralized Actions" Strategy assumptions lead to a 16% reduction of transport energy demand compared with the Reference Scenario, as a result of accelerated efficiency

TABLE 9.3 Impacts of Alternative Strategies on Final Energy Demand in 2030

	Final energy demand (Mtoe)		Changes from the Reference Scenario in 2030 (%)	
	Present situation (2010)	Reference Scenario (2030)	SEM–EU Cooperation Scenario	Decentralized Actions Scenario
Final demand	118.8	243.3	−6.3	−17.8
By sector				
Industry	30.3	59.1	−12.7	−16.2
Households	29.3	59.1	−5.2	−22.4
Services	8.0	21.4	−6.0	−17.2
Agriculture	5.6	10.6	−2.8	−17.2
Transport	45.6	93.1	−2.4	−16.0
By fuel				
Solids	1.2	2.3	−59.0	−61.7
Oil	68.0	111.1	−3.0	−17.5
Gas	20.1	54.9	−1.7	−27.1
Electricity	27	71	−11.9	−13.2
Biomass	3	4	0.0	0.0
Share of hybrids in car stock (%)	0	7.1	8.2	13.2

Source: MENA–EDS model.

improvements in conventional vehicles, faster turnover of vehicle fleet, and higher penetration of hybrid passenger cars which are projected to reach 13% of the SEM car stock in 2030.

Overall, the alternative strategies examined have significant implications for the evolution of energy intensity of GDP index, which has been characterized by a complete stagnation in the period 1990–2010. The Reference Scenario represents a slight improvement (−0.4% per annum between 2010 and 2030), whereas the Cooperation Scenario leads to annual improvements of the order of −0.7%. In the "Decentralized Actions" Scenario, an improvement of −1.2% is projected, which however is lower than the 2% annual reduction projected for the EU-28 region.

5.2 Power Generation

The impact of alternative strategies on power generation is projected to be particularly pronounced, due to the large-scale development of low-carbon technologies. Both alternative strategies show a similar trend away from fossil fuels and toward massive expansion of renewables, which is partly driven by higher fuel prices but also by proactive policies promoting investment in renewables. Although the mix of renewable technologies differs between the two alternative scenarios, they share common policies which support development of the power grid to facilitate expansion of renewables and the achievement of a stable business and financing environment allowing greater access to finance resources and security enabling increasing flows of foreign direct investment.

The two alternative strategies are assumed to follow contrasting pathways for the development of renewables. The "SEM–EU Cooperation" Strategy bases renewable development on electricity exports (generated mainly from large-scale centralized projects, such as large-scale CSP power plants and wind farms). Exports are addressed to the EU, which is assumed to absorb 235 TWh in 2030 via special DC interconnectors to be built by that time. Apart from exports, centralized production from renewables also covers domestic needs. For example, the penetration of CSP is projected to be significant, especially in countries with high CSP potentials such as Algeria (36%), Libya (35%), Morocco (23%), and Tunisia (21%).

The "Decentralized Actions" Strategy takes a different view regarding development of renewables. The orientation is assumed to develop highly decentralized renewable generation projects, mainly based on small-scale wind turbines and photovoltaics. Such development requires expansion and enhancement of domestic grids and allows larger domestic electrification. The scenario projects significant capacity for wind (30 GW higher than the Reference Scenario), especially in countries with high potentials (Egypt, Morocco, and Algeria). Photovoltaic generation also expands vigorously, but its contribution in total electricity needs remains rather limited (6.5% of generation in the SEM region compared with 1% in the Reference Scenario in 2030). In both strategies,

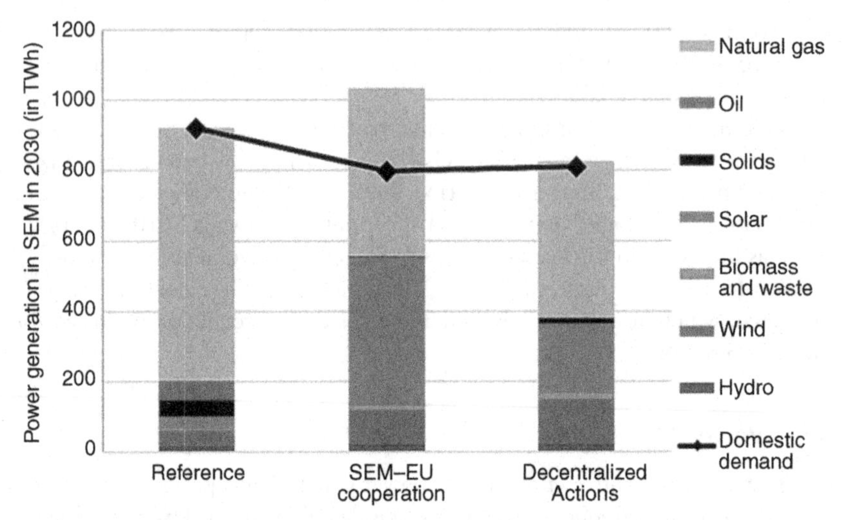

FIGURE 9.4 Impacts of alternative strategies on power generation structure in 2030. *(Source: MENA–EDS model.)*

total renewable generation reaches a particularly high share (42% in "Decentralized Actions" Scenario and 38% in "the SEM–EU Cooperation" Scenario).

Renewable development allows coal- and oil-based generation to virtually vanish in both scenarios by 2030. The share of natural gas in total generation is projected to reduce significantly to 55% in the Decentralized Actions Strategy (Fig. 9.4), down from 78% in the Reference Scenario. The average rate of use of gas capacities is also projected to reduce in the Decentralized Actions Strategy reaching 53% in 2030, down from 62% in the Reference Scenario. Gas capacities will have to remain in the high renewables scenario (Decentralized Actions) but produce less electricity. Instead gas will increasingly have to offer balancing and ramping (load-following) services to the system and therefore has to be remunerated accordingly.

5.3 Carbon Emissions

During the period 1990–2010, all countries of the SEM region have nearly doubled, or more, their energy-related carbon emissions, at a time when the EU managed to reduce theirs by about 10%. The Reference Scenario projects carbon emissions, per megawatt hour of electricity produced, to decline in all countries at varying rates depending on the penetration of renewables, substitution of oil with natural gas, and the introduction of more efficient thermal technologies, such as combined cycle gas turbines. Yet, overall energy-related carbon emissions in the SEM are projected to increase by 90% in the next two decades, following historical trends. Emissions per capita in the SEM would increase from 2.7 tCO_2 in 2010 to 4.1 tCO_2 in 2030, which is however still lower than 6.5 tCO_2 in Europe in 2010.

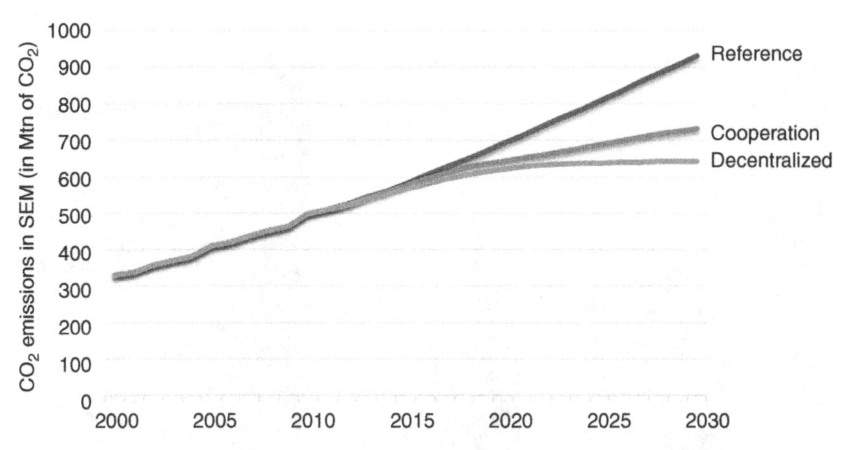

FIGURE 9.5 Carbon emissions in the SEM region. *(Source: MENA–EDS model.)*

The two alternative strategies project carbon emissions to remain below Reference Scenario projections (Fig. 9.5) due to increased penetration of RES technologies that substitute fossil fuels and energy efficiency improvements. In the "Decentralized Actions" Strategy, carbon emissions are projected to stabilize in the 2020s despite vigorous economic growth and increasing living standards, as a result of significantly accelerated energy efficiency improvements that exceed "SEM–EU Cooperation" levels, especially in households, and the higher penetration of hybrid vehicles in the road transport sector by 2030.

5.4 Power Generation Costs

The two alternative strategies imply drastic changes in the structure of power generation costs (Fig. 9.6). The substitution toward capital intensive RES technologies (CSP, photovoltaics, and wind turbines) implies an increase of 40–45% in overall cumulative investments costs relative to Reference Scenario levels (despite lower domestic electricity demand in the alternative scenarios[5]). On the other hand, the significant reduction in fossil fuel based power generation results in a sharp decline in cumulative variable costs, which include variable operation and maintenance costs and fuel costs but do not include carbon costs. The reduction in variable costs more than counterbalances the increased capital in terms of levelized annual costs[6]; thus, annual power generation total costs reduce between 12% and 13% by 2030 in the alternative scenarios compared with the Reference Scenario. The cost savings are sufficient to finance

5. The costs presented in the current section do not include CSP plants directed for exports in the "SEM–EU Cooperation" Strategy.

6. Power generation costs do not include the additional grid investments and storage costs required to integrate large amounts of intermittent RES in power supply systems.

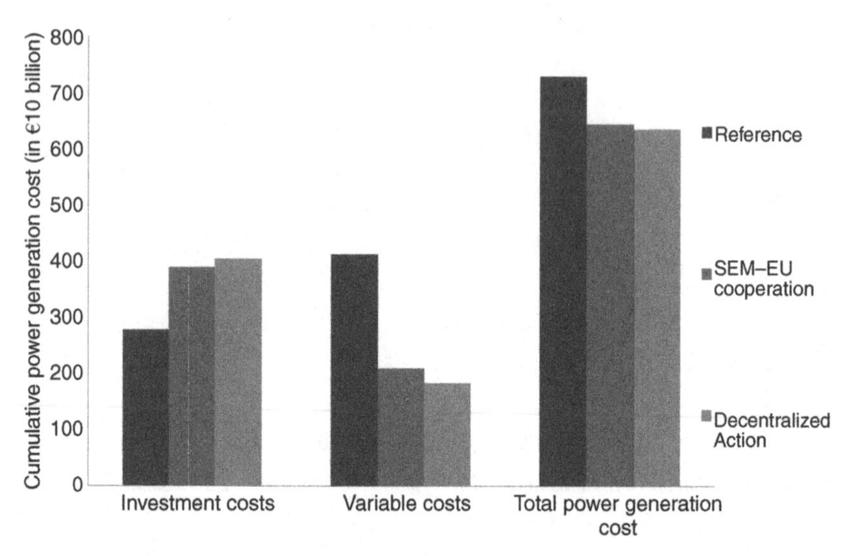

FIGURE 9.6 Cumulative power generation costs (2010–2030). *(Source: MENA–EDS model.)*

grid extensions and enhancements to allow for the development of renewables. In addition, it is assumed that the infrastructure for electricity exports to the EU, based on renewables and DC lines, is self-financed on the basis of a long-term power purchasing agreement covering all costs. It is estimated that a levelized price between 90€/MWh and 100 €/MWh (of exported renewable electricity) would be sufficient to recover all costs; such a price is obviously competitive when compared with the marginal costs of high renewable development domestically in the EU. This result supports the view that the SEM–EU Cooperation setting provides a win–win opportunity to both parties within the context of a strong decarbonization strategy for the EU.

5.5 Fossil Fuel Trade

Sustaining fossil fuel exports is important for the hydrocarbon-exporting economies of the SEM region, including Algeria, Egypt, and Libya, while the reduction in fossil fuel imports is equally important for the rest of the SEM countries which are hydrocarbon importers. The Reference Scenario projects oil production in the SEM region to increase from 192 Mtoe in 2010 to 276 Mtoe in 2030 (Table 9.4), mainly due to the major expansion in production in Libya, through enhanced oil recovery, the development of known fields (reserves), and parallel expansion of pipeline infrastructure. By contrast, Algerian and Egyptian production of oil is projected to follow a declining trend due to the maturity of oil fields and relatively limited undiscovered resources (BGR, 2009).

Production of natural gas in the SEM region is projected to expand under the Reference Scenario (from 143 Mtoe in 2010 to 322 Mtoe in 2030), due to expected developments in Algeria and Egypt which jointly account for about 80% of overall net increases (Table 9.5). In both these countries, increased gas

TABLE 9.4 Oil Primary Production and Exports in the SEM Countries (Mtoe)

		Algeria	Morocco	Tunisia	Egypt	Libya	Israel	Lebanon	Jordan	SEM
2010	Primary production	78.7	0.0	3.9	34.7	74.6	0.0	0.0	0.0	191.9
	Net exports	62.3	−10.7	0.0	−0.7	62.7	−10.7	−4.9	−5.1	93.0
Reference Scenario	Primary production	74.4	0.0	2.7	32.0	165.3	0.0	0.0	1.8	276.3
	Net exports	52.1	−16.0	−3.5	−22.1	148.5	−16.9	−4.2	−6.0	131.9
SEM–EU Cooperation Scenario	Net exports	52.6	−13.2	−3.2	−16.2	150.8	−16.7	−2.7	−5.3	146.2
Decentralized Actions Scenario	Net exports	55.7	−12.5	−2.5	−8.1	153.4	−15.2	−2.4	−4.4	163.9

Source: MENA–EDS model.

TABLE 9.5 Primary Production and Exports of Natural Gas in the SEM Countries (Mtoe)

		Algeria	Morocco	Tunisia	Egypt	Libya	Israel	Lebanon	Jordan	SEM
2010	Primary production	71.2	0.0	2.7	53.0	13.7	2.5	0.0	0.2	143.2
	Net exports	49.7	−0.4	−1.1	18.6	8.2	−1.7	−0.6	−1.6	71.0
Reference Scenario	Primary production	144.2	0.0	3.4	119.5	34.9	20.0	0.0	0.2	322.3
	Net exports	93.5	−9.9	−6.1	30.4	22.2	10.0	−3.8	−5.6	130.7
SEM–EU Cooperation Scenario	Primary production	128.8	0.0	3.1	93.9	30.4	20.6	0.0	0.2	277.0
	Net exports	90.6	−9.7	−4.0	26.0	21.6	9.8	−4.2	−4.7	125.3
Decentralized Actions Scenario	Primary production	149.9	0.0	3.0	115.0	35.0	22.6	0.0	0.2	325.7
	Net exports	116.6	−6.2	−3.3	54.1	26.3	13.0	−3.8	−4.4	192.2

Source: MENA-EDS model.

production is sufficient to meet the rapidly expanding local gas demand and, at the same time, nearly double exportable surplus. The Reference Scenario assumes rapid expansion of Israeli offshore fields and, by 2030, Israel is projected to export about 10 Mtoe of natural gas. The rest of the countries meet their expanding needs primarily from imports based on intraregional trade.

The two alternative scenarios assume the same evolution of primary oil production as in the Reference Scenario.[7] The "Cooperation" Scenario leads to a 10% reduction in primary oil consumption in the SEM region as a result of significant decline in oil-based power generation. The "Decentralized Actions" Scenario has a much wider scope for reduction in oil demand, mainly due to accelerated efficiency gains in final demand sectors and the higher uptake of hybrid vehicles in the transport sector. As a result, the exportable surplus of oil for the region as a whole is found to increase to 146 and 164 Mtoe in the "Cooperation" and in the "Decentralized Scenario," respectively, both above the 132 Mtoe in the Reference Scenario.

In the "SEM–EU Cooperation" Scenario primary gas production in the SEM region is projected to decline by 45 Mtoe compared with the Reference Scenario. This is due to lower demand for gas in Europe as renewable exports to the EU exert a crowding out effect in the EU power generation sector. Algerian gas production is also negatively affected by reduced gas consumption in the neighboring markets of Tunisia and Morocco. By contrast, the "Decentralized Actions" Scenario projects a significant increase of the region's gas exports from 131 Mtoe to 193 Mtoe in 2030, as a result of sharp reduction in domestic gas demand (efficiency gains and substitution toward renewables in electricity production). The exporting gas surplus in the "SEM–EU Cooperation" Scenario is comparable with the Reference Scenario despite the decrease in primary production, thanks to lower domestic demand for gas in this scenario. The "Decentralized Actions" Scenario predicts a significant increase in gas export surplus. Therefore, from the perspective of hydrocarbon exports by exporting countries and hydrocarbon consumption in importing countries, the alternative scenarios provide significant benefits to both groups of countries in the SEM region.

6 CONCLUSIONS

Assuming political stabilization and a return to sustained economic growth, the development of the energy, transport, and power sectors needs to be sustainable and affordable so as to improve economic competitiveness, financial stability, and living standards.

The Reference Scenario, which extrapolates current policies, finds them insufficient to achieve the required level of sustainability despite changes included in this scenario. The Reference Scenario assumes a gradual removal

7. The analysis assumes that international oil markets can absorb the quantity of oil produced in the SEM region with no effect on world oil prices.

of subsidies to energy products mainly in order to mitigate adverse effects on public finance, but this removal is partial and inefficiency persists. Despite the progress based on current policies, the Reference Scenario fails to tap into the vast potential of renewable energies and energy efficiency progress within the SEM region.

Further exploitation of these potentials, which is assessed in the context of alternative scenarios, is found to provide significantly larger benefits for consumers and the economy, such as wider access to electricity supply in rural and less developed areas, lower running costs for houses and industry due to energy savings, a higher surplus of hydrocarbons for exporting countries, and lower energy import bills for hydrocarbon-importing countries. In order to obtain these benefits, significant financial resources will be required, in addition to projected funding in the Reference Scenario, as both renewables development and energy efficiency improvements are capital-intensive investments.

Political and financial stability, as well as good governance and regulatory predictability, are the basic prerequisites for attracting foreign direct investment and for enabling private investment under favorable financing conditions. However, the amounts needed for investment in renewables and efficiency are particularly high and therefore it is worth examining concrete cooperation strategies with the EU in order to frame direct investment which will allow development of infrastructure and facilitate achieving favorable business conditions.

The analysis identified that a win–win cooperation situation can be achieved between the SEM region and the EU on the basis of two complementarity pillars: first, by extending EU-ETS to the SEM region with free allocation of allowances to this region for the power sector and energy-intensive industries; second, development of a large-scale renewable infrastructure in the SEM region to export electricity (mainly CSP and wind energy from centralized projects) to the EU through DC-based interconnections. The first pillar will make possible mitigation of the adverse effects of price reforms and promotion of renewables in the domestic power sector of the SEM countries, allowing lower domestic consumption of hydrocarbons. The second pillar can be self-financed by means of a power-purchasing contract at prices that can be more competitive than domestic renewables development in the EU. Model-based quantification of the SEM–EU Cooperation Scenario demonstrates that benefits (monetary and environmental) can be attained by both parties. The adoption of the 2030 EU energy and climate policy framework and the Market Stability Reserve will strengthen the EU-ETS price. In the context of cooperation between the SEM and the EU, the latter will benefit from lower costs for purchasing emissions allowances (lower ETS carbon price), which are projected to more than counterbalance the costs for the RES electricity exports region borne by the EU (by about €40 billion cumulatively by 2030). As the stringency of the European decarbonization effort will increase after 2030, resulting in further escalation of ETS prices, the benefits for the EU will be significantly higher if considered over a longer time horizon (by 2050). The introduction of ETS prices in the SEM region (with free allowances)

TABLE 9.6 Costs and Benefits in Alternative Strategies for the SEM Compared with the Reference Scenario

Costs and benefits (constant billion euros in cumulative terms until 2030)	SEM–EU Cooperation Scenario	Decentralized Actions Scenario
Oil exports	94.3	212.1
Natural gas exports	−14.7	168.0
Power generation costs	84.9	93.8
Value of carbon allowances	24.3	—
Energy efficiency investments		−90.0
Total	188.9	383.9
Total (% of GDP)	0.9	1.8

Source: MENA-EDS energy system model and Jablonski and Tarhini (2013).

will lead to a large reduction in oil-based power generation and thus to increased oil exportation surpluses (amounting to €94 billion), which combined with reduction in power generation costs and selling emissions allowances to the EU leads to an overall gain of €189 billion (Table 9.6).

An alternative strategy has also been examined, which assumes complete removal of current market and pricing distortions (energy subsidies) by 2020, a highly decentralized orientation of development of renewables, and the adoption of policies and standards for energy efficiency in all energy consumption sectors. This strategy is more aggressive than the SEM–EU Cooperation Scenario when it comes to modernization of the domestic energy system of the SEM region. However, its implementation requires significant incentives for investors in decentralized renewables and efficiency projects, and also very favorable financing conditions. It is probably less realistic than the "SEM–EU Cooperation" Scenario, but the benefits to be obtained for domestic SEM economies in the "Decentralized Actions" Scenario are found to be superior than those of the "SEM–EU Cooperation" Scenario. Final energy demand is projected to decline more sharply in the former strategy than in the latter, the contribution of renewables in power generation will be higher, access to electricity services in rural and less developed areas could be facilitated by the development of decentralized renewables and the overall costs can be lower. In addition, the former strategy allows significantly larger hydrocarbon exportation surpluses than both the Reference Scenario and the Cooperation Scenario for exporting countries and also significantly lower hydrocarbon imports for importing countries in the SEM region. The overall net benefits from the "Decentralized" Actions Scenario are projected to amount to 1.8% of the region's GDP in cumulative terms until 2030; this is twice as much as the net benefit from the "Cooperation" Scenario.

Implementation of the Cooperation Scenario strongly depends on the EU commitment to elaborate a concrete strategy in order to foster energy cooperation with the SEM region, effectively tackle the barriers to renewables deployment, and provide the required capital and financing mechanisms, policy frameworks, technical assistance, and know-how. The "Decentralized Actions" Scenario is inscribed in a framework of very optimistic hypotheses about overall political, economic, and social normalization of SEM economies and a significant improvement of financial and business conditions. The strategy assumes a favorable investment climate for both local investors and foreign direct investment, extensive reforms of energy subsidization schemes, investments in LNG terminals, and implementation of policies aimed at unlocking the potential of decentralized RES development, energy efficiency improvements, and hydrocarbon exploitation. Consequently, despite the higher economic benefits that emerge from the Decentralized Actions Scenario in terms of hydrocarbon surpluses and sustainable development objectives, the chances of materialization are considerably lower than with the "Cooperation" Scenario.

REFERENCES

BGR, 2009. Energy Resources 2009: Reserves, Resources and Availability of Energy Resources.

Blanc, F., 2012. Energy Efficiency: Trends and Perspectives in the Southern Mediterranean. MED-PRO Technical Report No. 21, CEPS, Brussels.

Coady, D., Gillingham, R., Ossowski, R., Piotrowski, J., Tareq, S., Tyson, J., 2010. Petroleum Product Subsidies: Costly, Inequitable, and Rising, IMF Staff Position Note, SPN/10/05. IMF, Washington, DC.

DLR (German Aerospace Center), 2005. MED-CSP: concentrating solar power for the Mediterranean region. Study prepared for the German Ministry of Environment, Nature Conversation and Nuclear Safety, Franz Trieb (Ed.). Available from: http://www.dlr.de/tt/med-csp

E3MLab, 2012. The MENA–EDS model. Available from: http://www.e3mlab.eu/e3mlab/MENA-EDS%20Manual/The%20MENA-EDS%20model.pdf

European Commission, 2014. A Policy Framework for Climate and Energy in the Period from 2020 up to 2030. Impact assessment accompanying the Communication, EC, Brussels.

Fattouh, B., El-Katiri, L., 2013. Energy subsidies in the Middle East and North Africa. Energy Strat. Rev. 2 (1), 108–115.

Folkmanis, A.J., 2011. International and European market mechanisms in the climate change agenda – an assessment of their potential to trigger investments in the Mediterranean solar plan. Energy Policy 39 (8), 4490–4496.

Fragkos, P., Kouvaritakis, N., Capros, P., 2013. Model-based analysis of the future strategies for the SEM energy system. Energy Strat. Rev. 2 (1), 59–70.

Fragkos, P., Kouvaritakis, N., Capros, P., 2015. Incorporating Uncertainty into World Energy Modelling. Environ. Model. Assess 20 (5), 549–569.

Hafner, M., Tagliapietra, S., 2013. A New Euro-Mediterranean Energy Roadmap for a Sustainable Energy Transition in the Region. MEDPRO Policy Paper No. 3, CEPS, Brussels.

Hafner, M., Tagliapietra, S., El Elandaloussi, E.H., 2012. Outlook for Oil and Gas in Southern and Eastern Mediterranean Countries. MEDPRO Technical Report No. 18, CEPS, Brussels.

High Representative of the Union for Foreign Affairs and Security Policy and the European Commission, 2011. Communication on A New Response to a Changing Neighbourhood: A Review of European Neighbourhood Policy, EC, Brussels.

IEA, 2012. World energy outlook 2012. Available from: http://www.worldenergyoutlook.org

IEA, 2013. Online database for energy subsidy data. Available from: http://www.iea.org/subsidy/index.html

Jablonski, S., Tarhini, M., 2013. Assessment of selected energy efficiency and renewable energy investments in the Mediterranean Partner Countries. Energy Strat. Rev. 2 (1), 71–78.

NEEAP, 2010. Lebanon National Energy Efficiency Action Plan (NEEAP) developed by the Lebanese Center for Energy Conservation (LCEC) 2011–2015.

OME, 2011. Mediterranean Energy Perspectives 2011. Mediterranean Observatory of Energy (OME).

Paroussos, L., Fragkiadakis, K., Charalampidis, I., Tsani, S., Capros, P., 2015. Macroeconomic scenarios for the south Mediterranean countries: evidence from general equilibrium model simulation results. Econ. Syst. 39 (1), 121–142.

Plan Bleu, 2008. The Blue Plan's sustainable development outlook for the Mediterranean. Available from: http://www.medsolutions.unisi.it/sites/default/files/UPM_EN.pdf

Schwanitz, V.J., Piontek, F., Bertram, C., Luderer, G., 2014. Long-term climate policy implications of phasing out fossil fuel subsidies. Energy Policy 67, 882–894.

Supersberger, N., Führer, L., 2011. Integration of renewable energies and nuclear power into North African energy systems: an analysis of energy import and export effects. Energy Policy 39 (2011), 4458–4465.

Trieb, F., Müller-Steinhagen, H., 2007. Europe–Middle East–North Africa cooperation for sustainable electricity and water. Sustain. Sci. 2 (2), 205–219.

United Nations, 2011. World Population Prospects: The 2010 Revision, Volume I: Comprehensive Tables. ST/ESA/SER.A/313, Department of Economic and Social Affairs, Population Division.

World Energy Council, 2010. Energy efficiency: a recipe for success, Annex 2, overview of energyefficiency policies measures. Available from: http://www10.iadb.org/intal/intalcdi/PE/2010/06689a03.pdf

Chapter 10

Benefits of Market Coupling in Terms of Social Welfare

Pedro Mejía Gómez
OMIE, Madrid, Spain

1 INTRODUCTION

In February 2011, the Heads of State and Government agreed that "The internal market should be completed by 2014 so as to allow gas and electricity to flow freely" (European Council, 2011). The European Council thereby reinforced the political support for an effective integration process, provided a specific date, and accelerated the implementation of the so-called "Third Package."

Within this context, and according to the cross-regional roadmaps set up by the Agency for the Cooperation of Energy Regulators (ACER), National Regulatory Authorities, and the European Commission, power exchanges have been working with the support of all the European stakeholders on the coupling of day-ahead markets across the European Union (EU).

On May 13, 2014 the southwestern Europe (SWE) and the northwestern Europe (NWE) day-ahead (DA) markets were successfully coupled. As a result, the SWE and the NWE projects, stretching from Portugal to Finland, now operate under a common DA power price calculation using the price coupling of regions (PCR) solution.

With the achievement of the full coupling of the SWE–NWE DA markets, the cross-border capacity of all interconnectors within and between the following NWE and SWE countries is now optimally allocated in the DA timeframe: Belgium, Denmark, Estonia, Finland, France, Germany/Austria, Great Britain, Latvia, Lithuania, Luxembourg, the Netherlands, Norway, Poland (via the SwePol Link), Portugal, Spain, and Sweden. Following the NWE–SWE full coupling, further extensions of market coupling (MC) with the PCR solution are envisaged. The combined DA markets of the NWE and SWE projects account for about 2400 TWh of yearly consumption.

The DA MC will bring significant benefits for end consumers derived from a more efficient use of the power system and cross-border infrastructures as a consequence of a stronger coordination between energy markets. This chapter

Regulation and Investments in Energy Markets. http://dx.doi.org/10.1016/B978-0-12-804436-0.00010-2

addresses the benefits of this key milestone on the road toward building an internal energy market and analyzes what this integration project means for European customers in terms of economic gains (Fisher et al., 2013).

The chapter also offers some thoughts on the impact this market-based approach to the management of electricity interconnections may have in neighboring Mediterranean countries.

2 DAY-AHEAD ELECTRICITY MARKETS IN EUROPE

The core activity of wholesale electricity markets in Europe is the DA market, where trading takes place on one day for the delivery of electricity the next day (Hunt, 2002). Market members submit their orders electronically to power exchanges, after which supply and demand are matched and the market price is calculated for each hour of the following day.

Power exchanges operate in an open trade context, so market agents can either bilaterally engage in any type of agreement for the delivery of energy (in the so-called bilateral market) and then declare their production/consumption schedule directly to the system operator at the market gate closure or submit bids for buying and selling power to power exchanges. These organized markets are optional, anonymous, and accessible to all participants satisfying admission requirements.

Ideally, the main objective of power exchanges is to ensure a transparent and reliable wholesale price formation mechanism in the power market, by matching supply and demand at a fair price, and to guarantee that the purchases made on the exchange are finally delivered and paid for.

Market equilibrium in the short term is illustrated schematically in Fig. 10.1, which is also useful for determining consumer surplus. This surplus is defined as total utility less total price paid (i.e., the area located between the demand curve and the equilibrium price). Similarly, the area to the right of the equilibrium point underneath the demand curve is the aggregate consumer utility that is not satisfied because marginal utility is lower than the equilibrium price. In a perfectly competitive electricity market, each producer's optimum output and price are defined by the marginal cost curve, and, in the aggregate, this ensures that demand is met by the most efficient units.

The price of electricity is becoming a major issue for society at large; not only for household users but also for business and industry due to the impact it has on their competitiveness in an open and globalized environment. This has been the backdrop to the launch of the EU's internal electricity market, with the first steps being taken in the 1990s, with the aim of passing on the benefits of the liberalization of the energy sector, involving better prices and services, to households and businesses alike (Boltz, 2013).

Those years saw the beginning of the creation of organized markets in all of Europe's regions. OMIE manages the spot market in the Iberian Peninsula, in much the same way as Nord Pool Spot does in the Nordic countries; EPEXSpot

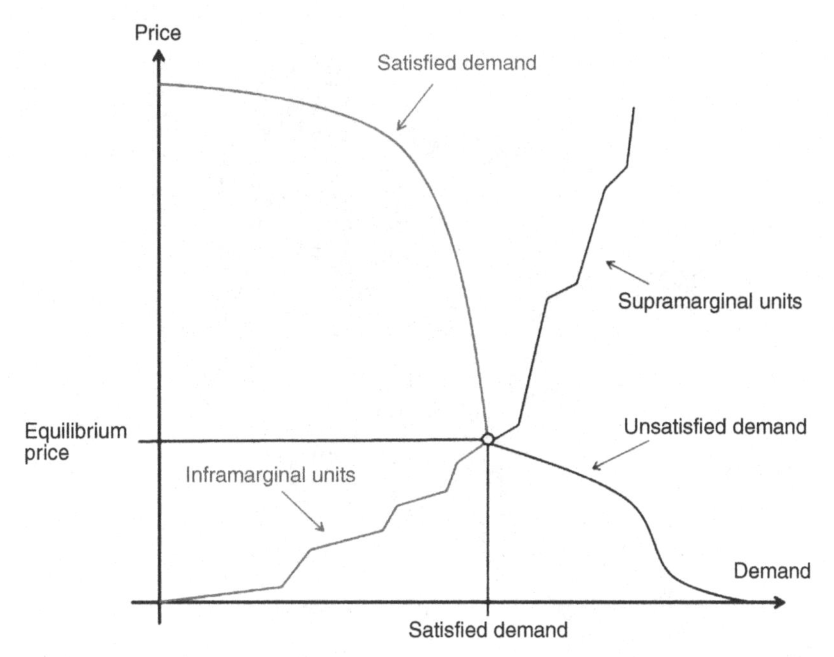

FIGURE 10.1 Short-term equilibrium between supply and demand in marginal price systems. *(Source: OMIE.)*

in France, Germany, and other Central European countries; and GME in Italy. Figure 10.2 displays different DA markets in the EU.

3 BENEFITS FROM ELECTRICITY CROSS-BORDER TRADING

A competitive wholesale electricity market at the EU level will be a reality as soon as cross-border electricity transactions between different member states flow in the most economically efficient direction (ERGEG, 2010), making the overall costs more efficient, and maximizing the welfare of both consumers and producers. International trading will lead to congestion on most of the EU borders and price spreads between different market areas will foster proper investments in more interconnections.

Therefore, the number of congested hours per year could be an indicator of performance, which reflects the level of cross-border trade. Indeed, the more integration we have the smaller the number of congested hours should be.

Another potential indicator reflecting the degree of cross-border trade over a specific border could be the congestion costs on this border. This indicator would be more complete than the previous one because it also takes into account the financial value of congestions. The congestion costs are the cost to society of having different prices in power markets. In other words, congestion costs are the loss in social welfare due to the congestion.

FIGURE 10.2 **DA markets in the EU.** *(Source: OMIE, PCR Project.)*

Figure 10.3 shows the net export curve (NEC) of interconnected markets A and B. For a given hour, the NEC of each market is constructed from the supply and demand curves of the market. Specifically, for each price *P* there is a given demand for imports (excess domestic demand) or supply of exports (excess domestic supply). These quantities represent the difference between offers and bids corresponding to each price (*P*). In other words, the NEC of a market gives, for each additional megawatt exported or imported by the market, the price that would be observed in this market.

As long as there is enough interconnection capacity between markets A and B, there will be a single clearing price for both markets, which is the intersection of NEC (A) and NEC (B). On the contrary, if there is insufficient interconnection capacity, markets A and B will be cleared at PB and PA, respectively. The foregone surplus (congestion costs) is represented by the striped area.

In this context, the implementation of congestion management mechanisms to optimize the use of current cross-border capacities, to ensure that cross-border

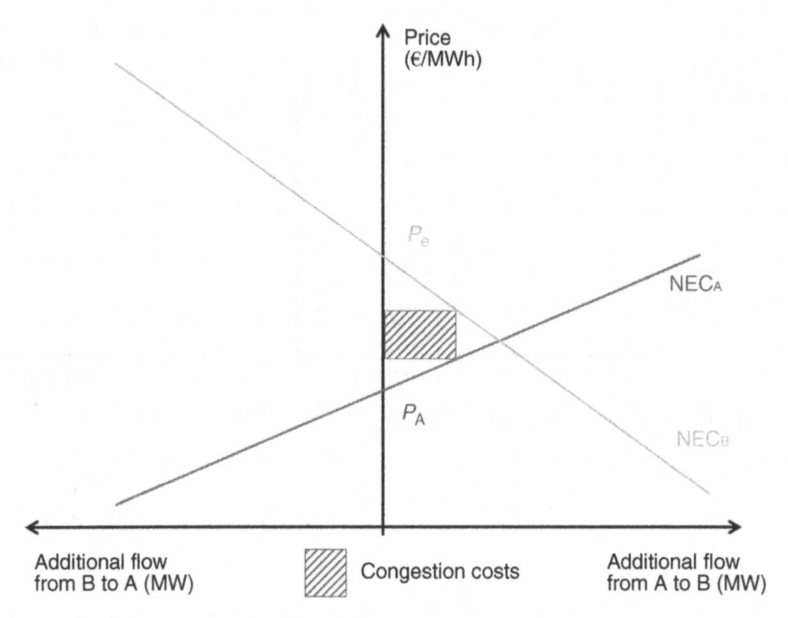

FIGURE 10.3 NECs of interconnected markets A and B. *(Source: OMIE.)*

flows go in the right economic direction and to promote more price convergence, has been a key priority of European regulators in the last decade.

4 DAY-AHEAD MARKET COUPLING

To solve the problems of congestion, insufficient utilization of the network, and the inefficient flow of energy between areas, an implicit auction system has been introduced.

This way of joining and integrating different energy markets into one cross-border market is called market coupling (MC) (Belmans et al., 2009). In a coupled market, demand and supply orders on one power exchange are no longer confined to the local territorial area. On the contrary, in an MC approach, energy transactions can involve sellers and buyers from different areas, only restricted by the constraints of the electricity network.

The main benefit of the MC approach lies in the improvement in market liquidity combined with the beneficial side effect of less volatile electricity prices. Additionally, MC is beneficial for market players too. They no longer need to acquire transmission capacity rights to carry out cross-border transactions, since these cross-border exchanges are given as a result of the MC mechanism. They only have to submit a single order in their market (via their corresponding power exchange), which will be matched with other competitive orders in the same market or other markets (provided electricity network constraints are respected).

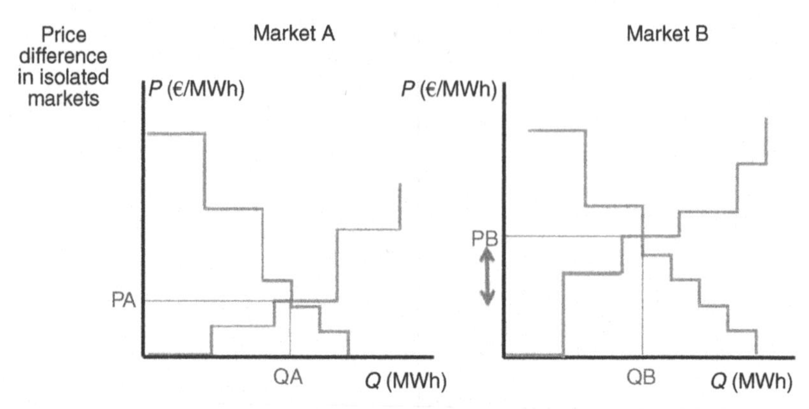

FIGURE 10.4 Not connected. *(Source: OMIE.)*

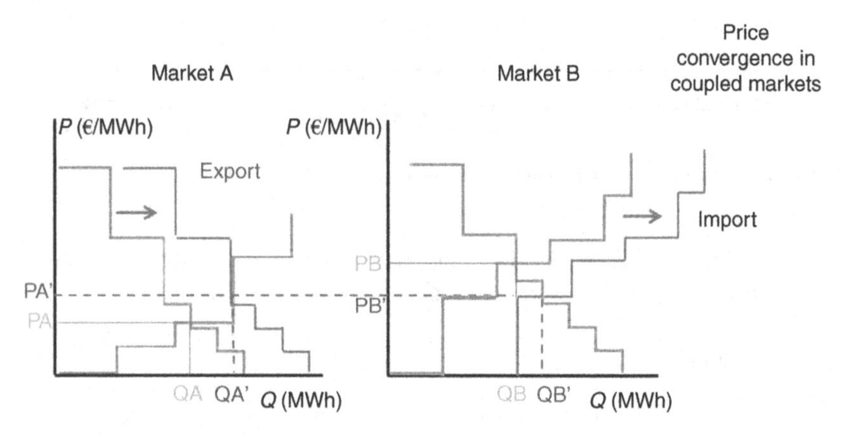

FIGURE 10.5 Connected. *(Source: OMIE.)*

The idea behind MC is to maximize the social welfare of both consumers and producers. As shown in the graph (Fig. 10.5), the clearing price at which the algorithm sets the price is the most efficient, creating the greatest area of welfare possible.

The first scenario in Fig. 10.4 shows two unconnected markets; both markets have different prices and therefore different welfare areas. The idea of connecting them both with the MC algorithm (Fig. 10.5) will benefit both as a whole. The cheaper market will increase the price and the most expensive will reduce it, but seen as a single market, the overall social welfare will increase, the combined market will be more efficient and beneficial to both consumers and producers.

4.1 EUPHEMIA Algorithm

EUPHEMIA (pan-European Hybrid Electricity Market Integration Algorithm) is an algorithm developed by European power exchanges to couple different

DA markets and handle standard and more sophisticated order types with all their requirements. Its aim is to rapidly find a good solution from which it continues to improve and increase overall welfare. This is the sum of consumer surplus, producer surplus, and congestion rent across the regions. EUPHEMIA is a generic algorithm: there is no hard limit on the number of markets, orders, or network constraints; all orders of the same type submitted by the participants are treated equally.

The development of EUPHEMIA started in July 2011 using one of the existing local algorithms as a starting point. The first stable version capable of covering the whole PCR range was internally delivered one year later (July 2012). Since then, the product has been evolving, including both corrective and evolutionary changes.

EUPHEMIA receives information about the power transmission network, which is then modeled in the form of constraints to be respected in the final solution. This information will be mainly provided by transport system operators (TSOs) as input to the algorithm.

The information received is divided into bidding areas and EUPHEMIA computes a market clearing price for every bidding area per period and a corresponding net position (calculated as the difference between matched supply and matched demand quantities belonging to that bidding area).

Bidding areas can exchange energy between themselves in an available transfer capacity (ATC) model, a flow-based model, or a hybrid model (a hybrid of the other two).

5 BENEFITS FROM PAN-EUROPEAN MARKET COUPLING

On May 13, 2014 in a landmark move toward an integrated European power market, full coupling of the SWE DA markets was successfully launched. As a result, the SWE and NWE projects, stretching from Portugal to Finland, now operate under a common DA power price calculation using the PCR solution.

Since that date DA transmission capacity on the French–Spanish border has been implicitly allocated through the PCR solution, replacing the previous daily explicit allocation. Full price coupling between the NWE and SWE projects allows simultaneous calculation of electricity prices and cross-border flows across the region. This will bring a benefit for end consumers derived from a more efficient use of the power system and cross-border infrastructures as a consequence of stronger coordination between energy markets.

Following the launch of the PCR on February 4, daily average cleared volume over these markets amounted to 3.2 TWh, with an average daily value of over €200 million. Following the NWE–SWE full coupling, further extensions of MC with the PCR solution are envisaged.

In the following sections, the chapter addresses the benefits of this project in the case of the interconnection between France and Spain.

5.1 Optimizing the Use of Existing Cross-Border Capacities

The most immediate effect after the implementation of implicit auctions across Europe is improvements in the use of existing interconnections. In the case of MIBEL (*Mercado Ibérico de la Electricidad*, Iberian Electricity Market), this was very important for French–Spanish interconnection in the first months of MC.

According to our own calculations for 2012 and 2013, the percentage of hours per year in which the cross-border capacity between France and Spain was not used at maximum capacity was around 55 and 47%, respectively. Therefore, during around 50% of the hours, the interconnection between the two countries was underused. This indicator was even worse in the first months of 2014, until MC was implemented. From January 2014 until May 13, 2014, the interconnection capacity was not used at its maximum for around 62% of the time.

After implementation of the PCR on May 13, 2014, the use of cross-border capacity improved significantly. In the period May 13–July 31, cross-border capacity was fully used 94% of the time and only 6% of the time was it not used at maximum, due to the fact that Spain and France had the same price levels.

In this context, another indicator used to evaluate optimization of the French–Spanish interconnection may be average capacity not used when a price difference exists. In 2012 and 2013, around 515 and 454 MW were not finally used in average terms for time with unused capacity.

In 2014, this indicator remained at the same level (around 453 MW) until May 13 when MC between SWE and NWE entered into operation. After implementation of the new scheme, average capacity not finally used at times when there was a price difference was zero.

5.2 Cross-Border Electricity Flows in the Right Direction

In 2010 the national energy regulators of the SWE electricity region started publishing regional reports on the management and use of interconnections. One of the most relevant findings of these reports was that important amounts of capacity were not being used in the price differential direction.

Figure 10.6 shows how many cross-border transactions at the French–Spanish border took place in the wrong direction due to explicit allocation of capacities before price formation in DA markets. On the contrary, Fig. 10.7 presents what is currently happening with cross-border transactions at the same interconnection. As can be seen, after implementation of MC there are no transactions in the antieconomic price direction. In other words, cross-border electricity flows go from lower to higher prices in all cases.

According to OMIE's estimations for 2012 and 2013, cross-border transactions were executed in the wrong direction for 17 and 13% of the time, respectively. In 2014 the same situation happened for 16% of the time (until MC was implemented). Afterward, this situation has not occurred again, as has

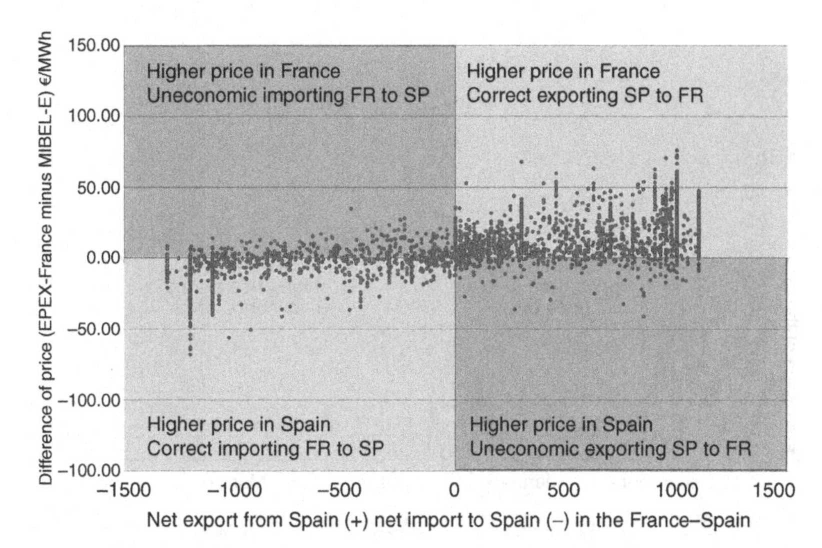

FIGURE 10.6 Difference price versus net export/import for the French–Spanish interconnection (until May 13, 2014). FR, France; SP, Spain. *(Source: OMIE.)*

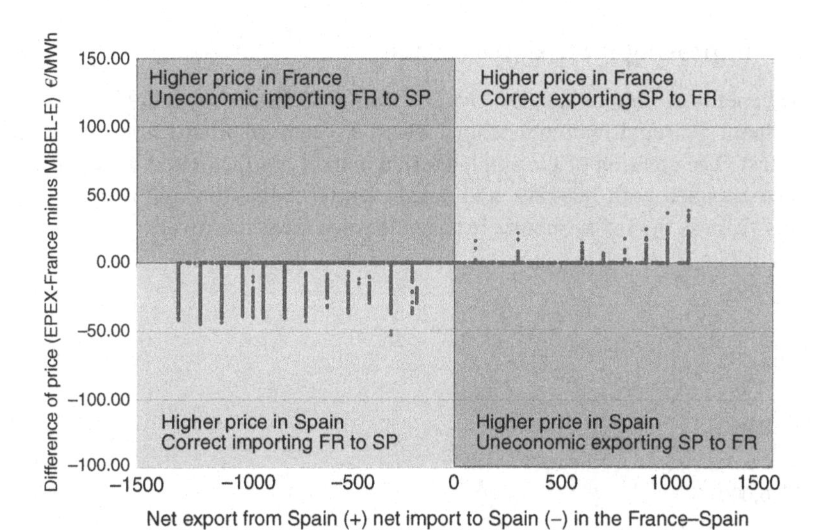

FIGURE 10.7 Difference price versus net export/import for the French–Spanish interconnection (from May 13 until December 31, 2014). FR, France; SP, Spain. *(Source: OMIE.)*

been the case with other European interconnections where MC has been implemented. Figure 10.8 shows how antieconomic, cross-border flows have also disappeared from the French–British interconnection.

As for average capacity used in the wrong direction, it stood at around 780 MW before MC (estimation for 2012, 2013, and January–May 2014).

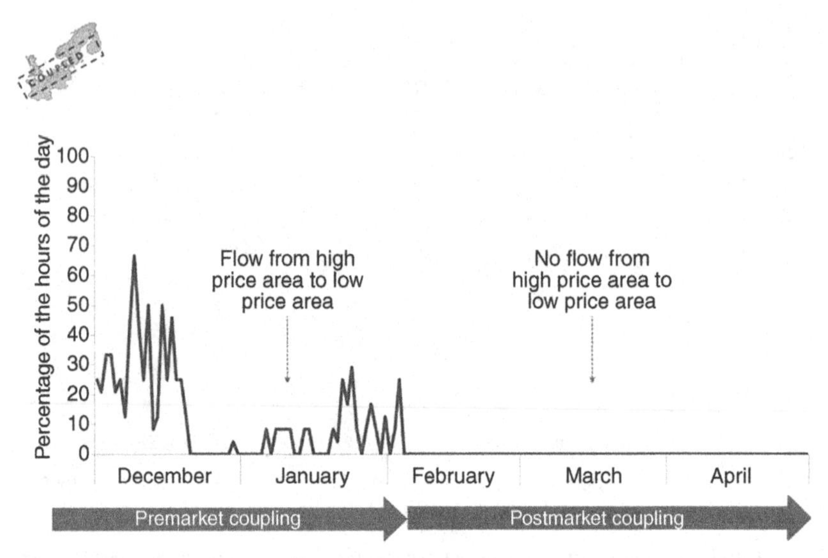

FIGURE 10.8 MC flows for IFA interconnector between France and Great Britain. *(Source: NWE Project, 2014.)*

5.3 Improving Price Convergence

The benefits of implicit auctions will be similar to those shown in Fig. 10.9 for the Spain–Portugal interconnection, where we can see what has happened in the past. The creation of the single Iberian market has facilitated price convergence between both markets, and consequently congestion rent decreased in these years as the infrastructure between the two areas improved.

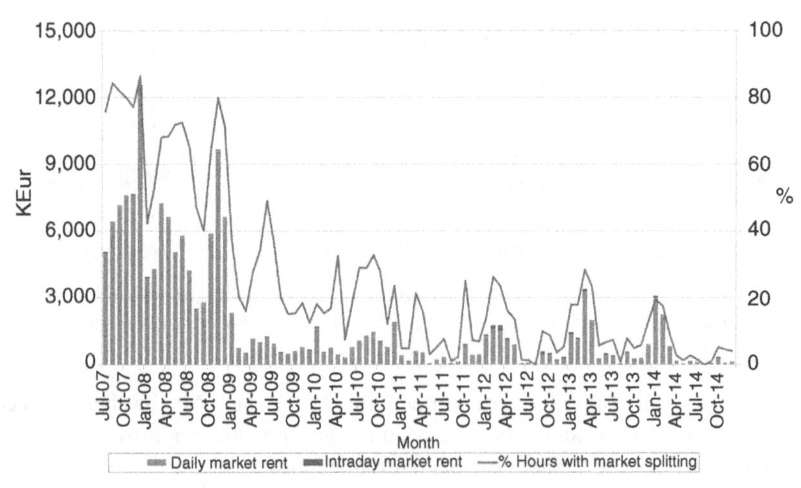

FIGURE 10.9 Congestion rent in the Portuguese–Spanish interconnection in daily and intradaily markets (July 2007–December 2014). *(Source: OMIE.)*

These same effects will be seen on the French–Spanish border in coming years; in particular, when the new interconnection comes into operation in the coming months. From May 14, 2014 until December 31, 2014 the price was the same for 8.4% of the time in the MIBEL market (Spain and Portugal) and the French area. This figure will certainly increase when the new line on the French–Spanish border begins operation in 2015.

In an attempt to show the possible effects on prices of the new 1400 MW interconnection between France and Spain, we have made a simulation of the results for 2013 and compared them with actual prices on the MIBEL market.

The data showed that the percentage of time with price differential <1€/MWh would increase from 5.8% to 30.8%, and the percentage of time with price differential <2€/MWh would increase from 10.4% to 33.2%.

This simulation was made with the old market algorithm that was used during 2013.

5.4 Gross Welfare of Market Coupling and Interconnectors

In order to complete the benefits of MC, we analyzed efficiency in terms of social welfare. Economic efficiency should allow lowest cost producers to cover the demand in neighboring areas. An interesting study of this issue is undertaken by ACER on a yearly basis (ACER, 2013).

This study compares for 2012, "*ceteris paribus*" the remaining data, the possibility of trading between different price areas while considering isolated markets unable to trade with each other. Another analysis is "incremental gain" as a result of enlarging the interconnection capacity of 100 MW on each border. Both indicators are presented in Fig. 10.10.

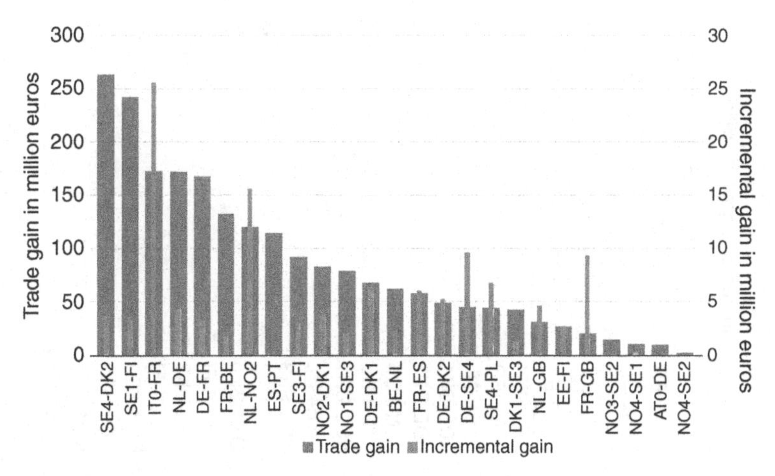

FIGURE 10.10 Gross welfare benefits from cross-border trade and incremental gain per border in 2012 (million euros). *(Source: ACER (2013).)*

The results of these simulations shed light on the benefits that MC has already brought to trading within the MIBEL market and also foretells the benefits that implicit auctioning with the rest of the continent will bring in terms of social welfare for both NWE and SWE areas.

In the first case, the Portugal–Spain (MIBEL) trading gain was the eighth most beneficial within Europe, estimated as it was at approximately €120 million. The France–Spain (SWE) real trading gain estimated up until 2015 was the 14th most beneficial, but it would have been the 7th most beneficial with a potential incremental gain of over €5 million (considering an increase in the size of the current interconnection of 100 MW).

Analogously, ACER also presented the results of incremental gains calculated for 2013 using the same methodology but considering the potentiality of the new PCR algorithm that is now being used to calculate European electricity prices (Fig. 10.11). The data showed that incremental gain on the France–Spain border would be close to €10 million, making it the fifth most beneficial within Europe with an increase of 100 MW in cross-border capacity. However, the Portugal–Spain border would receive the smallest benefit from an increase in capacity, which means that capacity at this border is already adequate.

In addition to this study by ACER, power exchanges in the NWE region have been publishing a monthly report on social welfare since February 2014 (Tennet, 2014) when the MC went live. In this report they calculate the gains that could be obtained in the event there were no network constraints in the central western European (CWE) region, in light of an infinitely available trading capacity. The simulation showed that from February until July 2014, the potential gain from having no network constraints in the CWE region would have been €61.8 million.

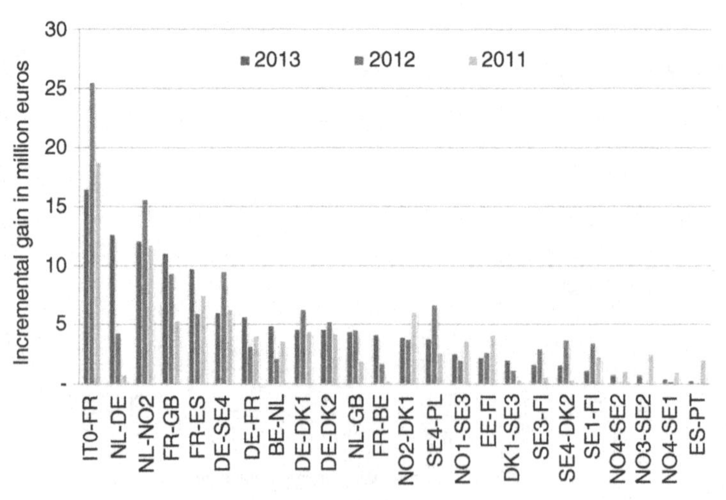

FIGURE 10.11 Incremental gain from PCR MC in different EU interconnections. *(Source: ACER, Florence Forum, November 2013.)*

Further studies have been done within the MIBEL market by OMIE. In this case, we have studied the price difference obtained using the interconnection with France compared with the scenario of having an isolated market. The results show that despite the interconnection with France still being very small, average improvement in price per hour is 1.7€/MWh.

6 SOME THOUGHTS ABOUT CROSS-BORDER TRADE BETWEEN THE IBERIAN ELECTRICITY MARKET (MIBEL) AND NORTHERN AFRICA

It is also worth noting that the European MC project will affect all surrounding countries. Although a market-oriented reform in neighboring countries is an obvious prerequisite to potentially extending the MC mechanism to these third countries, the new reality of DA markets in Europe will have immediate consequences for them.

In the case of cross-border trade between MIBEL and Northern Africa, the current situation is as follows.

In past years electricity flowed from Spain to Morocco, leaving import capacity (Morocco > Spain) completely unused. However, export capacity has been used 100% of the time with a use ratio of 64% of the available capacity in 2012 (total export to Morocco of 4909 GWh), 67% in 2013 (total export to Morocco of 5378 GWh), and in 2014, this increased to 77% of the capacity (total export to Morocco of 5928 GWh). These amounts are the maximum import volumes the Moroccan system can currently manage.

Although this interconnection is far from being congested, imports of electricity from Spain represent around 17% of Morocco's electricity consumption. Therefore, this is a significant percentage of total demand in this neighboring country – one that will have a more robust price reference than in the past. In other words, the more European price formation applies in electricity wholesale markets the more reliable the price will be for neighboring countries. Consequently, it will be up to them and their national energy policies to rely on more European imports or develop their own generation resources to compete with (and export to) Europe.

7 CONCLUSIONS

A competitive wholesale electricity market at the EU level will be a reality as soon as cross-border electricity transactions between different member states flow in the most economically efficient direction, maximizing the use of existing interconnections and fostering investments where needed. MC has been a key priority for European institutions and national regulatory authorities in accelerating the building of the long-awaited internal electricity market.

In this context, power exchanges and transmission system operators have worked together on the implementation of pan-European MC. This project was launched thanks to the initiative of the most relevant power exchanges in Europe (under the so-called Price-Coupling Region initiative) and now MC integrates the DA markets of 17 countries covering around 70% of European electricity demand.

Currently, the NWE and SWE regions are already fully coupled and this allows simultaneous calculation of electricity prices and cross-border flows across the regions. The direct consequences are:

- Optimization and more efficient use of existing cross-border capacities.
- Elimination of important amounts of capacity not being used in the price differential direction.
- Improvement of price convergence between different areas and a more reliable electricity price on the DA time horizon.

These benefits are already visible and will be further measured in a more precise way in the months to come.

As for neighboring countries, although it is not yet realistic to think about extending MC to them in the short term, they will benefit from a more robust price formation as MC is paving the way toward establishing (as soon as additional interconnections are built) a solid DA electricity price for Europe. It will be up to neighboring countries to increase imports or start developing their own new generation resources to compete with Europe.

REFERENCES

ACER, 2013. Annual Report on the Results of Monitoring the Internal Electricity and Natural Gas Markets. Agency for the Cooperation of Energy Regulators, Ljubljana. http://www.acer.europa.eu/official_documents/acts_of_the_agency/publication/acer%20market%20monitoring%20report%202013.pdf.

Belmans, R., Cole, S., Meeus, L., Vandezande, L., 2009. Market coupling and the importance of price coordination between power exchanges. Energy 34, 228–234.

Boltz, Walter., 2013, The Challenges of Electricity Market Regulation in The European Union in Evolution of Global Electricity Markets. (Chapter 8). E-control, Austria.

ERGEG, 2010. Incentive Schemes to Promote Cross-Border Trade in Electricity. ERGEG, Brussels.

European Council, 2011. Conclusions of the European Council on 4 February 2011. EUCO 2/1/11.

Fisher, L., Newbery, P., Noël, P., Strbac, G., Pudjianto, D., 2013. Benefits of an Integrated European Energy Market. Booz & Co, Amsterdam, London, Cambridge University, Singapore.

Hunt, S., 2002. Making Competition Work in Electricity. John Wiley & Sons, New York.

Tennet, 2014. Social welfare report. http://www.tennet.eu/nl/grid-projects/international-projects/market-coupling/downloads.html (accessed 25.11.2014.).

Chapter 11

Power Market Structure and Renewable Energy Deployment Experiences From the MENA Region

Emma Åberg*, Nurzat Myrsalieva, Tareq Emtairah†**

**Independent Consultant, Stockholm, Sweden; **Regional Center for Renewable Energy and Energy Efficiency (RCREEE), Cairo, Egypt; †International Institute for Industrial Environmental Economics (IIIEE), Lund University, Lund, Sweden*

1 INTRODUCTION

In many parts of the world, electricity sector liberalization has been a hot topic of debate during the last couple of decades. In developed and developing countries, sector reforms have been associated with high expectations regarding positive impacts on the economical and technical performance of the power industry (Dyllick-Brenzinger and Finger, 2013). However, due to an energy situation historically characterized by vast amounts of cheap hydrocarbons, such prospects have been less interesting for the Middle East and North Africa (MENA), and in almost all countries in the region power sectors are still characterized by a high degree of vertical integration and state control.

Since the early 2000s several countries have undertaken reforms in order to open up the power market to private sector involvement. The reforms relate to a number of aspects, which differentiate the individual markets and their level of liberalization: unbundling of the power sector; introduction of independent electricity regulators; and opening up the market to private power generation for different purposes. This chapter serves to assess to what extent such reforms have taken place in the different MENA countries (Sections 2 and 3) and how such reforms affect the achievement of renewable energy (RE) targets (Section 4). Section 5 reflects the overall development and provides some policy recommendations to improve private sector participation in RE generation. It should be noticed that, due to the relatively late electricity sector reforms, the amount of studies on progress and results are particularly scarce when it comes to the MENA region.

Regulation and Investments in Energy Markets. http://dx.doi.org/10.1016/B978-0-12-804436-0.00011-4

2 UNBUNDLING OF THE POWER SECTOR

2.1 On the Reform of the Power Sector

Power markets worldwide have traditionally seen a high level of state ownership, monopoly, and vertical integration. High vertical electricity sector integration refers to a situation in which all stages in the electricity value chain (generation, transmission, distribution, and retail) are owned and operated by the same actor. Contrariwise, *unbundling* of the power sector defines a condition in which different segments of a system are separated into clearly defined entities. Many actors, such as the European Commission, see a separation of the different activities of the electricity system as one of the key preconditions for increasing transparency and fair competition in the power market. Specifically, unbundling ensures the promotion of private generation investments in a nondiscriminatory way through fair access to transmission and distribution for all actors (Manoussakis, 2009).

Unbundling of the power sector can take different forms in relation to the degree of vertical separation. At a general level, it is possible to distinguish between ownership, legal, functional, and accounting separation. The highest degree of unbundling is *ownership separation*, which refers to a situation in which all network infrastructure (transmission and distribution) is fully separated from the generation stage. In the remaining forms of unbundling, network activities and generation are to different extents separated, although still operating under the same ownership. *Legal separation* is the second strongest form of unbundling and is similar to ownership separation in the way that power sector activities are managed under companies that have separate boards and reporting systems, but operate under the same ownership. *Functional separation* requires network activities and generation to have functionally independent management units in which the transmission entity works in a nondiscriminatory way toward all power generation companies. The weakest form of unbundling is *accounting separation*, which requires the costs and assets associated with different activities to be reported separately.

It is important to note that there are some major controversies surrounding the debate on the advantages and disadvantages of unbundling and specifically in relation to the degree and type of unbundling. In Europe, the debate is focused around the importance of separating transmission from all sorts of generation activities. The European Commission claims that legal or functional separation is not sufficient to stimulate an acceptable level of competition within the electricity market (Gugler et al., 2013). In the Commission's adopted package of energy proposals from 2009, ownership unbundling is highlighted as the preferred option of unbundling (Pollitt, 2008). The European debate to a large extent influences the MENA debate through various Euro-Mediterranean cooperation and integration initiatives. Nevertheless, distribution unbundling is likely to become an issue of debate in the near future, since the development of RE generation and smart grids raises the question of whether distribution unbundling favors or hinders innovation on this level (Meyer, 2012). Moreover, current disagreements and the fast development of RE and smart grids call for further and comprehensive discussions on what an unbundling of the electricity sector in the MENA region should actually look like.

2.2 Current Unbundling Situation in the MENA Region

In relation to RE deployment in the MENA region, there is general concern around the extent to which current distribution and transmission ownership arrangements can deliver transparent and nondiscriminatory access to the electricity grids. Some countries in the region have taken steps toward unbundling the electricity sector with different levels of ambition. Jordan is the country that is found to have come furthest in such development.

Since the establishment of General Electricity Law No. 64 in 2003, Jordan has put into operation a fully unbundled power sector characterized by ownership separation of generation, transmission, and distribution. One state-owned and three privately owned companies dominate the generation segment, in which 75% of total power generation is privately produced. Three private companies own and operate the distribution lines while the state-owned National Electric Power Company (NEPCO) owns and operates the transmission network (NEPCO, 2014). The General Electricity Law grants nondiscriminatory access to NEPCO's transmission lines and the Renewable Energy Law from 2012 further obliges this company, as the single buyer in the market, to purchase all electricity produced, although without any priority access or priority dispatch for RE (Dii, 2014a).

While Jordan has implemented full ownership separation, most other countries with ambitions to unbundle have taken steps toward a legal or functional separation of their transmission operators. The most common way to accomplish such separation is to create a state-owned holding company with subsidiaries that separately operate the different activities within the power sector value chain. This type of unbundling can be found in Abu Dhabi, Algeria, Egypt, and to some extent in Saudi Arabia.

Abu Dhabi is seen as a pioneer when it comes to power sector liberalization in the United Arab Emirates given that Reform Law 2 from 1998 brought far-reaching structural changes including unbundling of the sector. All network activities and one power generation company are gathered as legally separated entities under a state-owned holding company, Abu Dhabi Power Corporation (Dyllick-Brenzinger and Finger, 2013). What distinguishes Abu Dhabi from other countries in the region is that a majority of the power generation has been privatized since the reform. However, a generation company still operates under the same ownership as the network companies. For this reason, the Abu Dhabi power sector cannot be categorized as fully unbundled. Dubai, which has the other dominating power sector in the UAE, is still characterized by a high level of vertical integration and has not been unbundled (IRENA, 2014).

In Algeria, Law 02-01 from 2002 provides for full ownership unbundling of the electricity sector. Until now, the country has implemented a form of legal or functional unbundling where generation and transmission have been separated into different branches under the state-owned holding company Sonelgaz Group. What makes Algeria different from other countries is its structuring of network activities. Transmission ownership and transmission operation have been separated and the distribution companies operate as single buyers. The transmission organization lays the foundation for a specific unbundling model, the Independent System Operator (ISO) model, which allows for a large number of transmission owners. However, studies of the Algerian power sector indicate that the process toward a higher degree of unbundling, including the creation of a market operator and an ownership separation, has halted. The law guarantees access to the grid without any priority access for RE projects. Once an RE project has been connected to the grid, priority dispatch is guaranteed by decree No. 06-429 from 2006.[1]

The Egypt Reform Law was adopted in 2001 and opens up a fully unbundled power sector. Today, the reform has reached an implementation level characterized by a form of legal or functional unbundling in which the Egyptian Electricity Holding Company (EECH) gathers all state-owned companies: one transmission company, nine distribution companies, and six large electricity generators. The transmission company operates as the single buyer and, even if not required by law, has in practice committed itself to purchasing all RE-generated electricity. In the network access contracts between the transmission company and the private power producer it is specified that RE electricity will have priority dispatch. Priority access and dispatch are foreseen to become a statutory reality in the new electricity law being drafted.

Saudi Arabia created the legal context for unbundling the power sector in its Electricity Law established in 2005. Except for a number of independent power producers (IPPs), all power actors are gathered under the Saudi Electricity Company (SEC). This is a joint stock company with a majority of its shares held by the Saudi government. Unbundling of the sector is progressing slowly but as an important first step the transmission company has been legally separated from any other power sector activities. The remaining power activities under SEC have reached the stage of accounting separation.

Another territory worth mentioning in this context is Palestine. Due to its particular situation, it imports most of its electricity needs from Israel directly via the distribution networks that are owned and operated by a handful of different municipally owned companies. There are no high voltage transmission lines and only one conventional power plant, which is not operating at the moment due to fuel shortage and its location in war-affected Gaza (Åberg, 2014). Nevertheless, there are prospects for high-voltage transmission lines – the Palestine Electricity Transmission Line Company (PETL) was introduced in 2010 – and power

1. Décret exécutif no 06-429 du 2006.

generation in the future. With such prospects in mind, the General Electricity Law No. 13 that was adopted in 2009 includes the legal context for unbundling the power sector (IRENA, 2014).

All countries in the MENA region that have committed to unbundling reforms are still in the process of implementing transparent and functional organizations and it is therefore a complex task to assess the extent to witch separate entities actually operate independently. The implementation of rather complex sector models takes time and the coming decade should be seen as a transitional period. For the enforcement of the legal provisions that are the basis for an open market, independent regulators will play a crucial role. All countries that have legally undertaken unbundling of the power sector have also established regulatory agencies. Agencies independent from political power and ministry control are generally recognized as a guarantee of regulatory commitment of a country, something that will promote investor confidence (Cambini and Franzi, 2013). Among the MENA countries, only the Jordanian Electricity Regulatory Commission (ERC) can be considered as somewhat independent. ERC is responsible for the enforcement of unbundling and can make decisions regarding tariff setting and third-party access of the grids without receiving approval or informing the government or parliament of its work (Cambini and Franzi, 2013). In contrast, the level of political independence of the regulatory bodies in Algeria, Egypt, Palestine, and Saudi Arabia has been shown to be low with the agencies mainly operating as advisory bodies to the governments which takes the final decisions on licensing, tariff setting, etc. (Cambini and Franzi, 2013; Åberg, 2014; Dyllick-Brenzinger and Finger, 2013) The regulatory body in Abu Dhabi has government representatives on its board and, for this reason, cannot be seen as administratively independent; however, it is said to have far-reaching regulatory powers in licensing and monitoring activities (Dyllick-Brenzinger and Finger, 2013).

As for the rest of the MENA countries and their power sectors, it is likely that some of them would to a certain extent fall under the category of accounting separation, at least in practice. Nevertheless, as long as no laws or electricity regulators have been established, these types of separations will not be considered unbundling (Table 11.1).

3 RENEWABLE ENERGY AND PRIVATE SECTOR PARTICIPATION

Another important aspect of a liberalized power market is the level of private participation. While private participation in transmission and distribution remains very restricted, many MENA countries have opened up for private engagement in the power generation sector. Private participation in power generation activities exists in various forms, which are most often determined by law. It is possible to distinguish between private power generation for utility supply, third-party sales, direct export, and self-consumption. With the exception of

TABLE 11.1 Status of Power Sector Unbundling in MENA countries, 2014

Country	Unbundling status	Electricity regulatory agency
Jordan	Ownership separation	Electricity Regulatory Commission (ERC)
Palestine	Ownership separation	Palestinian Electricity Regulatory Council (PERC)
Egypt	Functional or legal separation	Egyptian Electric Utility and Consumer Protection Regulatory Agency (EgyptERA)
UAE (Abu Dhabi)	Functional or legal separation	Abu Dhabi Regulation and Supervision Bureau (RSB)
Saudi Arabia	Functional or legal separation	The Electricity and Co-Generation Regulatory Authority (ECRA)
Algeria	Functional or legal separation	Commission de Régulation de l'Electricité et du Gaz (CREG)

the latter, which is sometimes referred to as autoproduction, all types of private power generation fall under the concept of IPPs.

IPPs typically build, own, and operate power plants to sell electricity either to the utility or to a third party. While the first option is a common and sound means for private actors to operate, third-party sales offer additional incentives for investors (Dii, 2013). A third-party supply option can be particularly appealing for larger industrial and commercial actors that have specific electricity needs that can be best satisfied by a third party. In any case, IPPs are particularly dependent on some sort of insurance, which guarantees that their generated electricity is being purchased to a fixed and long-term price (IRENA, 2014). A means by which states can provide such a guarantee and enable IPPs investment in RE projects are through public competitive bidding processes, direct proposal submissions, or feed-in tariffs, which all result in power purchase agreements (PPAs) with the state-owned utility (IRENA, 2014). In the case of third-party sales, a purchase guarantee needs to be assured via a bilateral agreement directly between an IPP and one or several large electricity consumers (Dii, 2013). The same type of bilateral agreement would apply for the unusual situation of direct export of electricity to another country. In addition, all IPPs need nondiscriminatory access to the grid in order to transport their electricity to their customers; such conditions need to be assured through legislation and regulated grid-transporting tariffs. Consequently, the overall number of IPPs in the market usually reflects the level of liberalization and competition as well as access to the market for private developers (RCREEE, 2013).

Private actors that generate electricity primarily for self-consumption (autoproducers), typically at large industrial sites, have often been seen as exceptions under traditionally strict state electricity monopolies. In the MENA region, autoproduction companies have often been state owned and not connected to the grid. Therefore, early autoproducers should not be seen as a sign of liberalization in the same way as IPPs (Dyllick-Brenzinger and Finger, 2013). Today, self-generation of RE can be an attractive way for households, private industries, and commercial actors to protect themselves from rising electricity prices and to a certain extent also against power outages. In order to encourage self-generation, the regulatory framework should allow RE-generating companies to feed any potential electricity surplus to the grid under some sort of net metering scheme. In addition, power-generating actors should be allowed to locate their RE generation plants outside their premises with the right to use the grid for *wheeling* of electricity to any area of use.

Most MENA countries allow for some sort of private participation in power generation activities and 17 out of 18 investigated countries have adopted legislation authorizing IPPs and/or autoproduction. Libya, which is the only country with a fully closed electricity sector, has a new electricity law underway which will open up for private participation also in this country. However, private sector participation in power generation remains limited in relation to RE-sourced electricity. While many countries have IPPs producing conventional electricity, Morocco and Abu Dhabi are in fact the only countries in which private actors own and currently operate renewable power plants (IRENA, 2014) (in Morocco in the form of a 50-MW wind park that has been in operation since early 2000 and in Abu Dhabi in the form of a 100-MW concentrated solar power (CSP) plant in operation since 2013 (MEMEE, 2009)). However, a number of RE IPP projects are now in the pipeline in Egypt, Jordan, Abu Dhabi, and Morocco.

In order to increase the share of private participation in the electricity sector, both overall legal structures and support mechanisms are needed. Previous studies of large-scale private participation show that support mechanisms directly affect the amount of private capacity being installed, while overall legal structures determine whether to invest or not in the first place (Fox-Penner et al., 1990).

3.1 Renewable Energy for Utility Supply

The majority of MENA countries authorize private power generation for utility supply (i.e., selling electricity to a single buyer). While direct proposal submissions are allowed only in 2 countries, 14 of the 18 MENA countries have introduced public competitive-bidding processes. Six of these 18 had officially announced tenders for some sort of RE project by the first quarter of 2014 (Fox-Penner et al., 1990).

What characterizes the MENA region regarding the possibility for private actors to produce electricity for utility supply is a lack of clear signals from governments. In most of the countries, private actors are fully dependent on government announcements of tenders and as a general rule competitive-bidding processes are only planned for one or two projects. To increase certainty for investors there need to be clear targets and predictability around the total number of RE projects to be developed through PPAs.

Egypt, Morocco, and Saudi Arabia are the only countries that have set targets for total installed capacity of RE to be developed through the IPP competitive-bidding approach. Morocco is well on track toward meeting its targets, which are to develop 1000 MW of solar and 1000 MW of wind power through competitive bidding (RCREEE, 2013). Some 350 MW of wind power and 160 MW of CSP have been awarded PPAs and the right to sell electricity to the utility ONE. As a second phase of both the solar and wind programs, another 850 MW of wind power and 300 MW of CSP are now in the tendering process.

Compared with Morocco, Egypt has seen large delays in its plans to develop 2500 MW of wind power through competitive bidding. Tenders were planned to be launched in blocks of 250 MW and the first of these tenders was issued in 2009. The project saw 10 bidders prequalify, but was later interrupted due to events linked to the revolution. In the last couple of years, the National Renewable Energy Authority (NREA) has talked about a relaunch of the 250 MW projects, as well as tenders for other PPA projects. Despite these ambitions, no announcements of competitive biddings have been made; instead, Egypt has launched an auction for the concession of land designated for private wind power projects. However, the auction differs from the previous projects in the way that it does not allow the sale of any electricity to the utility EETC. As a complementary measure to increase power generation for utility supply, Egypt moved forward with its long-awaited feed-in tariff scheme in 2014. The scheme will be applicable for projects up to 50 MW and allow both households and commercial actors to sell electricity to the utility companies.

Comparable with the Egyptian case, announcements of planned tenders have also been delayed in Saudi Arabia. In 2012, the country launched very ambitious RE targets that were followed up by a White Paper specifying the framework for three tendering rounds for PPAs with the state-owned Sustainable Energy Procurement Company (SEPC). The tendering rounds would allow for the development of between 5500 MW and 7800 MW of RE. The first introductory round was supposed to be launched in late 2013 but no official announcements have yet been made.

In addition to these countries, Jordan, UAE, and Tunisia are interesting cases despite the fact that they have not announced any targets for their competitive-bidding processes. Jordan encourages private participation in producing electricity for the grid, both through competitive biddings and direct proposal submissions for PPA. Jordan and Palestine are in fact the only MENA countries allowing direct proposal submissions. In Jordan, a total of 252 MW of solar and 90 MW of wind power projects have been awarded PPAs with utility companies. At the beginning of 2014, a 400-MW combined solar and wind tender was launched, but later interrupted due to grid problems and a rejected grid improvement grant from one of the Gulf Cooperation Council (GCC) countries. A relaunch can be expected as soon as solutions to the grid problems are found. In the UAE, Abu Dhabi was the first Emirate to announce and award a PPA tender for a RE power plant. Since 2013, a 100-MW CSP plant has been operating successfully; new tenders and a feed-in tariff scheme are expected in the near future. The last year witnessed promising steps toward possibilities for IPPs to produce electricity for the grid in Tunisia. The new RE electricity law that was adopted in September 2014 opens up the way for competitive bidding for private projects larger than 10 MW and for a feed-in tariff support for projects less than 10 MW. However, after being adopted the law was overridden by the constitutional court on the grounds of violations of certain constitutional articles. Nevertheless, there seems to be no disagreement on the actual content of the law, which is expected to be slightly revised and adopted again.

3.2 Renewable Energy for Third-Party Sales

In countries that allow for third-party sales and the possibility of establishing bilateral agreements between IPPs and large electricity consumers, new innovative business models have emerged and overcome some important barriers for both small and large-system RE deployment (Kollins et al., 2010). Yet, a very limited number of countries in the MENA region authorize private participation in the electricity generation sector for this purpose. Despite a more liberal approach toward private participation in general, the issue of third-party sales seems to be a sensitive political topic in many of the countries in the region.

Today, Algeria, Egypt, Morocco, Saudi Arabia, and Syria* authorize such a power generation option. In Morocco, Law 13-09 allows RE IPPs to sell electricity directly to large consumers and bypass the single buyer ONE. However, this option is to some extent being hindered by the fact that the medium-voltage grid has not yet been fully opened to third-party access. Despite these obstacles Morocco was able to apply this model in practice. In January 2013, the NAREVA Holding Company commissioned three wind projects with a total capacity of 200 MW to supply power to large industrial customers (Rouaud, 2013).

Egypt is relatively close to applying third-party sales in practice. Egyptian law allows for third-party sales from IPPs and the New and Renewable Energy Authority (NREA) has, in the first half of 2014, announced and awarded concessions for land in the Gulf of Suez earmarked for 600 MW of wind power. A private actor who intends to sell electricity to third parties was awarded the contract. The third-party option in Egypt seems to have been developed as an alternative to competitive-bidding plans that have been heavily delayed. Considering the low electricity tariffs in the country, it is unclear if the project will receive any type of economic support from the government.

Third-party sales would theoretically be possible in Saudi Arabia; however, this option is not part of the RE agency K.A.CARE's plans and it is perceived to be difficult to develop projects outside this plan (Dii, 2014b). Similarly to the Moroccan and Saudi Arabian cases, third-party sale is legally addressed in Algeria (Law 02-01), but legislative availability has not led to any project so far.

*Due to the ongoing armed conflict, all RE development in Syria has stalled.

3.3 Renewable Energy for Direct Export

A hot topic of debate during the last five years has concerned the export of RE electricity, particularly to Europe. Various cooperation projects on the interconnection of countries have taken place, but power export and import activities remain highly unregulated in MENA countries. As a result, the possibility for private actors to produce RE for export to other countries is very limited and all electricity-exporting activities that currently occur in the region are performed by national utilities (Dii, 2013). The region sees only few countries currently addressing RE electricity export in their legislation. These countries have a rather restrictive approach and at present not enough details have been specified to make export possible in practice.

The countries that legally address RE electricity exports are Morocco, Jordan, Algeria, and Tunisia. Law No. 13-09 in Morocco allows for export of RE-produced electricity by using the national grid and interconnections. Morocco foresees that any IPP that aims to export will be subject to technical approval by the state-owned utility ONE; however, details are still missing and transmission costs for exporting have not been specified. In Jordan, the General Electricity Law specifies that import and export is to be handled on a case-by-case basis relying on authorization by the Council of Ministers. In Algeria, the regulatory framework allowing RE electricity export exists but is not fully developed regarding the authorization procedure and other specifications. The national regulator will be the authorizer with the right to deny any approval if local electricity demand is not fulfilled. Tunisia has specified the conditions for RE-sourced electricity export in its new electricity law that is waiting to be adopted. According to Tunisia's National Agency for Energy Conservation (ANME) export will be allowed under the new law but grids have to be built by the developer. The requirement for private actors to build their own grids will most likely constitute a great barrier to deployment of RE for export. The remaining MENA countries do not seem to foresee private export of RE as an option in the near future.

3.4 Renewable Energy for Self-Consumption

Actors generating RE electricity for their own consumption depend on some sort of regulatory possibility to feed surplus electricity to the grid whenever their electricity load profiles differ from their generation profiles. While many countries authorize self-generation in general, this chapter only considers countries that have schemes in place that allow feeding excess electricity to the grid under some sort of net metering scheme. Today, six countries, Jordan, Egypt, Lebanon, Tunisia, Palestine, and the UAE, have such schemes, which have seen different levels of success.

Jordan allows for self-generation and net metering through Law No. 13 adopted in 2012. The scheme states that projects up to 5 MW can be connected to the grid and use the net metering scheme to offset electricity consumption from the utility. Any projects with a larger capacity are handled on a case-by-case basis. The net metering scheme allows for any excess electricity at the end of the billing period to be purchased at a preferential price. Such monetary purchase of electricity usually provides an extra incentive for private investors; however, Jordan has set a restrictive cap on installed capacity and this results in very small quantities of excess electricity available for purchase. In a detailed case study of the Jordanian net metering scheme performed by the Regional Center for Renewable Energy and Energy Efficiency (RCREEE) in cooperation with Lund University, it was shown that the regulation fails to specify important details, which has resulted in difficulties and misinterpretations. A major uncertainty concerns how the industrial sector's time-based electricity tariffs will be taken into account in the offsetting calculations. Presently, some 300 solar PV systems have been installed under the net metering option in Jordan.

Egypt adopted a net metering decree at the beginning of 2013. However, this scheme only applies to smaller scale solar PV projects connected to the low-voltage grid and no projects have yet been implemented. No mechanism that allows for larger autoproducers has been established, although the first RE self-producer project, an Italgen wind farm in the Gulf El Zayt near the Red Sea, is currently under construction. This wind farm with a capacity of 120 MW is expected to generate 500 GWh/year, and will satisfy approximately 35% of Suez cement factory electricity needs (RCREEE, 2013). It is unclear how this project is being supported by the government and if it will be allowed to sell excess of electricity to the grid.

In Tunisia, the overall details for self-generation with the possibility to feed excess electricity to the grid is determined in Law 2004-72 and its modification found in Law 2009-7 for energy conservation. It allows private actors to feed surplus electricity to the low-, medium-, and high-voltage grids under a net metering scheme. The scheme targeting the residential sector has been very successful in stimulating deployment of RE systems for self-generation; more than 2600 solar PV systems had been installed as of the end of 2013. However, the net metering scheme targeting the commercial and industrial sector has not been sufficient to stimulate investments even though it allows for monetary compensation of surplus electricity within the limits of 30% of what is being produced annually.

In territories with very high electricity tariffs, such as Palestine, a net metering scheme could be a very good way to incentivize private investments. This has been recognized by one of the distribution companies, Jordanian Electric Power Company (JEPCO), which has introduced its own net metering possibility for self-generation. No support mechanism for self-generation exists on a state level at the moment. However, Palestine's interruption of its feed-in tariff scheme has led to revision of the country's strategy to stimulate private RE generation, and a net metering scheme can probably be expected in the near future.

Morocco is the only country with large-scale, grid-connected RE autoproducers in practice: 32 MW belonging to Lafarge Ciments and 160 kW belonging to a shrimp-processing factory. These autoproducers were established as a result of a specific investment program launched by the country's utility operator ONE in 2006. The EnergiPro program allowed industrial groups to produce their own electricity up to a capacity of 50 MW and the utility company guaranteed the purchase of excess electricity at preferential rates equivalent to 20% above the peak tariff (RCREEE, 2013). This specific program ended in 2012 and autoproducers are no more eligible to benefit from these incentives. However, a net metering scheme is under development at the time of writing this chapter.

4 RENEWABLE ENERGY SHARES AND TARGETS

The overall share of RE-sourced electricity in the MENA is relatively low in comparison with the region's RE generation potential. By the end of 2014, total regional share of RE-installed capacity in MENA reached about 5%. The great majority (4%) of installed RE capacity comprises hydropower, which saw large deployment some decades ago. The total share of both wind and solar power is negligible compared with hydropower, although it does have significant growth targets attached. The only countries with any considerable, yet very small, amounts of installed wind and solar power are Egypt, Morocco, and Abu Dhabi. With the inclusion of Jordan, these countries also have the largest amount of RE capacity in the pipeline. While almost all countries have some sort of strategy for the development of RE, only a few have officially adopted RE targets. Among the countries that have the most ambitious targets and plans are Egypt, Morocco, and Jordan. These are all net importers of energy and have been heavily affected by rising fuel prices and increasing domestic electricity demand (Table 11.2).

To increase the installed capacity of RE, most countries are dependent on a combination of both public and private investments. Many MENA countries have laid the foundation for private participation in conventional power-generating activities and should be able to create significantly better conditions for RE private generation without making any major structural changes in the short term. Under current market structures, all MENA countries need to promote a higher degree of legal certainty through further developing, implementing, and enforcing the existing legal frameworks. This includes the development of detailed secondary regulation to support current general electricity laws as well as enacting new specified RE laws. In the MENA, where power sectors are still characterized by a high level of vertical integration, it is particularly important to regulate RE grid access details and transmission costs. The countries that have come farthest oblige the single buyer to purchase all RE electricity produced; however, such a guarantee is not enough to stimulate investment certainty. The nature of solar and wind technologies in combination with few storage options requires that developers can feed their electricity to the grid at the time of generation; this can only be assured by regulating priority dispatch. The obligation for the single buyer to purchase RE electricity should in turn

TABLE 11.2 MENA Renewable Energy Targets, 2014

	RE targets							
						Total		
	Wind (MW)	PV (MW)	CSP (MW)	Biomass (MW)	Geothermal (MW)	MW	%	Target date
Algeria	50	280	325	0	0	660	6	2015
	270	800	1,500	0	0	2,570	15	2020
	2,000	2,800	7,200	0	0	12,000 × 25,000	40*	2030
Bahrain	0	0	0	0	0	0	0	None
Egypt	7,200	1,320		0	0	11,320**	20*	2020
	0	700	2,800	0	0	3,500		2027
Iraq	50	200	50	0	0	300	2	2017
	–	–	–	–	–		5	2030
Jordan	1,200	500	100	50	0	1,850	10[†]	2020
Lebanon	60–100	10	0	15–25	0	125–165[‡]	12	2015
Libya	260	124	0	0	0	384	3	2015
	600	344	125	0	0	1,069	7	2020
	1,000	844	375	0	0	2,219	10*	2025
Morocco	2,000	2,000		0	0	6,000[§]	42[¶]	2020
Palestine	44	45	20	21	0	130	10*	2020

(Continued)

TABLE 11.2 MENA Renewable Energy Targets, 2014 *(cont.)*

	RE targets							
						Total		
	Wind (MW)	PV (MW)	CSP (MW)	Biomass (MW)	Geothermal (MW)	MW	%	Target date
Syria	1,000	2,000	1,300	250	0	4,550	30	2030
Sudan	680	667	50	68	54	1,582[††]	11[¶]	2031
Tunisia	1,500	1,900	300	300	0	4,000	30[¶]	2030
Yemen	400	8.25	100	6	160	674.25	15[¶]	2025
UAE – Abu Dhabi	–	–	–	–	–	1,500	7	2020
UAE – Dubai	–	–	–	–	–	3,000	15	2030
Saudi Arabia	9,000	16,000	25,000	3,000[‡‡]	1,000	54,000	30	2032
Kuwait	–	–	–	–	–	–	15	2030
Qatar	–	–	–	–	–	1,800	20	2024

**Electricity generation.*
***Including current installed capacity of hydro.*
†Primary energy.
‡Including 40-MW hydro.
§Including 2000-MW hydro.
¶Installed capacity.
††Including additional 63-MW hydro.
‡‡Waste to energy.
Source: RCREEE (2013).

always be combined with priority access, which ensures that RE projects are granted priority if several actors are requesting access to the grid in a certain location. If not, there is a potential risk that RE projects that request approval for grid connection are being delayed. To further increase legal certainty, an electricity regulator with responsibility for tariff setting, license issuance, and overall power sector monitoring should be established. It has been shown to be challenging for a number of MENA countries to transfer responsibility to the regulator and this might have to be done gradually.

5 CONCLUSION – POLICY IMPLICATIONS

In addition to increased legal certainty, most MENA countries need to improve and diversify the possibility for the private sector to participate in RE power generation activities. As shown in this chapter, most countries in the region allow private participation through utility supply but very few enable RE power generation for third-party sales, export, or self-generation. The focus on PPAs for utility supply, mainly through competitive-bidding processes, has put the private sector in a situation in which it is heavily dependent on the government to take initiatives for tenders. With that being said, it should be stressed that PPA tendering is an important and good way to stimulate large-scale RE projects, given that the processes are effectively executed. Conversely, the situation today shows that tendering processes in many countries have proved to be complex and lengthy, causing great delays and uncertainty among investors. This situation can be improved through streamlined tendering processes instead of a case-by-case approach. More streamlined and predictable tendering processes will hopefully encourage governments to announce more tenders, something absolutely necessary to meet ambitious RE targets. Good experiences from RE tendering for PPAs can be found in Morocco and this knowledge must be spread to the rest of the region.

As for other options for private actors to participate in RE power-generating activities, self-generation and alternative utility supply options seem to be most relevant in the near future. Export alternatives have been given a lower priority due to increasing domestic electricity needs and lower interest from European countries to import RE-sourced electricity from MENA. Likewise, third-party sales have been given a lower priority due to the fact that enabling this option seems to be a rather sensitive topic even among countries that have come far in their power market liberalization process. However, an alternative that becomes particularly interesting with rising electricity prices is self-generation under a net metering scheme. Net metering schemes with the possibility to sell surplus electricity to the grid are suitable ways of encouraging both small- and large-scale electricity consumers to become RE producers. It is however a somewhat complex task to design net metering schemes suitable for different contexts and consumer segments. Experiences in Jordan and Tunisia show that regulated grid access details and standardized authorizing and licensing procedures are crucial. A handful of MENA countries have experience in the net metering field, and sharing that experience within the region could be a very good start. A complementary option should be to proceed with alternative utility supply options such as direct proposal submissions and feed-in tariffs.

In addition to the measures mentioned to strengthen current structures, it is important for all countries to focus on how their power markets can become more open and transparent. Some actors would argue that full unbundling of the power sector would be needed to guarantee transparency and nondiscriminatory access to grids. However, it is recommended to investigate at the regional level how and if the MENA region would benefit from different levels of power market unbundling. As shown earlier, there are major disagreements on the topic of unbundling and most recommendations done in the academic literature and by the European Commission have not taken into account the fast development of RE and smart grids. Before moving on with unbundling, these aspects should be investigated.

REFERENCES

Åberg, E., 2014. Solar Power in the MENA Region: A Review and Evaluation of Policy Instruments for Distributed Solar PV in Egypt, Palestine and Tunisia. Lund University, Lund, Sweden.

Cambini, C., Franzi, D., 2013. Independent regulatory agencies and rules harmonization for the electricity sector and renewables in the Mediterranean region. Energy Policy 60, 179–191.

Dii, 2013. Desert Power: Getting Started. Dii, Munich, Germany.

Dii, 2014a. Jordan: regulatory overview (1–10). Dii, Munich, Germany. Available from: http://www.dii-eumena.com/fileadmin/Daten/RegulatoryOverview/Regulatory Overview Jordan.pdf

Dii, 2014b. Regulatory overviews. http://www.dii-eumena.com/de/unsere-arbeit/regulatory-overviews/algeria.html (accessed 19.11.2014.).

Dyllick-Brenzinger, R.M., Finger, M., 2013. Review of electricity sector reform in five large, oil- and gas-exporting MENA countries: Current status and outlook. Energy Strat. Rev. 2 (1), 31–45.

Fox-Penner, P.S., Tolley, G., Phillips, A., Hall, G., Moss, B., Mroz, T., Hu, S.D., 1990. Regulating Lessons of the PURPA Approach: Producers Policies Act of 1978, or The passage of the Public Utility Regulatory Prior to this PURPA, heralded a new era in electric power regulation. legislation,' it was difficult for a firm to generate its own, 12, pp. 117–141. Available from: http://ac.els-cdn.com/016505729090045K/1-s2.0-016505729090045K-main.pdf?_tid=03188a12-698a-11e4-996e-00000aab0f01&acdnat=1415700400_8c1279b4c4c0c789c5afda9392b9163a

Gugler, K., Rammerstorfer, M., Schmitt, S., 2013. Ownership unbundling and investment in electricity markets – a cross country study. Energy Econ. 40, 702–713.

IRENA, 2014. Pan-Arab Renewable Energy Strategy 2030. Roadmap of Actions for Implementation. IRENA, Abu Dhabi, United Arab Emirates.

Kollins, K., Speer, B., Cory, K., 2010. Solar PV Project Financing: Regulatory and Legislative Challenges for Third-Party PPA System Owners, revised February 2010.

Manoussakis, S., 2009. Liberalization of the EU Electricity Market: Enough to Power Real Progress? Analysis of Ownership Unbundling and the Project for Liberalization of the European 2 (2), 227–240.

MEMEE, 2009. Stratégie energétique nationale. Horizon 2030.

Meyer, R., 2012. Vertical economies and the costs of separating electricity supply – a review of theoretical and empirical literature. Energy J. 33 (4), doi: 10.5547/01956574.33.4.8.

NEPCO, 2014. Electricity sector structure. http://www.nepco.com.jo/en/electricity_sector_structure_en.aspx (accessed 08.11.2014.).

Pollitt, M., 2008. The arguments for and against ownership unbundling of energy transmission networks. Energy Policy 36 (2), 704–713.

RCREEE, 2013. Arab Future Energy Index (AFEX): Renewable Energy. RCREEE, Cairo, Egypt, http://www.rcreee.org/sites/default/files/reportsstudies_afex_re_report_2012_en.pdf.

Rouaud, P., 2013. Nareva wants to become a major energy player in Morocco. Usine Nouvelle. Available from: http://www.usinenouvelle.com/article/nareva-veut-devenir-un-des-acteurs-majeurs-de-l-energie-au-maroc-selon-son-pdg-ahmed-nakkouch.N202432

Soares, I., Sarmento, P., 2010. Does unbundling really matter? The telecommunications and electricity cases.

Chapter 12

Northern Perspective: Developing Markets Around the Baltic Sea

Riku Huttunen
Energy Department, Ministry of Employment and the Economy, Finland

1 INTRODUCTION

The idea of this article is to give an outsider's view on issues relevant to the regulation of Mediterranean energy markets. Electricity and gas markets surrounding the Baltic Sea make an interesting comparison. The economic and political environments are naturally different, but many features are also common to Mediterranean and Baltic markets. Therefore, conclusions drawn from North European experience may inspire Mediterranean actors too.

In both cases, the sea divides electricity and gas markets. At the same time, it links energy markets, especially in the case of oil products and LNG. Submarine cables and pipelines necessary for developing the markets are in some cases the reason for strong political disagreement due to their importance for security of supply and also because of wider geopolitical considerations.

Inside both regions and also in neighboring areas there are remarkable gas and oil producers. In Northern Europe, the Russian Federation and Norway are the most important energy-exporting countries. For both of them, oil and gas production represent an extremely important part of GDP, tax basis, as well as export income.

Across the Mediterranean, there are countries and economies facing very different stages of development and in some cases also political unrest. Around the Baltic Sea, the situation has in general been more stable, but the Russia–Ukraine crisis and the following international reactions have once again uncovered some potential problems. In both regions, energy and geopolitics are highly topical issues.

Each of the sea regions is very important for the European Union (EU), both politically and economically. The Baltic Sea is surrounded by eight EU member states. Energy trade with Russia, especially gas trade, is crucial. In both regions, the EU has a strong interest, as a big consumer of imported energy, to ensure reliable supply from energy-producing countries. This means, for example,

Regulation and Investments in Energy Markets. http://dx.doi.org/10.1016/B978-0-12-804436-0.00012-6

active involvement in transit negotiations. The Energy Charter Treaty is an international instrument designed to bind parties to follow similar principles regarding security of investments, transit of energy, etc. The energy community, for its part, is a way to widen EU internal market rules in electricity and gas to third countries, be they EU candidate countries or other neighboring partners. The interest in setting these kinds of binding rules is shared by many countries in both sea regions.

2 POLITICAL AND ECONOMIC INTEGRATION IN THE BALTIC REGION

Economic integration of the Baltic Sea Region has been a step-by-step process, especially regarding European integration. Germany (West) was one of the original six states founding the European Coal and Steel Community (ECSC), European Economic Community (EEC), and European Atomic Energy Community (EURATOM) in the 1950s. Denmark joined the communities in 1973. The EU memberships of Finland and Sweden in 1995 meant that the Baltic Sea (and the Northern Dimension) gained more importance in European affairs. Norway declined membership but is an active member of the European Economic Area (EEA) and in this way the European internal market, especially in the field of energy (Fig. 12.1).

The biggest change, however, was launched by the collapse of the Soviet Union and the Eastern Bloc. The eastern part of Germany was naturally immediately integrated into the Union. Poland and the three Baltic States, Estonia, Latvia, and Lithuania, joined the EU almost 15 years later, in 2004. After that, the only shores round the Baltic Sea not belonging to the EU were the Leningrad Oblast (the province surrounding St. Petersburg) and the Kaliningrad region belonging to the Russian Federation. Belarus is normally also considered part of the Baltic Sea Region but it has no direct connection to the sea.

The general, political development has also meant a new phase in energy cooperation. With Russia, it still typically takes the form of bilateral trade. The geopolitical dimension is important and the EU strives to speak with a single voice toward Russia. This has been more and more evident after the first Ukrainian gas crisis in 2006 and especially during the recent crisis (starting in 2014). In any case, Russia has not ratified the Energy Charter Treaty, although it has signed the treaty and promised politically to follow its principles. This means, however, that there are no effective, legally binding arrangements in place tailored for energy trade. A recent sign of lacking cooperation is the fact that Russia did not sign the new International Energy Charter (Hague II) in May 2015. Russia's recent membership of the World Trade Organization (WTO) has some positive effects. Still, the WTO does not give very precise rules addressing specific issues in the energy field and its application is open to potential disputes as well as long dispute settlement processes.

FIGURE 12.1 **The map of Baltic Region.**

The room for genuine multilateral dialog and cooperation is small, the Baltic Sea Region Energy Cooperation (BASREC) being a rare example. It has been organized under the umbrella of the Council of the Baltic Sea States (CBSS), which also includes Russia. BASREC's field of activity is quite limited and its future seems somewhat unclear at the moment. Its emphasis has been on the promotion of energy efficiency, the use of renewable energy, and other sustainable supply sources.

TABLE 12.1 Nordic Generation Capacity (MW) by Power Source, 2013

	Denmark	Finland	Norway	Sweden	Nordic region
Installed capacity (total)	14,681	17,300	32,879	38,273	103,313
Nuclear power	–	2,752	–	9,531	12,283
Other thermal power	9,863	11,135	1,040	8,079	30,117
Condensing power	1,929	2,465	–	1,375	5,769
CHP, district heating	7,372	4,375	–	3,631	15,378
CHP, industry	562	3,180	–	1,498	5,240
Gas turbines, etc.	–	1,115	–	1,575	2,690
Hydro power	9	3,125	30,900	16,150	50,184
Wind power	4,809	288	811	3,745	9,653
Solar power	563	0	N/A	43	606

Source: Nordic Market Report 2014 (NordREG), modified by author.

3 NORDIC ELECTRICITY MARKET – A SUCCESS STORY

The Nordic electricity market has a relatively long history and has been the prime example of a liberalized international power market. The roots of the Nordic wholesale market are in the 1990s when four Nordic countries liberalized their power markets and brought them together in a common market.

Deregulation, in the form of a decreasing government role, was undertaken to create a more efficient market, with increased security of supply. Available power capacity can be used more efficiently in the larger region, and integrated markets enhance productivity and improve efficiency. Power from many different sources enters the grid as shown in Table 12.1. In the Nordic countries, production is dominated by hydro in Norway, hydro and nuclear in Sweden, and coal and wind in Denmark. In Finland, the energy mix is wider, with a relatively big share of nuclear and especially wood-based production, both in forest industry burning wood residuals and CHP district heating plants.

Nord Pool Spot (NPS) is the Nordic power exchange. It is owned by the Nordic and Baltic transmission system operators. The majority of the volume handled by NPS is traded on the day-ahead market (ELSPOT) in hourly trade. There is an intraday market (ELBAS) supplementing the day-ahead market and helping to secure the balance between supply and demand in Northern Europe. Being located in Oslo, NPS is licensed and regulated by the Norwegian energy regulator, NVE, working where relevant in cooperation with other Nordic and Baltic regulators.

Despite its general success, the Nordic electricity market is not without problems. Bottlenecks exist both between countries and inside them. For the latter reason, Sweden was, in 2011, divided into four bidding (price) zones and Norway has five. Limited transmission capacity causes price differentiations between bidding zones, depending on power supply and demand. For example, connections between Sweden and Finland are currently often crowded, even with their high capacity (being at the moment 2700 MW).

There has also been a request from Nordic ministers to the Nordic energy regulators (NordREG) to draw and elaborate a roadmap for creating regional retail markets for electricity. However, national decision-making in Nordic countries has not been concerted and only small steps toward implementation have been taken so far.

Estonia, Latvia, and Lithuania deregulated their power markets a couple of years ago. The Baltic States are now also being integrated into the Nordic wholesale market, step by step, but remain physically part of the synchronous Russian (ex-Soviet) grid, UPS/IPS. It has been suggested many times, especially by Baltic politicians, that this synchronization be replaced by a close connection to Central European power grids. At the moment, there is no concrete plan for that kind of arrangement, which would be a costly and time-consuming exercise. The European Council in October 2014 hinted at supporting *inter alia* synchronization of Baltic electricity grids with continental power networks.[1] Realistically speaking, more concrete arrangements are only likely to take place in the 2020s. The Nordic electricity market now forms part of the northwestern Europe (NWE) day-ahead price-coupling area launched in February 2014. The NWE region, stretching originally from Finland to France, and later in 2014 to Spain and Portugal, operates under a common day-ahead power price calculation. The intraday coupling of European markets has not progressed as smoothly for a number of reasons, including undefined roles and the competences of relevant actors (power exchanges, transmission system operators, and regulators). In any case, intraday coupling would be very welcome considering the high volume of intermittent production (wind and solar) in Europe.

4 GAS – WEAKENING RUSSIAN DOMINANCE

Traditionally, Russia has dominated the gas supply in the Baltic Sea Region. The gas pipeline infrastructures of Finland and the three Baltic States are only connected to the Russian (Gazprom, Russian export monopoly) network and practically all natural gas has been imported from Russia. Poland has also been dependent on Russian gas deliveries.

1. "Preventing inadequate interconnections of Member States with the European gas and electricity networks and ensuring synchronous operation of Member States within the European Continental Networks as foreseen in the European Energy Security Strategy will also remain a priority after 2020." European Council conclusion of October 23/24, 2014, paragraph 4.

Motivated by the Russia–Ukraine crisis, the European Commission has conducted stress tests on the short-term resilience of the European gas system (European Commission, 2014). The focus was on two regions, the Baltic States and Finland forming one and the southeast region of the EU forming the other. The idea was to look at the consequences of a 6-month Russian gas supply disruption. It was no surprise, regarding the Baltic Region, that supply shortfalls would be most serious in Finland and Estonia. Under cooperative conditions, however, Estonia could rely on the large Inčukalns Underground Gas Storage Facility in Latvia.

Finland is a better example of the fact that security-of-supply issues are multidimensional and cannot be assessed too mechanically. At first sight, regarding gas supplies, Finland is the most vulnerable EU member state with no alternative suppliers and practically no storage capacity. However, the share of protected customers being supplied by gas is minimal. Due to effective preparedness measures being in place (i.e., obligatory fuel-switching arrangements and high alternative fuel stock obligations, 5 months), gas volumes could be replaced fairly well, but with notable additional costs of course.

The gas supply situation in the Baltic Region is changing rapidly at the moment. There are many infrastructure projects – pipelines and LNG terminals – with the aim of enabling alternative import sources. Besides intensified competition, security of supply is the main motivation for these projects in EU member states. At the EU level, the original idea behind the Energy Union was to address gas issues, in particular, in order to increase the security of supply. However, the Energy Union has been constructed to include other aspects of energy policy as well.

Two large LNG terminals finalized in 2014–2015 are bringing more competition and security of supply to the region. They are located in Klaipeda (Lithuania) and Śvinoujście (Poland). There are also terminal projects in Finland and Estonia and a plan to build a submarine pipeline connecting Finland and Estonia (Balticconnector). A very important gas connection is to be constructed between Poland and Lithuania in the next 4 years.

High-capacity LNG terminals and storage facilities can be used as hubs to serve the fast-growing number of smaller LNG terminals, in this way developing an off-grid gas supply, etc. In the Baltic Region, construction of small LNG terminals is accelerated by the international obligation (MARPOL 73/78, the International Convention for the Prevention of Pollution from Ships) to limit emissions, especially SO_x, from ships.

National political considerations are sometimes very strong in the case of infrastructure projects. A good example of that is the Nord Stream, a 1222-km submarine (twin) pipeline connecting Russia and Germany via the Gulf of Finland and the Baltic Sea. It has high transit capacity (2×27.5 Mm^3/year). The Nord Stream project was strongly advocated by political leaders in Russia and Germany. Political and other concerns of neighboring EU member states did not affect realization of the project to any great degree. The outcome is a gas connection bypassing Finland, the Baltic States, and Poland, which does not really increase the security of supply in those countries. The integration of regional markets has to be realized by other means (as briefly described earlier).

5 INCREASING THE ROLE OF THE EUROPEAN UNION

The role of the EU in energy market integration has already been touched upon. It is still worth looking at the issue more closely. One of the top priorities in EU energy policy is to integrate European electricity and gas markets, finally including energy community partners. A critical part of that work was to create the necessary infrastructure. The original political goal was to connect energy islands (isolated areas) to wider EU markets by 2015. This is not going to be realized as such, but many important steps have now been taken.

Infrastructure projects with a strong European dimension are known as projects of common interest (PCIs). The idea is that PCIs are provided with fast track licensing procedures in member states. EU part-funding might be available through the Connecting Europe Facility (CEF).

The Baltic Energy Market Interconnection Plan (BEMIP) is an initiative launched to find ways of connecting Lithuania, Latvia, and Estonia to wider European energy networks. As mentioned earlier, there are many links missing such as creating a "Baltic Ring" for electricity.

In October 2014, the first CEF financing decisions were made. Altogether, a total of more than €500 million was granted to BEMIP projects (gas €339 million, electricity €168 million). This shows that many project and investment plans have progressed well. In June 2015, the European Commission and the Baltic Sea Region countries signed a Memorandum of Understanding modernizing and strengthening BEMIP.

The starting point is that investment projects should be standalone, economically rational, and viable. In some cases, however, infrastructure investments with positive externalities may need this kind of support in order to launch. As for energy production, subsidies might have even more serious distortive effects. The examples of subsidies for electricity produced using renewable energy sources in Spain, Germany, etc. show that overscaled operating aid may cause serious distortions in relevant markets. In the EU, the European Commission has the power to assess member states' aid schemes.[2] At the moment, more stringent guidelines are in place for renewable energy source (RES) subsidies. More restrictive rules are also being developed for capacity mechanisms (i.e., aid for basic power production), which can be seen as a consequence of very generous subsidies for wind and solar power production. All this is welcome

2. "Save as otherwise provided in the Treaties, any aid granted by a Member State or through State resources in any form whatsoever which distorts or threatens to distort competition by favouring certain undertakings or the production of certain goods shall, in so far as it affects trade between Member States, be incompatible with the internal market." Article 107 of the Treaty on the Functioning of the European Union.

because it is timely to create a clear and limiting state aid regime in order to enable well-functioning electricity markets.

6 CONCLUSIONS

The lessons to be learned from the Baltic Sea energy markets can be summarized as follows.

- Integrating energy markets requires huge investments in production, transmission, etc. More generally, the International Energy Agency (IEA) has estimated the amount of energy investment needed in Europe alone as US$5 trillion in 2014–2035 (International Energy Agency, 2014). Security-of-supply considerations, due to recent political turmoil, intensify the demand for new capacity.
- Investments in interconnections are often very costly and require cost allocation and political agreements between countries. However, the potential benefits from the wider market are typically notable. Effective market integration generates welfare through optimization of production and enables more effective price signals. It also promotes security of supply.
- Stability and predictability are necessary in order to realize large investments, which is a remarkable challenge anywhere, let alone in Europe. Regulation should be clear and focused. Political (national) considerations may cause suboptimal results for integrated markets. This should be avoided.
- In some cases with high positive externalities, subsidies might be necessary in order to get large infrastructure investments launched. However, operating aid is highly distortive and subsidies should be granted very cautiously. The aim should be to develop energy-only markets, where public support, price regulation, etc., are exceptions, not the rule.

REFERENCES

European Commission, 2014. Communication from the Commission to the European Parliament and the Council on the Short Term Resilience of the European Gas System: Preparedness for a Possible Disruption of Supplies From the East During the Fall and Winter of 2014/2015. COM(2014) 654 final, Brussels.

International Energy Agency, 2014. World Energy Investment Outlook Special Report. IEA, Paris.

Part III

Investments for Grids and Generation Projects

Chapter 13

Private Participation in Energy Infrastructure in MENA Countries: A Global Perspective

Ernesto Somma, Alessandro Rubino
DISAG Department of Business and Law Studies, University of Bari Aldo Moro, Bari, Italy

1 INTRODUCTION

The global electricity landscape is rapidly changing, creating challenges even for the most advanced nations. Consumption of electricity continues to grow, and yet over 1.3 billion people still do not have access to electricity, most of whom live in rural areas. Universal access, volatile prices for fossil fuels, and climate change make the energy sector a key priority for the global development community. Lack of available electricity hinders progress on many other development fronts, and high prices are able to dampen, if not halt, progress and innovation in large sectors of the economy.

Other chapters in the book analyzed the expected trends for energy-related supply and demand. Chapter 15 shows us that significant additional investment will be required in the near future in the region to match growing demand. A significant portion of this additional capacity will be provided by renewable energy source (RES) generation in Middle East and North African (MENA) countries (see Chapters 3, 4, and 5). We have also learned from Chapter 1 that creating the right condition to foster investment in energy infrastructures (in both generation and transport) requires the existence of a number of preconditions that need careful consideration when planning policy interventions. Chapter 14 highlights how a National Regulatory Agency could play a significant role in promoting investment in additional generation capacity, and shows a strong and positive correlation with other institutional dimensions measuring enforcement of the rule of law and of the control of corruption.

At the same time, severe budget constraints will not allow public investment alone to close the gap between growing demand and projected supply.

Regulation and Investments in Energy Markets. http://dx.doi.org/10.1016/B978-0-12-804436-0.00013-8

Therefore, new business models need to be developed to allow active private sector participation, as underlined in Chapter 17. In light of the evidence collected in the literature and recalled in this book, it is useful to understand how public and private partnership (PPP) in investment is faring in countries belonging to the MENA region, and how they compare when we look at investment in the energy sector on a global scale.

We will look at this evidence by providing information on the evolution of PPP and by looking at private participation in infrastructure (PPI). Private sector participation in the provision of infrastructure services began in the early 1990s in a few pioneering countries including Argentina, Chile, Malaysia, and Mexico (Panayotou, 1998). It has since spread throughout the developing world: 137 developing countries have implemented infrastructure projects with private participation in at least one sector since 1990 (Private Participation in Infrastructure Database, 2015). Whereas governments remain the main source of infrastructure financing in developing countries, providing around 70% of the funds necessary, the private sector is also a key source, contributing 22% – well beyond the 8% provided by official development assistance.

Large deficits remain, with current investment in infrastructure meeting less than half the needs in developing countries. These deficits have motivated many governments to view private participation in the provision of infrastructure services as an integral part of their development strategy. This is particularly true with reference to the current needs and the status of the energy sector in the Mediterranean region.

The Mediterranean energy sector presents a high degree of interdependence, both for electricity and gas. Despite its long history of interaction, the Mediterranean today is a highly fragmented region that is facing unprecedented challenges in its social, economic, and political dimensions. The events connected with the so-called "Arab Spring"[1] that started in 2011 are deeply connected with a process of democratization of civil societies and with the unbalanced distribution of wealth and opportunities at national and at regional levels. The region, in fact, is characterized by uneven distribution of wealth, ranging from the affluent north basin[2] to areas of deep poverty and scarcity of resources, despite which the population is increasing rapidly in the south and in the east of the basin.[3] Strong asymmetries in terms of socioeconomic development in the region are confirmed when we look at the distribution of gross domestic production (GDP) and the population today as well as in projections for the region. The analysis highlights a number of remarkable diverging scenarios.

1. See Roy (2012) for a comprehensive review and analysis.

2. For the purposes of our analysis, we follow OME classification and include as north Mediterranean countries (NMCs) both EU countries (Cyprus, France, Greece, Italy, Malta, Portugal, Slovenia, Spain) and non-EU countries (Albania, Bosnia and Herzegovina, Croatia, Romania, Serbia).

3. Algeria, Egypt, Libya, Morocco, Tunisia, Turkey, Israel, Jordan, Lebanon, Palestine, Syria.

South and East Mediterranean Countries (SEMCs) are expected to grow at a faster rate than North Mediterranean Countries (NMCs). Also population trends will describe a strong upward curve in countries located in the SEMCs. As underlined by Cambini and Rubino (Capter 8), this will mean that most of the population will be based in the MENA region by 2030.

This unbalanced situation has profound implications for energy balance and trade too. In fact, even at current energy availability and consumption we can observe a great disparity between north and south. While it is expected that this gap will be reduced in the coming 20 years, NMCs will still be consuming twice as much as the south on a *per capita* basis. All this takes on even greater emphasis when we consider that the Mediterranean region, as a whole, imports nearly half of its energy needs (over 90% for NMCs). Energy consumption, and in particular electricity demand, in the Mediterranean is closely linked to economic development and population growth. Thus, demographic trends and economic development expected in the south imply a significant increase in demand, whereas in NMCs economic maturity and weak population trends contribute to declining electricity consumption. However, despite these expected trends, regional discrepancies between NMCs and SMCs will continue.

In a situation where persisting discrepancies are expected to last for decades to come, although with a declining gap, there is certainly economic justification for the promotion and development of crossborder electricity and gas transmission infrastructures (L'Abbate et al., 2014). Crossborder interconnection is expected to play a positive and significant role in reducing the energy gap among subregional markets (Rubino, 2014). Moreover, the significant role that renewable energy generation will play in the Euro-Mediterranean region requires national electricity systems to become highly interconnected and robust to growing demand. In addition, they need to be able to accommodate the presence, in the generation mix, of large quantities of, or intermittent, wind and photovoltaic generation. According to the Mediterranean Observatory of Energy (OME) projection (Observatoire Méditerranéèn de l'Energie, 2011) over €700 billion will be needed by 2030 to ensure the additional generation required, with an additional 3000 MW of north–south interconnections in the Mediterranean basin, requiring investment of the order of €20 billion up to 2020 (Med-TSO, 2013).

Although state-level energy policies are still dominant in the energy sector, it is indisputable that MENA countries will not be able to deliver investment of this variety and size via the public budget only. Therefore, it is important at this stage to shed light on how regional markets are scoring in terms of capability to attract private investments, and their abilities to cooperate with the private sector to deliver the necessary investment required.

This chapter initially discusses how infrastructure investments are evolving worldwide; it will then look at the performance of the energy sector in the various

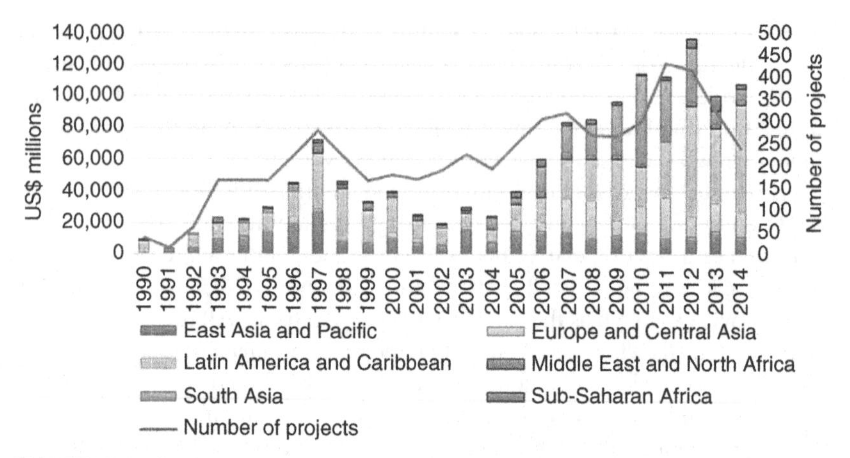

FIGURE 13.1 Total investment in energy, transport, and water by region (in current US$ million). *(Source: World Bank and PPIAF, PPI Project Database (http://ppi.worldbank.org accessed 06.08.2015).)*

regions. Subsequently, we look at the MENA region's performance. This overview will allow, in the final section of this chapter, some conclusions to be drawn.

2 GLOBAL OVERVIEW

The focus of this analysis is to evaluate the performance and trends of PPP in infrastructure. Infrastructure investment is considered to entail public–private participation when a public–private partnership establishes a contract "[...] for providing a public asset or service, in which the private party bears significant risk and management responsibility" (World Bank, Asian Development Bank, Inter-American Development Bank, 2014). In addition, the remuneration allowed to finance the infrastructure needs to be linked to performance, in order for the investment to qualify as a PPP. This proposed definition, adopted by the World Bank, allows inclusion in the analysis of a wide spectrum of investment projects, including new assets and services and different levels of private participation and engagement in projects. We will consider these types of investments further in the chapter.[4]

Total PPP investment[5] in infrastructure[6] was US$109.4 billion in 2014, compared with US$102.7 billion in 2013. The level of total investment is still below the record values of 2012, but registers a +7% increase from 2013 (Fig. 13.1).

4. Public private partnership (PPP) and private participation in infrastructure (PPI) will be used interchangeably in this chapter.

5. "Investment" refers to investment commitments at the time of financial closure or in the case of brownfield concessions, contract signing.

6. Infrastructure refers to energy, transport, and water projects, excluding oil and gas extraction but including natural gas transmission and distribution.

TABLE 13.1 Investment Committed by Sector, 2014

Sector	Number of projects	Average investment (US$ million)	Total investment (US$ million)	Total investment (%)
Energy	157	30,613	48,063	45
Transport	49	112,802	55,273	51
Water and sewerage	33	12,424	4,100	4
Total	239		107,436	100

Source: Authors' elaboration based on World Bank and PPIAF, PPI Project Database (http://ppi.worldbank.org accessed 07.11.2015).

The largest number of new projects were in energy (157), followed by transport (49), and finally water and sewerage (33). Although the energy sector had the most new projects, the sector with the greatest investment was transport, receiving US$55.2 billion, or 51% of total global investment. The energy sector accounted for US$48 billion, or 45%, and the water and sewerage sector had just above US$0.4 billion, which was 4% of total investment committed (Table 13.1).

When we look at performance per region, there is a significant drop in the level of investment in Sub-Saharan Africa (SSA) on a year-on-year (YoY) basis (from 9.2% to 2.5%). At a more general level, MENA countries and SSA together represent 5% of the level of global investment against 64% of Latin America and the Caribbean and, respectively, 11 and 13% of the East Asia Pacific region and Europe and Central Asia.

Of total cumulative PPI of US$107.4 billion, the ranking of regions by order of volume in 2014 was (1) Latin America and the Caribbean, (2) Europe and Central Asia, (3) East Asia and Pacific, (4) South Asia, (5) Middle East and North Africa, and (6) Sub-Saharan Africa (Table 13.2).

In this framework, investment in MENA is slowly recovering after the complete halt registered in the region in 2011, following political turmoil during the Arab Spring. Historically, the MENA region has been unable to attract investment commitments to private infrastructure projects, representing only 3% of total PPP investment committed since 1990.[7] It is extremely sensitive to political instability. If we look at private participation as a percentage of GDP

7. The largest recipient of PPP investment, in the period 1990–2015, are Latin American and Caribbean countries (41%), East Asian and Pacific countries (21%), South Asian countries (19%), European and Central Asian countries (13%), Sub-Saharan and African countries (3%), and finally Middle East and African countries (3%).

TABLE 13.2 Investment Committed by Region, 2014

No.	Region	Total investment (US$ million)	Total (%)
1	Latin America and the Caribbean	69,113	64
2	Europe and Central Asia	14,254	13
3	East Asia and Pacific	11,514	11
4	South Asia	6,743	6
5	Middle East and North Africa	3,302	3
6	Sub-Saharan Africa	2,510	2
	Total	107,436	100

Source: Authors' elaboration based on World Bank and PPIAF, PPI Project Database (http://ppi.worldbank.org accessed 07.11.2015).

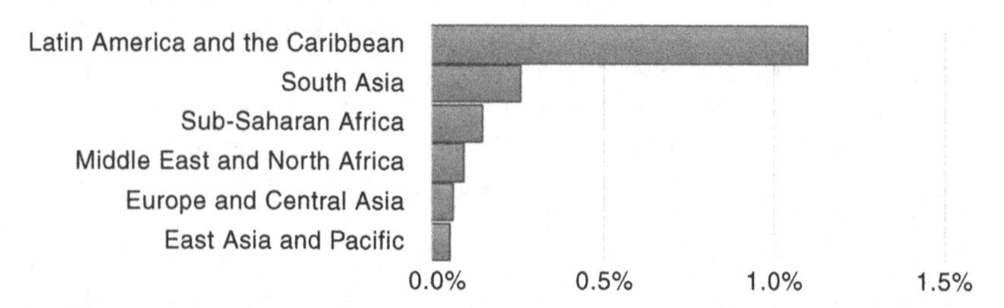

FIGURE 13.2 PPI as %GDP by region. *(Source: World Bank and PPIAF's PPI Project Database.)*

(Fig. 13.2), we find that Latin American (LAC) countries are those best suited to attract private investments (1.09%), compared with the size of their economy. The small amount of private participation is striking when we look at EAP countries. Although the trend is stronger for China, it remains true throughout the rest of the region.

The top six countries engaged in PPI in the last 3 years (2012–2014) were: (1) Brazil, (2) Turkey, (3) Peru, (4) Chile, (5) Mexico, and (6) Morocco. These six countries attracted US$141 billion of investment, representing 72% of all the PPI commitments in the developing world in the considered period. Brazil drew the highest volume of investment (US$70.6 billion), followed by Turkey (US$32.2 billion), and Peru (US$12.5 billion).

Private participation in the largest PPI market – Brazil – continued to show strength by attracting 36% of PPI in the 3-year period considered.[8] The largest

8. PPI were large in Brazil for the closure of a number of significant projects, mostly on energy and transport, such as the US$15 billion Belo Monte hydro project in 2012.

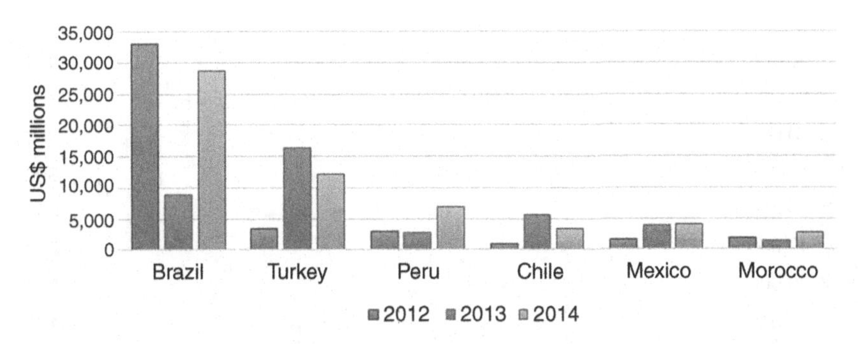

FIGURE 13.3 PPI in the top six countries. *(Source: Authors' elaboration based on World Bank and PPIAF, PPI Project Database (http://ppi.worldbank.org accessed 07.11.2015).)*

global investment, a US$10.0 billion airport project in Rio de Janeiro, alone accounted for 20% of global totals. Galeao International Airport, originally built in 1952, will receive another major upgrade to accommodate the upcoming 2016 Olympics in Rio de Janeiro. Sao Paulo's Orange Line Metro was the second largest overall investment at US$3.8 billion.

Turkey was able to rank second because of a number of transport projects – the US$1.05 billion Istanbul Salipazari Cruise Port and the US$2.9 billion Third Bosporus Bridge and Northern Maramara Highway Project. Growing consumerism continues to drive investment in Turkey, and the expansion of the Salipazari passenger terminal on the west coast of the Bosporus is no exception as tourism in the country increases. In addition, energy demand in Turkey continues to rise faster than most OECD countries. With an expected annual growth rate of 4.5% from 2015 to 2030, electricity demand should grow even faster. To meet these needs, Turkey is rapidly becoming an energy transit hub between Europe, Russia, and the Middle East by being an integral part of the oil and natural gas supply movement (Fig. 13.3).

3 ENERGY INVESTMENT

While the global value of infrastructure investment is in line with levels registered in 2010 (if not describing an upward trend), the energy sector has failed to recover the level recorded in 2010, in terms of both volume and numbers of projects. This was due to the dramatic drop in investment in India and Brazil, where the energy sector is facing multiple challenges. Total number of PPI[9] in the energy sector in developing countries in 2014 was 157, a 44% decrease from 2013 (Fig. 13.4). Average project size was US$306.1 million suggesting a significant increase in size compared with 2013 (US$259.2 million).

9. Projects are considered to have private participation if a private company or investor is at least partially responsible for operating cost and associated risks. The projects tracked are those that have at least 25% private equity or, in the case of divestitures, at least 5% private equity.

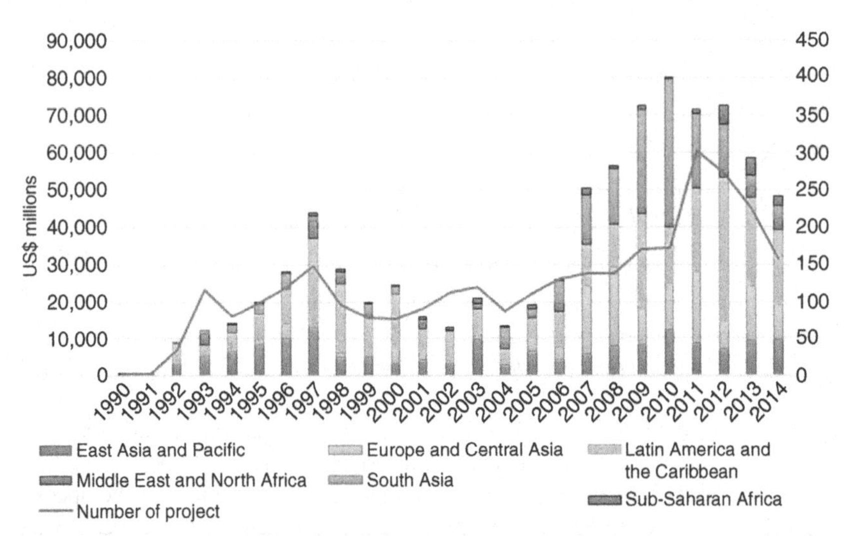

FIGURE 13.4 Total investment in energy by region. *(Source: Authors' elaboration based on World Bank and PPIAF, PPI Project Database (http://ppi.worldbank.org accessed 07.11.2015).)*

Moreover, despite the reduction in investment registered in two large countries, there was significant movement in regional rankings. The traditionally strong South Asia (SA) region dropped from second to fifth place in 2013, and fourth place in 2014 as the Indian PPI market faced multiple challenges. Europe and Central Asia moved from fourth place in 2012 to second in 2013 and third in 2014 registering a more than twofold increase in overall investment levels (in 2013). This was largely fueled by a spike in Turkish PPI. MENA and the SSA region consistently registered low levels of investment throughout the period (Fig. 13.5).

The top six countries in PPI for energy in the period 2012–2014 were: (1) Turkey, (2) Brazil, (3) Chile, (4) Thailand, (5) Morocco, and (6) Mexico. These six countries attracted US$106 billion, representing 67% of all PPI commitments in the developing world for the energy sector. Turkey attracted the highest volume of investment (US$23 billion), followed by Brazil (US$21.7 billion) and Chile (US$9.3 billion). Thailand saw the largest energy PPI for the SA region, attracting US$5.9 billion out of a total of US$18.9 billion. There was little correlation between number of projects and total energy PPI levels, with Brazil, India, and China having the greatest number of projects and India and China occupying secondary positions for PPI levels. On the other hand, Turkey came just below China for number of projects, but had a greater average project size at US$527 million. Morocco had the largest average project size of US$1.4 billion from just four projects in the period (Table 13.3).

Private participation in Turkey increased sixfold in 2013, rising from US$2.2 billion in 2012 to US$13.2 billion in 2013. This was mainly driven by Turkey's energy privatization drive, which resulted in a series of brownfield

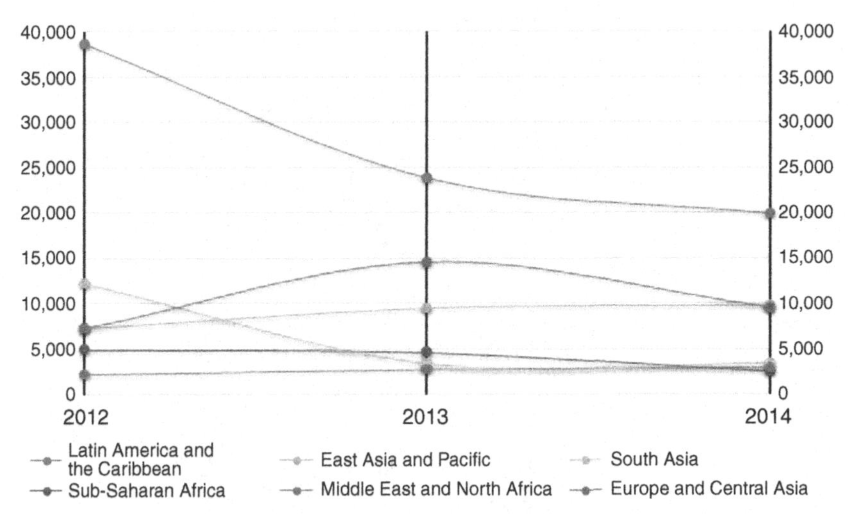

FIGURE 13.5 Total investment in energy by region (2012–2014). *(Source: Authors' elaboration based on World Bank and PPIAF, PPI Project Database (http://ppi.worldbank.org accessed 07.11.2015).)*

concessions for state-owned and operated power distribution projects. This constituted four out of six projects of more than US$1 billion in size, with a total investment of US$7.2 billion. Chile also did well in 2013, attracting 380% more PPI than in 2012. On the other hand, investment in Brazil slowed dramatically, dropping 69% from US$19.3 billion in 2012 to US$220 million in 2013. A major reason for this drop was the completion of large investments in 2012, such as the US$15 billion Belo Monte mega hydro project.

There was significant investment in renewable energy in the period considered. Renewable energy investment in 2013 totaled US$21.4 billion, which comprised 38% of total PPI in the energy sector. Within this, the top three PPI regions for renewable energy were LAC (66%), followed by EAP (12%) and SSA (11%). The top five renewable energy PPI destination countries reflected the regional investment distribution, with Brazil coming in top, followed in sequence by Chile, China, South Africa, and Mexico. However, overall PPI levels in renewable energy contracted by more than half as compared with 2012 levels. While all regions experienced drops, the global decrease was largely due to a US$15 billion drop in the LAC region.

4 REGIONAL OVERVIEW – THE MENA REGION

The MENA region moved from sixth position to fifth in 2014 increasing slightly from the prior year and comprising just 6% of global energy sector PPI at US$3 billion. Between 2012 and 2014 only Jordan and Morocco managed to commit PPI investments in energy projects for a total of US$7.7 billion (Table 13.4).

TABLE 13.3 Investment Committed by Country in the Energy Sector (2012–2014)

							Grand total*	Total*	Total (#)	Percentage of total
Countries	2012*	#2012	2013*	#2013	2014*	#2014	1990–2014	2012–2014		
Turkey	2,223	11	13,246	20	7731	13	49,940	23,200	44	21.9
Brazil	19,312	34	220	35	2171	31	68,033	21,703	100	20.5
Chile	885	9	5092	13	3338	12	17,885	9315	34	8.8
Thailand	1,280	9	1251	12	3410	15	177,156	5941	36	5.6
Morocco	1,867	2	1438	1	2600	1	15,317	5905	4	5.6
Mexico	1,326	3	1179	10	2749	8	14,968	5254	21	5.0
Grand total	42,711	271	32.969	226	30,333	157	1,466,774	106,013	654	67.27

In US$ million.

Source: Authors' elaboration based on World Bank and PPIAF, PPI Project Database (http://ppi.worldbank.org accessed 07.11.2015).

TABLE 13.4 Investment Committed in the MENA Region by Country in the Energy Sector (2012–2014)

	2012*	2013*	2014*	Grand total (1992–2014)*	Total (2012–2014)*
Morocco	1,867	1,438	2,600	13,017	5,905
Jordan	350	1,102	371	2,812	1,823
Algeria	0	0	0	2,462	0
Egypt, Arab Rep.	0	0	0	1,092	0
Tunisia	0	0	0	291	0
Total MENA	2,217	2,540	2,971	19,674	7,728
Grand total (global PPI)	42,710	32,969	28,643	1,224,927	104,322

**In US$ million.*
Source: Authors' elaboration based on World Bank and PPIAF, PPI Project Database (http://ppi.worldbank.org accessed 07.11.2015).

If we look at the longer history of PPI energy investment in the region, we notice that investment in the MENA region only represents 2% of global investment in the sector, making it the worst performing region globally. Investment in the MENA region is recovering from a 15-year investment low. In 2014, two projects reached financial or contractual closure, no new projects have been operational since 2009. The number of transactions is still below 2009 levels, and is struggling to recover after the financial crisis.

Several drivers appear to dominate the energy scenario which can be broadly summarized as: (1) continued political instability in the region, (2) strong demographic growth, and (3) increased penetration of nonprogrammable renewable energy sources. These all have a number of relevant repercussions on the investment outlook in the region.

Country risk (Chapter 17), which takes into consideration, among other indicators, political unrest in the region, is likely to require a significant markup in the return required by investment remuneration in the region, in particular when characterized by long lead times. Finally, the great emphasis currently placed on investment in RES is likely to require governments to be adept at adapting the existing energy paradigm, mostly based on fossil fuel generation, to a new and diverse model, thereby showing a certain capacity to evolve to the changing needs emerging in the new energy scenario.

As discussed in Chapter 14, there is robust evidence that suggests a positive correlation between the presence of an appropriate institutional framework in each country and the level of investment realized. It is therefore not surprising that Morocco and Jordan register a (relatively) high score in terms of political

TABLE 13.5 Political Stability and Absence of Violence/Terrorism (2012–2014)

	2011	2012	2013
Morocco	2.11	2.04	2.00
Jordan	1.98	1.98	1.88
MENA average			1.71
Tunisia	2.13	1.76	1.59
Algeria	1.14	1.18	1.33
Egypt, Arab Rep.	1.05	1.04	0.88

0, weak; 5, strong.
Source: Authors' elaboration from World Bank. WGI, Worldwide Global Indicators.

TABLE 13.6 Rule of Law (2012–2014)

	2011	2012	2013
Jordan	2.76	2.87	2.89
Tunisia	2.37	2.35	2.30
MENA average			2.31
Morocco	2.28	2.29	2.25
Egypt, Arab Rep.	2.10	2.04	1.90
Algeria	1.71	1.73	1.82

0, weak; 5, strong.
Source: Author's elaboration from World Bank. WGI, Worldwide Global Indicators.

stability[10] (PS) and for Rule of Law[11] (RoL). These two factors (see Tables 13.5 and 13.6) play a significant and positive role in providing an environment conducive to better engagement of private investments. PS signals the existence of sufficient social and political stability that, among other benefits, contributes to lowering the returns on investment required by investors (via a lower country risk). The RoL score, on the other hand, influences the growth rate of developing countries by restraining the government from intervening with *ad hoc* actions and offsetting private initiatives. This improves investors' confidence about expected return on investments and provides legal protection to business

10. "Reflects perceptions of the likelihood that the government will be destabilized or overthrown by unconstitutional or violent means, including politically-motivated violence and terrorism" from the Worldwide Governance Indicators Project.

11. "Reflects perceptions of the extent to which agents have confidence in and abide by the rules of society, and in particular the quality of contract enforcement, property rights, the police, and the courts, as well as the likelihood of crime and violence" from the Worldwide Governance Indicators Project.

activities. Morocco and Jordan register the highest scores in both indicators, and not surprisingly they are the only two countries that have been able to attract a significant number of PPP investments in energy infrastructures.

5 CONCLUSIONS

PPP is a relatively new feature of infrastructure investments in many low- and middle-income economies. Since 1990, a growing number of countries are experiencing increasing participation of the private sector in supporting the development and diffusion of effective infrastructure endowment. PPP can be beneficial for public service provision for at least a couple of economic reasons: first, it enhances the economic efficiency of service provision (de Bettignies and Ross, 2004) and, second, allows governments to define the characteristics of the service they need and the business model around which this service has to be provided without the need to own (entirely) the assets (Kirkpatric et al., 2006).

Theoretical studies and empirical evidence have illustrated how a number of factors coalesce to play a relevant role in determining the level and magnitude of private engagement in PPI. According to Mengistu (2013), these factors can be grouped into three main categories: (1) factors that determine governments engaging the private sector in infrastructure financing; (2) the underlying context in terms of the overall macroeconomic environment, which drives to some extent respective motivations of the public and private sectors; and (3) factors that affect the incentive and motivation of the private sector to enter into a PPP with the government.

The factor included in the first group has been discussed in other chapters of this book. The evidence collected indicates that available public budgets are not able to match the investment needs of the region (Chapter 15) and that an enhanced level of private participation in infrastructure investment is needed (Chapter 8) in most MENA countries. Chapter 1 also discussed how some of the most recent energy policy initiatives (e.g., RES targets and feed-in tariffs) have been specifically set up to encourage private participation in energy investment. Therefore, it appears that the factors inducing governments to engage the private sector in infrastructure financing are present in the MENA region and currently represent one of the main drivers in recent policy dynamics.[12]

The second group of factors, those that determine the general framework under which infrastructure investment takes place and define the conditions that both private and public investors need to consider, have been discussed in

12. Countries in the region are striving to attract investment from private and public funds in strategic economic sectors. A recent example includes the organization of an economic development conference in Egypt. The Egypt Economic Development Conference (EEDC) was a 3-day event that took place in Sharm el-Sheikh. On March 13, 2015, the EEDC attracted pledges of $12.5 billion in assistance from Egypt's allies in the Gulf Cooperation Council (GCC), $33.2 billion in investment agreements, and a further $89 billion in Memorandum of Understandings. See, for example, Wall Street Journal (2015).

Chapter 17. Chapter 17 concluded that country risk is always considered when planning long-term investment. Countries that are perceived less financially stable are obliged to grant investors higher returns to offset country risk, thus increasing total investment costs.

Finally, the third group of factors, which takes into consideration those aspects considered critical by private investors, include adequate regulatory frameworks and proper enforcement of laws, independence of regulatory institutions and processes, access to credit, consumer ability to pay for services, government effectiveness and responsiveness and political stability, and favorable public opinion on private provision of infrastructure services. Chapter 14 has carefully explored the role that institutional endowment plays in favoring investment in the energy sector. It emerges that the presence of a National Regulatory Agency (NRA) determines a positive impact on investments.

In our qualitative analysis, we have turned our attention to how PPP investment evolved in low- and middle-income countries at the global level and looked for possible explanations for the observed trend. Analysis, performed looking at the data made available by the PPI Database,[13] has shown that the region was able to attract most investment was Latin America and the Caribbean collecting more than 64% of the global investment in 2014. MENA countries only represented 3% of global investment in the same year. Historically, MENA countries have been unable to attract an adequate level of investment compared with the GDP level of the region. When we focus on the regional level we find out that only Morocco and Jordan have been able to attract PPP investments, with Morocco alone representing almost 6% of PPP investment in the energy sector. Moreover, the social and political unrest experienced in 2011 stopped this already weak trend, reducing the level of PPP investment in the region to zero. The region has been struggling to recover ever since.

Whereas the list of factors deemed to play a significant role in this dynamic is long, we have focused our analysis, following Mengistu (2013), on some of the dimensions that appear to be more significant when we look at the institutional endowment of recipient countries. As illustrated in Section 3, the indicators "Political Stability" and "Rule of Law" are positively correlated with the level of investment. The greater the political (and financial) stability, the lower the perceived country risk, thus the lower the return required over the investment. A high score related to the "Rule of Law" indicator implies greater certainty on the judicial and legal system, thus improving the level of contract enforcement in the country considered. The analysis performed shows significant disparity both within MENA countries and among the different regions

13. The PPI Database contains 25 years of data on private participation in infrastructure in 137 countries. The data set includes information on more than 4300 infrastructure projects that have reached financial closure. It covers projects in transport, energy, telecommunications, and water and sanitation. The database highlights contractual arrangements, committed investment, and key information on investors (http://ppi.worldbank.org/).

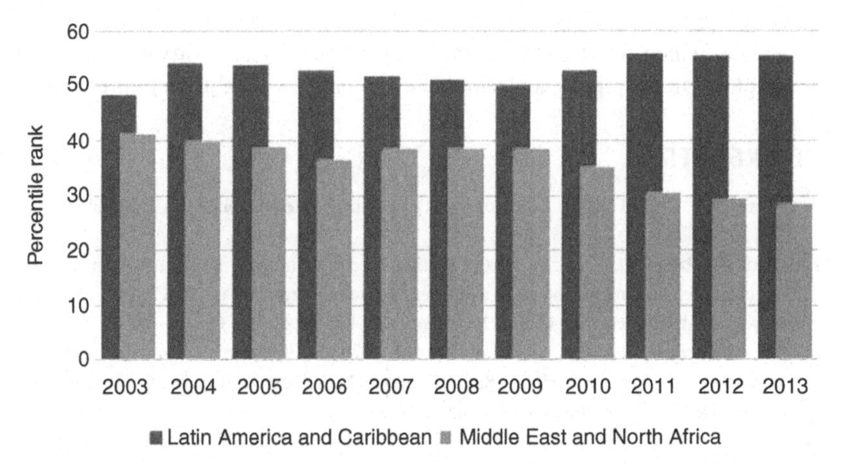

FIGURE 13.6 Political stability and absence of violence/terrorism (2003–2013) in MENA and LAC. Percentile rank 0–100 indicates world ranking (0 corresponds to the lowest and 100 to the highest). *(Source: World Bank. WGI, Worldwide Global Indicators.)*

globally. These differences yield a similar disparity in the level of investment. In this respect, our intraregional analysis shows that Jordan and Morocco are better endowed compared with other countries in the region (Algeria, Egypt, and Tunisia). When we compare the political stability score in MENA countries with that of LAC, we found that there is a persistent and increasing gap (see Fig. 13.6), which is able to partially explain the different performance observable in these two regions. MENA countries register the lowest score for political stability among the considered regions (with the sole exception of SA).

PPP is projected to play a growing role in infrastructure investment. This is particularly so in energy investment among low- and middle-income countries, as recent policy development suggests. It is a shortcut to modernizing the energy sector and provides the much-needed infrastructure. MENA countries show a significant gap in the level of PPP committed when compared with other regions globally. Among the long list of institutional dimensions predicted to play a role in this negative trend, political instability and rule of law seem to penalize MENA countries heavily. Our qualitative analysis suggests that MENA countries should look at the most successful experience worldwide to adopt international best practice and transition their institutional and regulatory dimension to become "investment friendly." While a more stable political and institutional environment undoubtedly favors a better investment climate, the MENA region should provide investment incentives that are robust to the existing financial and institutional uncertainty, which most Mediterranean countries experience these days. While this chapter has analyzed and described a number of interesting trends taking place in the region, further work should be carried out to explore the way in which international best practices can be adapted to the specific MENA situation. In addition, the role that financial markets can play in attracting PPP in a region dominated by radical uncertainty is a promising area of analysis which could bring together the diverse interest of private and public stakeholders in the region, and should be explored thoroughly in future work.

REFERENCES

de Bettignies, J., Ross, T., 2004. The economics of public private partnerships. Can. Public Policy 30 (2), 135–154.

Kirkpatric, C., Parker, D., Zhang, Y., 2006. Foreign direct investment in infrastructure in developing countries: does regulation make a difference? Trans. Corp. 15 (1), 143–171.

L'Abbate, A., Migliavacca, G., Calisti, R., Brancucci Martinez-Anido, C., Aymen, C., Fulli, G., 2014. Electricity exchanges with North Africa at 2030, the European and the Italian approach. In: Cambini, C., Rubino, A. (Eds.), Regional Energy Initiatives: MedReg and the Energy Community. Routledge, London, pp. 225–245.

Mengistu, T., 2013. Emerging Infrastructure Financing Mechanism in Sub-Saharan Africa. Pardee RAND Graduate School/RAND Corporation, Santa Monica.

Observatoire Méditerranéen de l'Energie, 2011. Mediterranean Energy Perspective. OME, Paris.

Panayotou, T., 1998. The Role of the Private Sector in Sustainable Infrastructure Development. Harvard Institute for International Development, Cambridge, MA.

Private Participation in Infrastructure Database, 2015. Public–Private Infrastructure Advisory Facility (PPIAF). http://www.ppiaf.org (accessed 07.07.2015.).

Roy, O., 2012. The transformation of the Arab World. J. Democr. 3 (23), 5–18.

Rubino, A., 2014. A Mediterranean electricity cooperation strategy: vision and rationale. In: Cambini, C., Rubino, A. (Eds.), Regional Energy Initiatives: MedReg and the Energy Community. Routledge, London, pp. 31–44.

Wall Street Journal, 2015. Egypt's Sisi Closes Economic Conference with Call for Further Investment. http://on.wsj.com/1JrSotp (accessed 15.03.2015.).

World Bank, Asian Development Bank, Inter-American Development Bank, 2014. Public–Private Partnerships. World Bank Publications, Washington, DC, Reference Guide. Version 2.0.

Chapter 14

Investment and Regulation in MENA Countries: The Impact of Regulatory Independence

Abrardi Laura, Cambini Carlo, Rondi Laura
Politecnico di Torino, DIGEP Department of Management, Torino, Italy

1 INTRODUCTION

In recent years, the south Mediterranean region has been acquiring increasing relevance in the energy sector. The reasons are manifold. First, on the supply side, the area is rich in natural resources: mainly oil, which represents 90% of the energy produced in the Middle East and North African (MENA) countries, but also gas and renewable energies, both showing increasing trends in generation, to the point that gas is expected to reach oil share levels by 2035. Second, on the demand side, the consumption of energy in MENA countries is growing steadily and rapidly, and at even higher rates than the global demand for energy. New investments are therefore necessary to exploit the resources of the region and reliably meet the increased demand, both locally and globally. Third, the south Mediterranean area has a convenient geographical location as a result of its proximity to markets and demand of the European region. Finally, policy makers of the energy sector hold considerable interest in achieving a transparent and harmonized landscape of regulation across the Mediterranean Basin.

With the twofold aim of increasing infrastructure investments and converging toward a homogeneous and harmonized regulatory framework with Europe, national regulatory agencies (NRAs) have recently been set up in several Mediterranean countries. The rationale behind the establishment of independent authorities lies in the attempt to avoid the limitations and inefficiencies related to the lack of commitment, time inconsistency, uncertainty, conflicting interests, and political interference, typical of situations in which regulation is carried

Regulation and Investments in Energy Markets. http://dx.doi.org/10.1016/B978-0-12-804436-0.00014-X

out by governmental entities.[1] Overall, the presence of independent regulatory bodies is generally recognized as conducive to higher effectiveness and efficiency of regulation with, ultimately, an expected positive influence on investment decisions. Our study aims at empirically investigating the impact of the presence of independent regulatory bodies on the incentives to invest in electricity capacity in a sample of 12 MENA countries (Algeria, Djibouti, Egypt, Iran, Iraq, Jordan, Lebanon, Libya, Morocco, Syria, Tunisia, Yemen) over the period 1990–2011.

The presence of NRAs apparently represents a necessary condition for the creation of a favorable regulatory framework, but it is not, in itself, sufficient to actually achieve the best outcomes. The reason lies in the effectiveness of the regulation which also depends on the characteristics of the environment in which NRAs operate: namely, on the set of political and social institutions of the country. Indeed, factors such as executive–legislative–judiciary relations, the bureaucratic system, the level of political stability, the degree of conflict, and the arbitrariness or scale of corruption have a significant influence on regulatory performance by determining the receptivity of the environment to regulation activities. In turn, this has predictably a strong impact on private investments.

Indeed, institutions influence the effectiveness of regulation in a deeper way other than by simply shaping the environment in which regulators operate. Institutions have a central role in the definition of the regulatory framework, which is indeed generated by a political process. In particular, institutions define the type of regulation and determine the level of technical expertise, resources, and degree of autonomy of regulatory bodies. Even more importantly, institutions affect the degree of flexibility, independence, commitment, and discretion of regulation.

Institutions then have a double impact on the effectiveness of regulation. They not only create the social, political, and economic environment in which regulators operate, but they are also pivotal to the formation and characteristics of the regulatory framework itself. Therefore, our study also aims at providing an assessment of the degree of interdependence between regulatory and institutional frameworks, evaluating the effect of this relationship on the infrastructural investments in the electricity sector of our selected sample of MENA countries.

The chapter is organized as follows. In Section 14.2 we briefly describe the interrelation between political institutions and the regulatory process. In Section 14.3 we provide an overview of the presence and characteristics of NRAs and of the institutional endowment in MENA countries. In Section 14.4 we present a model that we use to derive a testable hypothesis regarding the relationship between NRAs, the institutional context, and investments. Concluding remarks are in Section 14.5.

1. Quoting Stigler (1971, p. 3): "the political process defies rational explanation: 'politics' is an imponderable, a constantly and unpredictably shifting mixture of forces of the most diverse nature, comprehending acts of great moral virtue (the emancipation of slaves) and of the most vulgar venality (the congressman feathering his own nest)."

2 THE ESTABLISHMENT OF REGULATORY AUTHORITIES: PITFALLS OF THE INSTITUTIONAL ENDOWMENT OF COUNTRIES

In line with the experience of the United States (Geradin 2004; Joskow 2007), the EU makes the establishment of independent NRAs one of the pivotal elements to the competitiveness of utilities in regulated sectors. In particular, in Europe, the United Kingdom was the first country to adopt NRAs (Cambini et al., 2012; Saal 2002). At the EU level, the Directive 2003/54/EC carefully defined the institutional design of regulatory bodies and provided a first framework for pan-European coordination among regulators through the European Regulators Group for Electricity and Gas (ERGEG) setup. Then, the Directive 2009/72/EC, part of a third package of directives aimed at utility liberalization and energy market integration, further stressed the role of agencies, their duties, and the need for their effective independence; strengthened coordination at the EU level through the Agency for the Cooperation of Energy Regulators (ACER) was affirmed. Thus, NRAs gradually emerge in the EU regulatory experience as pivotal instruments for the electricity sector liberalization.

Based on the European experience, the model of independent national regulators is the one the EU promotes in Mediterranean neighboring countries through partnership programs and cooperation initiatives. In this regard the 2007 Euro-Mediterranean Ministerial Conference provided, for the first time, an assessment of the regulatory framework for the electricity sector at the MENA country level, including the role of existing regulatory agencies and the degree of diffusion of EU regulatory standards.

In determining whether a regulated energy sector is preferable to an unregulated one, the EU's promotion activity assigns a central role to NRA independence, as pointed out in the following paragraph.

2.1 Restructuring Utility Industries: The Role of National Regulators

The rationale behind the creation of a regulatory authority lies in the attempt to insulate regulators from political interference intended at influencing the investment or price-setting decisions of regulated firms. Specifically, the establishment of NRAs addresses two obstacles to regulatory action (Levi and Spiller, 1994): time inconsistency and regulatory commitments/credibility issues. Electricity is one of the sectors in which time inconsistency problems arise in association with different and often contending social interests. In democratic contexts, it is the legislative–executive dynamic, as well as the alternation of parties in power, that reveals such contending interests. Thus, delegating the rule implementation phase to technical agencies reduces a specific source of instability of the regulatory framework of a country, the one associated with the possibility of having the government replaced by other parties with different preferences and representing different social interests. As for the majority of MENA countries, long-lasting regimes show that the "risk" of being replaced by democratic alternation of parties in power has been almost absent for the past 20 years. Compared to most OECD or EU countries, for example, the institutional endowment of MENA is characterized by fewer checks and balances between domestic institutions, poorer enforcement of the Rule of Law and, at the same time, stronger incumbents and bureaucracies largely dependent on the ruling elites (Franzi, 2013). Paradoxically, in such contexts, the setup of NRAs may even reinforce the power of the incumbent, typically state-owned operator because its tight long-standing relationships with bureaucrats and government staff may weaken NRA independence in decision-making while at the same time reinforcing the capacity to infiltrate bureaucracy by elites in power (Gilardi, 2005). Similarly, the lingering relation between incumbents and bureaucrats strengthens bureaucratic elites, making bureaucrats one of the most relevant players in the region. Since MENA is mainly comprised of publicly driven economies, bureaucrats are influential actors in reforming processes involving the utility sector. To a certain extent, bureaucrats may infiltrate the elite in power, having developed that knowledge and technical expertise necessary to influence the implementation of rules and reforming projects; they are the actors who may assure continuity in sector management and stability of the regulatory framework in the case of unexpected regime changes, such as the ones that occurred during 2011–2012.

The second reason for the setup and independence of NRAs is related to regulatory commitment/credibility. Regulatory credibility is the sole insurance against the risk of administrative expropriation; when such credibility is lacking, it signals that political commitments toward sector liberalization are missing and the regulatory environment of the country is not transparent. The stability of authoritarian and monarchical regimes of the last 20 years have not been capable of generating new investment in electricity directed at improving both crossborder and MENA–EU power exchanges, the latter being limited to the interconnection between Spain and Morocco (EU Commission, 2010). As Levy and Spiller state (1994), the credibility of regulation in the utility sector is higher in countries in which executive and legislative discretions are reciprocally counterbalanced than in countries where such a counterbalance does not exist or is weak (see also Bortolotti et al., 2013). Missing executive–legislative counterbalance, every form of regulatory intervention may be easily knocked over. In this case, administrative expropriation is a serious risk for foreign investors interested in securing a fair return on their investments. Considering the scenario of MENA countries, a functioning judiciary power remains the sole body capable of assuring the degree of regulatory credibility necessary for spurring new investments. Judiciary power, when independent, works as a restraint to the discretion of incumbents. Thus, the higher the degree of judiciary independence the less serious the regulatory commitment problem. In the MENA region, such independence is undermined by scarce financial resources, arbitrary decisions on judges' appointment and dismissals, career improvements, and interference by incumbents in the administration of justice when verdicts mainly refer to regime opponents (Freedom House, 2011).

In order to effectively ensure the independence of regulatory agencies, it is crucial that the legal and institutional settings assure the separation of the agency from executive power, its accountability to elected bodies, and its operating autonomy in terms of both financial and occupational independence (i.e., staff and expertise recruiting). These characteristics are required if the purpose is to reduce the risk of capture and mitigate asymmetric information issues (Larsen et al., 2006). If these conditions are met, NRAs are independent in every respect, and this allows a more stable regulatory environment, with a positive impact on the investment decisions of public utilities. This result finds validation in empirical evidence from European (Cambini and Rondi 2011), Latin American, and Caribbean countries (Andres et al., 2006, 2007, 2008; Correa et al., 2006; Gutiérrez, 2003). The experience of Chile has been widely examined; being crucial both for reform sequencing (Newbery, 2001) and their value, the institutional context of the country has been assessed to be the main influential factor when reforming the electricity sector (Gutiérrez, 2003; Levi and Spiller, 1994; Zhang et al., 2005). Trillas and Montoya (2011) present an analysis of the evolution of regulatory independent agencies for 23 Latin American and Caribbean countries in the telecommunications industry. By defining agency independence in terms of the political vulnerability of regulators, the authors show that a higher degree of authority independence is associated with higher network penetration. Cubbin and Stern (2006) show that in those countries where an independent agency has been set up generation capacity has improved, confirming the relation between performance of the utility sector and the governance of regulatory institutions.

Because independence is one of the key features that the EU promotes abroad due to its relevance, its dimensions need to be clearly identified. Independence is the result of many institutional factors that can be summarized by:

- available resources, in terms of financial and human capital;
- regulatory tools; and
- accountability.

With respect to the first point, independence itself could be intended as a resource that supports regulator access to information, especially in contexts affected by information asymmetries (Armstrong and Sappington, 2006). Moreover, available economic resources and, consequently, the possibility of recruiting experts and properly managing relationships with stakeholders (i.e., energy industries, consumers) are dimensions of independency that should be opportunely considered when designing an optimal regulatory policy for the energy sector. At the same time, obtaining information, as well as remunerating expertise for coding this information, has its own cost. A proper budget ensures the possibility of paying competitive salaries and financing agency costs, two of the main structural conditions for maintaining independency.

As for regulatory tools, instruments available to regulators are pivotal to bringing about effective regulation. In fact, *if the regulator is not authorized to compel the firm to report data on its operations, the regulator will find it difficult, if not impossible, to make informed policy decisions* (Armstrong and Sappington, 2006, p. 337), thus compromising the activities of tariff setting and defining incentive mechanisms. Furthermore, constraints to regulator actions limit its independence by increasing the risk of capture, thus exposing its actions to external interests. For example, in the absence of sufficient resources (such as budget, jurisdiction, or technical expertise), the regulator may depend on the information provided by market operators that often are state-owned companies whose leaders are appointed by political authorities. A form of political capture may especially emerge when the issues at stake are related to tariffs. In this situation, consumer opinion deeply affects incumbents, whose main interest is not to lose voter support and maintain a sort of social order and stability. Even in this case, company managers and politicians share the same interest: *under-pricing to stimulate demand and secure political support* (Joskow, 2005, p. 9). Finally, a weak regulator working in close cooperation with the firm supposed to be regulated originates a discriminatory regime toward other companies. As we will see in Section 14.3, this problem is common to those MENA countries that maintain discriminatory regimes toward foreign investors despite the liberalization process undergone in the past.

A further form of influence is given by misunderstood *accountability* requirements and their translation into, for example, the obligation for regulators to require state permission for any kind of decision, thus leaving regulators a mere consultative role. Conversely, if correctly implemented, accountability provisions entail regulators working within rules codified by elected bodies – the Parliament – in front of which they have to justify their actions without renouncing their operational independence.

Finally, independence must entail the capability and power of regulators in defining the correct policy in terms of tariffs and incentive mechanisms.

2.2 The Role of the Institutional Framework

The degree of discretion accorded regulators (and consequently their performance) is tightly linked to the institutional framework in two ways. First, institutions are responsible for the formation and design of regulatory bodies and their functioning by assigning jurisdiction, instruments, and resources. Second, institutions are responsible for the characteristics of the economic and social environment in which regulators operate. Indeed, regulatory practices and rules are implemented differently depending on the institutional endowment of countries (North, 1990). In this regard Levi and Spiller (1996) highlight how the judiciary independence – a functioning checks and balances system, the presence of veto players and contending social interests, as well as the administrative capabilities of a country – are *exogenous* factors directly impacting regulatory restraints of countries and the independence of regulatory agencies. As a result, analyzing the effectiveness of the regulatory process necessarily requires considering the institutional background of a country.

Regulation is defined as a *design problem* that has two main components: governance – the mechanisms that constrain regulatory discretion – and incentives – rules on pricing, subsidies, network access, etc. The choice between specific governance mechanisms and incentive structures are *variables in the hand of policy-makers* (Levi and Spiller, 1994, p. 205), who choose on the basis of the *institutional endowment* of a country (North, 1990). In detail, the dimensions of the institutional endowment of a nation can be defined as the set of:

- executive and legislative institutions;
- judicial institutions;
- informal norms;
- contending social interests and their balance; and
- administrative capabilities.

Executive and legislative institutions refer to the formal mechanisms for appointing legislators and decision makers, law design and implementation, and the way executive–legislative relations are regulated. Judiciary independence from political power is one of the main dimensions of the Rule of Law; it makes potential investors confident of the seriousness of law implementation. In this respect, the mechanisms for judges' appointment and for assuring their impartiality need to be considered. Moreover, informal norms and customs must not be neglected as they are *generally understood to constrain the action of individuals or institutions* (Levi and Spiller, 1994, p. 206). In addition, different interests within society, their balance, and the role of ideology exert substantial influence on policy definition. Finally, the administrative capabilities of a country are also relevant for an effective implementation of regulation.

The conditions for positive institutional endowment can be obtained through: (1) explicit separation and counterbalance between executive, legislative, and judicial powers; (2) a written constitution that limits the first two powers and is enforced by the latter; (3) two elective chambers, elected using different voting rules; (4) an electoral system allowing opposition parties to emerge and counterbalance the governing party; and, finally, (5) a federal structure of the state or strong decentralization of powers.

Given the five dimensions of institutional endowment, the mechanisms and preconditions for an effective regulatory framework can be classified as:

- restraints on regulator discretion written in the regulatory system;
- restraints on the possibility of regulatory systems to evolve; and
- independent judiciary power.

The first aspect takes into account whether explicit, specific rules exist to define regulatory boundaries and ensure their credibility (thus requiring strong administrative capabilities) or a more flexible process is adopted, with informal norms and administrative laws.

The second mechanism refers to the flexibility of the regulatory system and its potential to evolve and adapt to changes in society and industry. The degree of flexibility is a delicate issue and requires careful consideration by policy makers, as it may come at the price of lower credibility of regulation. Indeed, the same mechanisms that constrain regulator arbitrariness to enhance the credibility of regulation may also make it difficult to adapt the rules to evolving circumstances. In this regard the political system of a country should be analyzed considering the distinction between situations in which a credible regulatory commitment may be achieved through legislation from those in which such credibility may be derived by licensing and contractual arrangements. In the former case, the relative rigidity of the legislative system limits the flexibility of the regulatory action, and the introduction of reforms may have to await drastic changes in the political system. In the latter, administrative discretion is constrained by formal regulatory contracts (licenses), based on upholding a system of property rights (Levi and Spiller, 1994). Electoral, executive, and legislative institutions are the core aspects to analyze to bring about this aim.

The third mechanism refers to the existence of a functioning judiciary system, characterized by independence from political power; thus, judiciary capabilities of refraining from administrative arbitrary action would be the main aspect to consider.

To the extent that the EU has recognized the importance of the institutional framework for positive reform of public utilities, EU programs have been directed at state capacity building through the promotion of good governance and the Rule of Law. First, the Euro-Mediterranean Partnership (1995) and, second, the European Neighbourhood Policy (2004) foster sustainable economic growth and market integration at the Euro-Mediterranean level (mainly defined by shared regulatory standards). Improvement in MENA Rule of Law and good governance is pursued through technical issues of cooperation. NRAs have to be viewed in the wider context of EU Rule of Law and good governance promotion, since they are indicative of transparent regulatory practices of a country.

3 THE REGULATORY AND INSTITUTIONAL LANDSCAPE IN MENA COUNTRIES

The MENA region hosts a population of about 398 million individuals, 85% of whom live in middle-income countries, 7% in low-income countries, and 8% in high-income countries. Economic and social welfare in the region is superior to that of the sub-Saharan region, but rapid population growth (at a rate of 19% in the last 10 years), modest development of the private sector, limited intensity of innovation, and finally difficult access to credit markets have led to the highest levels of unemployment in the world, especially among the young and women.

The region was severely affected by the global crisis. The recovery is at present hesitant and erratic (especially for net importers of oil products), the fiscal deficit has worsened, and investments have generally decreased (in particular, Egypt, Syria, Tunisia, and Libya experienced negative growth in foreign direct investment in 2011). Overall, macroeconomic disequilibria, social and political unrest, and the vast unfinished list of reforms are preventing the region from achieving its potential rate of GDP growth.

As for the demand for energy, at a global level it has grown by 35% from 2001 to 2013, despite the recessive effect of the crisis worldwide. Demand in the MENA countries exhibited an exceptional growth rate, even higher than the global average, with an increase of 73.3% from 2001 to 2012. The high level of consumption in the MENA region can be explained both by considerable population growth and by a generous subvention policy directed at keeping energy prices at accessible levels (Fig. 14.1).

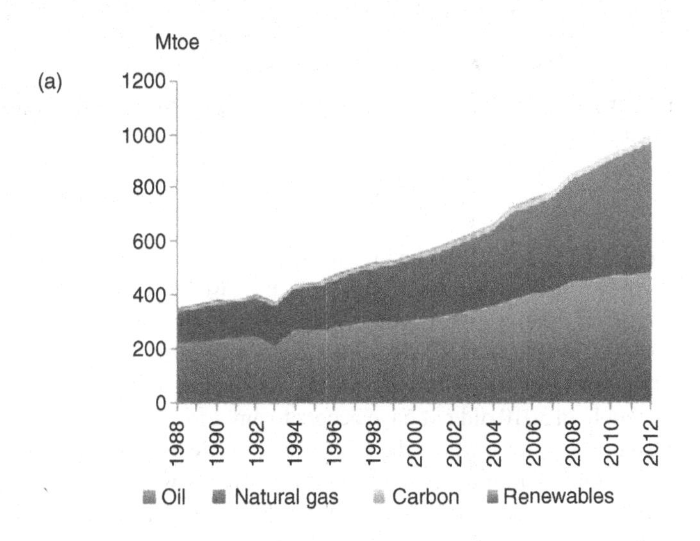

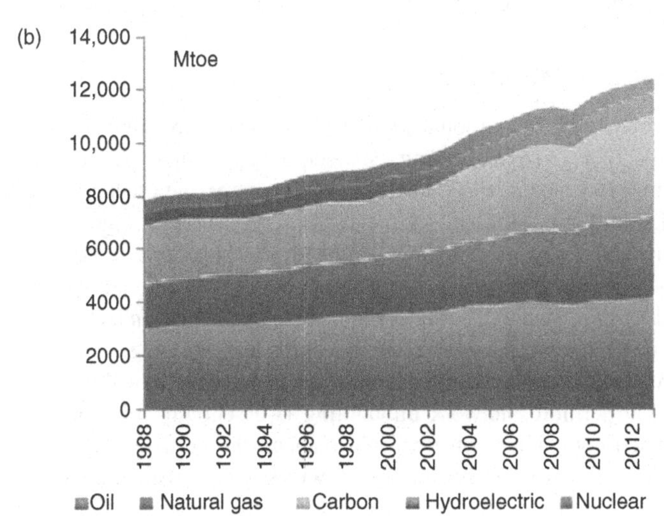

FIGURE 14.1 Demand for energy (Mtoe) in MENA countries (a) and worldwide (b). *(Source: BP, Statistical Review of World Energy, 2014.)*

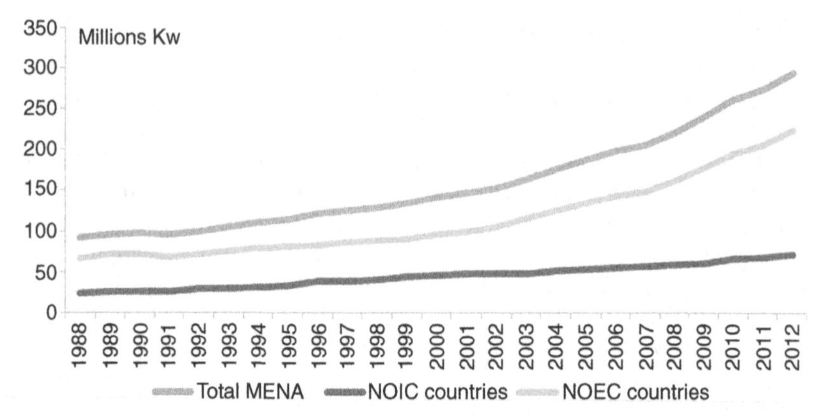

FIGURE 14.2 Evolution of the total electricity installed capacity in MENA countries. *(Source: EIA statistics.)*

In the electricity sector the increase in demand was actually accompanied by growth of installed capacity (Fig. 14.2). In the last 14 years, total installed electricity capacity has grown by 93%, mainly driven by investments in net oil-exporting countries (NOEC). Most MENA countries present high rates of electrification. Nonetheless, and despite the abundance of natural resources, 20 million people in 2010 did not have access to any form of electricity, mostly concentrated in rural areas, where the electrification rate in 2000 – according to Energy Information Administration (EIA) estimations – was on average equal to 79.9% in North Africa and 76.6% in the Middle East. Yemen and Djibuti's electrification rate is exceptionally low (36.9 and 50%, respectively).

3.1 NRAs in MENA Countries

In the MENA region, national regulatory agencies have recently been set up (see Table 14.1) in several countries. According to a survey conducted in 2012 (Cambini and Franzi, 2013), analyzing the progressive creation of NRAs (in particular in Egypt, Jordan, Morocco, Tunisia, Turkey, Algeria, Israel, and Lebanon) and their independence has been assessed along three main dimensions: decisional autonomy, organizational autonomy, and agency responsibility. Results show that the regulatory authorities in the region lack real (*de facto*) independence and act mainly in a consultative role for the government. In fact, executive power holds any decisional right about tariffs or third-party access (TPA). Furthermore, network unbundling is essentially functional, and fully state-owned companies still own and manage the network. Overall, political interference over regulatory activities is widespread and encompasses the main activities of regulatory agencies. Table 14.1 gives information about NRAs in MENA countries.

TABLE 14.1 National Regulatory Agencies in the 12 Selected MENA Countries

Country name	NRA	Executive branch commission	Year set up	Normative source
Algeria	Electricity and Gas Regulatory Public Utility Authority (CREG)		2002	Law No. 02-01 of February 5, 2002
Djibouti		Ministry of Energy and Natural Resources (MENR)		
Egypt	Egyptian Electric Utility and Consumer Protection Regulatory Agency (EgyptERA – EEUCPRA)	Ministry of Electricity and Energy (MoEE)	2000	Presidential decree
Iran		Energy Department (Ministry of Energy)		
Iraq		Ministry of Electricity		
Jordan	Energy and Minerals Regulatory Commission (EMRC); Jordanian Electricity Regulatory		2001	Temporary General Electricity Law No. (64) of 2002
Lebanon		Ministry of Energy and Water (MEW)		
Libya		The Energy Council (EC)		
Morocco	Request for its creation approved in 2014	Ministry of Energy, Mines, Water and Environment (MEMEE), Electricity and Renewable Energies Department,		
Syria		Ministry of Electricity; PEEGT and PEDEEE are responsible for regulating their areas of activity		
Tunisia	The National Agency for Energy Management	Ministère de l'Industrie, de l'Energie et des Petites et Moyennes Entreprises (TMIE); The Directorate General for Energy of the Ministry of Industry and Energy	2004	Law No. 2004-72 of August 2, 2004
Yemen		Ministry of Electricity and Energy (MEE)		

Of the individual regulatory agencies operating in MENA countries, only Jordan's Energy and Minerals Regulatory Commission (EMRC) appears to be (at least) formally regarding decision-making autonomy in terms of tariff setting, network planning, TPA, license issues, service quality, dispute settlement, and consumer protection. In particular, Jordan's authority is competent for those aspects identified as pivotal to sector liberalization: unbundling, TPA, and tariff setting.[2]

The most sensitive aspects of the electricity sector in Mediterranean countries without NRAs are comanaged by public companies and central administration apparatus. Thus, when looking at those issues on which credible regulatory commitments and independence of regulators are measured in the study – unbundling, TPA, and tariff setting – it is possible to conclude that NRAs in the Mediterranean area are not truly autonomous and independent in the decision process.[3] In terms of organization autonomy, the competence to decide the internal organization of regulators is shared between the regulator and the legislative power in Egypt and Jordan, while in Tunisia and Morocco NRAs are fully independent regarding their decision process. Similarly, personnel policy is a competence that NRAs share with the legislative power in Egypt and Jordan only. Finally, MENA countries rank generally low in terms of budget autonomy (especially with budgets defined for 1 year).

As for NRA accountability, the regulator does not typically require approval from the executive power for taking decisions on regulatory aspects and is autonomous in using regulatory tools. The only exception is Egypt, where the regulator is an officer within the executive. However, in some countries regulators have obligations to executive and legislative powers. The Egyptian agency has to submit an annual report to the executive for approval.

Overall, as reported in Cambini and Franzi (2013), considering the three dimensions of independence (decision-making, organizational, and accountability), the MENA countries in general do not perform particularly well, with the exception of Jordan. Despite being a monarchy where the executive has strong powers in the energy sector, Jordan's NRA is the only case where the regulator has decisional power on issues such as tariffs. Moreover, unlike other MENA countries, courts in Jordan may intervene in cases of appeals against regulatory decisions, confirming a better functioning judiciary power in restraining executive and regulator administrative discretion than seen in other MENA countries.

3.2 Political and Legal Institutions in MENA Countries

From an institutional point of view, all examined constitutions formally recognize the separation between executive, legislative, and judiciary powers.

2. However, some commenters note that the expansion of competencies to a new industrial sector (mines) in late 2014 also led to a reduction in the monitoring powers of the NRA.

3. It is worth pointing out that even in countries that declare to have *legal* capability on price-setting decisions, this does not imply that these regulators might not be influenced by external (i.e., government) pressure.

Moreover, with the sole exception of Israel, all countries have a written constitution. Two legislative chambers with different voting systems exist in the two monarchies, Morocco and Jordan, and in Tunisia, Algeria, and Egypt. However, in countries where a second chamber exists, it is elected partially by voters and partially by the Prime Minister – in Egypt – or formed by members of local councils or of professional chambers – as in Algeria and Tunisia. In practice, the rules for second-chamber elections are directed at consolidating monarchical regimes: in Jordan it is entirely the remit of the King; in Morocco it is formed only by people having certain attributes.

No MENA country has a federal structure of power, although there is a decentralized level of administration in Egypt and Morocco in which local authorities exert power over tax and spending legislation.

In terms of mandate termination and number of consecutive mandates given to a Prime Minister, only the Algerian and Israeli constitutions allow for a second mandate, as opposed to the Tunisian and Lebanese constitutions, which explicitly exclude such a possibility. In the Egyptian case the possibility of a second mandate has been introduced in the constitutional text adopted in March 2011 in the aftermath of the people's uprising.

Despite such provisions, presidential regimes have been characterized by long-lasting leaders. Excluding the two monarchies in which Prime Ministers are nonelected, the majority of presidential and semipresidential regimes show considerable political continuity, measured by the number of years Prime Ministers and their parties have been in power. On the contrary, the parliamentary republics – Israel and Lebanon – have been characterized by high government instability. The higher degree of stability registered by presidential regimes compared to parliamentary countries is evident also in terms of regularity of Parliamentary elections: it is higher in the former than in the latter case (the exception is Libya where there are no elections).

As for checks and balances between political forces, the MENA countries (especially monarchical, presidential, and semipresidential regimes) show a division of power in which the Government and Parliament are squeezed by strong executives, endowed with absolute majority and veto powers. Presidential and monarchical regimes favor a higher degree of discretion on the part of incumbents than parliamentary regimes, especially when majority opposition party dynamics are not at work. The parliamentary regimes of some MENA countries, such as Algeria and Tunisia, are instead characterized by a more genuine check-and-balance system and alternation of parties in power. At the same time, they also typically present a high level of system instability and weak majorities (e.g., Lebanon). As for administrative capabilities, pivotal to effective implementation of regulatory policies, bureaucracy quality is perceived to be high only in Israel (World Bank, 2013).

Similarly, in terms of the judiciary's independence and powers, the only country to perform well is Israel. The other countries show a low degree of confidence in the way the ruling system functions, in the capabilities of contractual

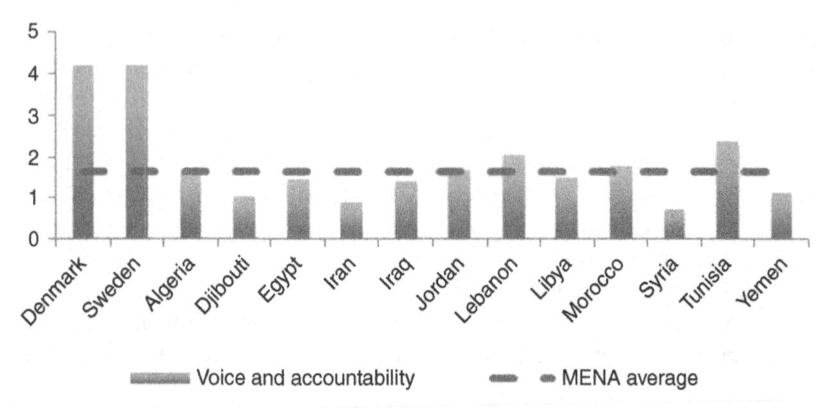

FIGURE 14.3 Voice and accountability in 2013 (0, weak; 5, strong). *(Source: World Bank's Worldwide Global Indicators.)*

enforcement, in the system of property rights, and in the correct functioning of the courts – all relevant dimensions when describing the investment environment of a country and its potential threats such as nationalization and administrative expropriations.

In our empirical analysis we control for a number of political and institutional characteristics, which may have influenced investment growth in the sector:

- voice and accountability;
- political stability;
- type of regime;
- Control for Corruption;
- checks and balances;
- Rule of Law.

"Voice and accountability" capture perceptions of the extent to which a country's citizens are able to participate in selecting their government, as well as freedom of expression, freedom of association, and a free media.[4] Figure 14.3 reports performance in terms of voice and accountability in the 12 MENA countries constituting our panel. As a reference measure, the figure also displays a simple average calculated on the whole set of MENA countries and includes the performance of Denmark and Sweden as a European benchmark. In terms of voice and accountability, all countries considered in the panel (with the exception of Tunisia, Lebanon, and Morocco) present an indicator value below the MENA average.

Political stability measures perceptions of the likelihood that the government will be destabilized or overthrown by unconstitutional or violent means, including politically motivated violence and terrorism.[4] Political stability creates an environment favorable to investment and as such is a factor of economic

4. Definition from www.theglobaleconomy.com

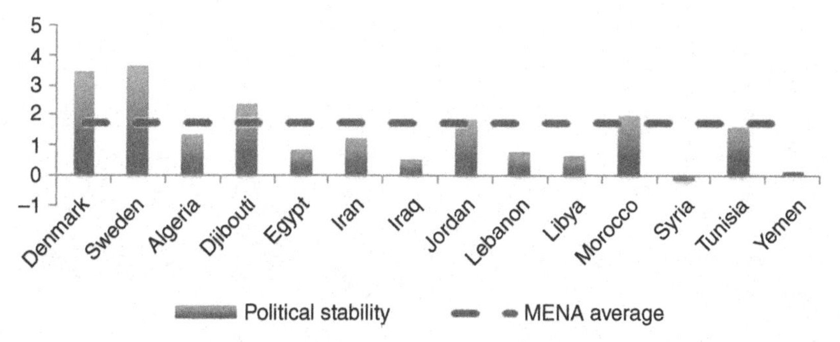

FIGURE 14.4 Political stability in 2013 (0, weak; 5, strong). *(Source: World Bank's Worldwide Global Indicators.)*

growth. In turn, positive macroeconomic performance is conducive to political stability, thus strengthening the interconnection between the two. Political instability creates uncertainty and volatility and shortens the time horizon of policy makers, leading to short-run, inefficient decisions (Fig. 14.4).

The type of regime takes into account the system of property rights and the level of democracy. Democracies have a controversial effect on economic growth. On the one hand, they apply pressure for immediate consumption, which has the effect of reducing investment in the long run. On the other hand, they have an interest in maximizing GDP growth, which has the effect of autocrats lacking credibility and time consistency.

Control of Corruption assesses the likelihood of encountering red tape, corrupt officials, and other groups.[4] It captures the perception both about the extent to which public power is employed to obtain private gain, and the extent to which the state is captured by private powers. Widespread corruption generally hinders investment (by decreasing the rate of return) and consequently economic growth, especially in small economies. Moreover, it prevents efficient allocation of resources and production decisions, hampers innovation, and wastes resources in rent-seeking activities (Fig. 14.5).

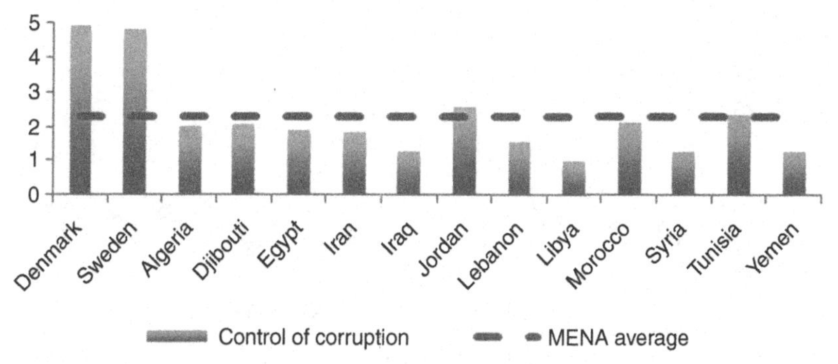

FIGURE 14.5 Control of Corruption in 2013 (0, weak; 5, strong). *(Source: World Bank's Worldwide Global Indicators.)*

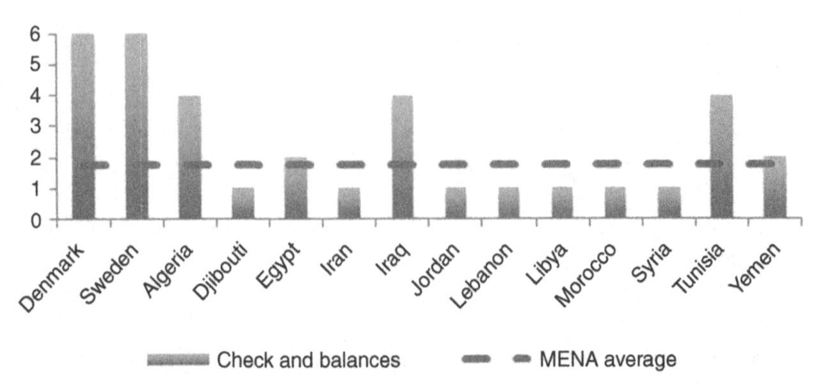

FIGURE 14.6 Checks and balances in 2012 (0, weak, 6, strong). *(Source: DPI (2012).)*

Checks and balances refer to the set of institutional mechanisms aimed at maintaining the equilibrium and separation between executive, legislative, and judiciary powers. An effective system of checks and balances lends credibility to public administration by imposing constraints on executive arbitrary activity (Keefer and Stasavage, 2003; North and Weingast, 1989), which is favorable to long-run economic growth (Mehlum et al., 2006; Robinson et al., 2006) (Fig. 14.6).

Rule of Law relates to the extent to which agents have confidence in, and abide by, the rules of society; in particular, the quality of contract enforcement, property rights, the police, the courts, and the likelihood of crime and violence.[4] The Rule of Law influences the growth rate of developing countries by restraining the government from intervening with ad hoc actions and offsetting private initiatives. This improves investor confidence about expected return on investments and provides legal protection for business activity (Fig. 14.7).

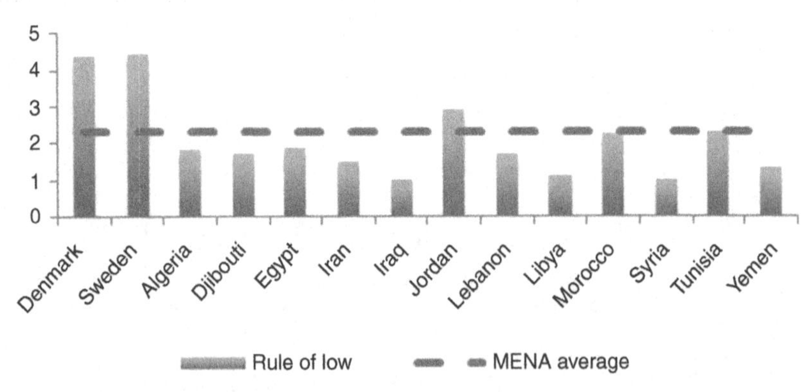

FIGURE 14.7 Rule of Law in 2013 (0, weak; 5, strong). *(Source: World Bank's Worldwide Global Indicators.)*

4 EMPIRICAL ANALYSIS

4.1 Research Design

The purpose of econometric analysis is to explore the role played by NRAs in the institutional framework when determining investment decisions in the energy sector.

We start by investigating the relationship between an NRA and investment in the generation capacity of MENA countries. As a result of limited data, the level of investment is approximated by the growth in *installed generation capacity*. However, as described in Section 14.2, the establishment of an NRA can be related to the quality of the institutional endowment (see Cambini and Rondi, 2011 and Bortolotti et al., 2013 for empirical analyses of a sample of European countries). With this aim in mind, we test the following hypothesis.

Hypothesis 1: The establishment of an NRA positively affects investment in installed capacity.

To test Hypothesis 1, we estimate a simple investment model that accounts for the traditional acceleration principle, linking investment to (electricity) demand growth and to other potential country-level determinants of capacity growth related to economic and energy systems. The generation of network *capacity growth* used in the model is estimated as the log difference of capacity installed in two adjoining years. The variable *NRA* is a dummy assuming the value 1 when a regulatory agency is in place.

The set of country-specific controls includes: 1 and 2-year lags in *electricity consumption growth*, measured by the rate of growth in demand for electricity; the log of *GDP*, to control for the different size and growth of economies; *population density*, which increases the demand for energy and at the same time leads to concentration of generation plants in densely populated areas; and *distribution losses* expressed as a percentage of total generated energy, which increases the energy requirement given the levels of actual consumption. We then add country dummies to account for residual time-invariant heterogeneity across countries and time dummies to control for common effects over time. Our baseline specification is as follows:

$$\begin{aligned}
&CapacityGrowth_{i,t} = \alpha_0 + \alpha_1\, ElConsGrowth_{i,t-1} + \alpha_2\, ElConsGrowth_{i,t\text{-}2} \\
&+ \alpha_3 \,\mathrm{Log}\, GDP_{i,t} + \alpha_4\, NRA_{i,t} + \alpha_5\, AV_{i,t-1} + \alpha_6\, PopDensity_{i,t-1} \\
&+ \alpha_7\, DistrLoss_{i,t-1} + \lambda_t + \mu_i + \varepsilon_{it}
\end{aligned}$$

where λ_t and μ_i are time and country dummies; and ε_{it} is the error term.

As suggested in Section 14.2, the decision to establish an NRA is likely to be influenced by the institutional endowment of the country (i.e., by the political and legal institutions that preside over and contribute to the functioning and efficacy of the government). These institutions also influence the extent of powers delegated *de jure* to the NRA and somewhat condition the degree of genuine (*de facto*) independence of the authority from the government. This reasoning leads to Hypothesis 2.

Hypothesis 2: The likelihood of an NRA being set up is affected by the political and institutional features of MENA countries.

To test Hypothesis 2 we estimate a simple logit model in which the probability of establishing an NRA is regressed on a set of political and institutional characteristics of the MENA countries, as described in Section 14.3.2.

Furthermore, we also consider that *NRA* might actually encompass and capture the impact of the institutional endowment on investment. In order to test this conjecture, we exclude *NRA* from the investment equation and check whether the institutional variables alone do influence capacity growth (i.e., have their own influence on investment). Finally, as a further robustness check, we test whether the effect of *NRA* survives when we include institutional characteristics in the investment model, in order to check that the impact of *NRA* does not disappear when we control for institutional features.

Before proceeding with econometric analysis, it is important to note that the discussion will relate to regulatory *governance* rather than to the content of regulatory *interventions* such as pricing policies or unbundling decisions.

4.2 Data

The data relate to 12 countries (Algeria, Djibouti, Egypt, Iran, Iraq, Jordan, Lebanon, Libya, Morocco, Syria, Tunisia, Yemen), a 22-year period (1990–2011), and a total of 263 observations. Energy data are publicly available from the website of the US Department of Energy, International Energy Agency (IEA), for all countries since 1990. Unfortunately, separate data on investment for the transmission and distribution sectors do not exist. Therefore, in our models we refer only to capacity at the *generation* level. However, in all these countries electricity companies are state controlled, vertically integrated in most cases (and where legally separated[5] they still keep the same owner), and, given the substantial absence of market competition, they are the only operating companies in the country. This implies that regulatory interventions on tariffs (both retail and wholesale) have an indirect effect on generation capacity too (see also Pargal, 2003; Cubbin and Stern, 2005; Zhang et al., 2005). Table 14.2 lists all variables used in the empirical analysis and describes their sources. In Panel A of Table 14.3 we report the variables used to measure the characteristics of the energy sector, the economy, and the population. Panel B gives the political and institutional characteristics. Basic descriptive statistics for all variables are given in Table 14.3.

5. Vertical unbundling between generation and transmission/distribution segments has only been implemented in Algeria since 2002, Egypt since 1996, Iran since 2001, and Jordan since 2002.

TABLE 14.2 Variable Description

Variable	Description	Unit of measure	Sources
Panel A: Energy and economic indicators			
Capacity	Annual installed capacity of the electricity generation system	Millions of kW	EIA statistics
Capacity-Growth	Rate of growth of annual installed capacity	%	EIA statistics
ElCons	Annual consumption of electricity	Billions of kW	EIA statistics
ElCons-Growth	Rate of growth of annual consumption of electricity	%	EIA statistics
GDP	Gross domestic product	Current US$	World Bank data
NRA	Presence of an independent regulatory agency	Dummy	Regulator websites
AV	Added value by industry as percentage of GDP	%	The global economy
PopDensity	Percentage of urban population over total population	%	World Bank data
DistrLoss	Distribution losses	Billions of kWh	EIA statistics
Panel B: Institutional indicators			
Checks	Checks and balances	Ordinal scale: 1 (unlimited power) –18 (subordination of the executive)	Database of Political Institutions (DPI)
Polity	Type of regime	Ordinal scale: 0 (full autocracy) –20 (full democracy)	Polity IV
PolStab	Political stability	Ordinal scale: 0 (weak) – 5 (strong)	World Bank data (WGI)
Voice&Acc	Voice and accountability	Ordinal scale: 0 (weak) – 5 (strong)	World Bank data (WGI)
Rule	Rule of Law	Ordinal scale: 0 (weak) – 5 (strong)	World Bank data (WGI)
Control-Corr	Control of Corruption	Ordinal scale: 0 (weak) – 5 (strong)	World Bank data (WGI)
Tenure	Government tenure How long the country has been autocratic or democratic?	Years	World Bank data (WGI)

TABLE 14.3 Summary Statistics of the Main Macroeconomic Variables

Variable	Observations	Mean	Standard deviation	Min	Max
Panel A: Energy and economic indicators					
Capacity	264	7.790	10.879	0.09	65.31
CapacityGrowth	252	0.040	0.101	−0.571	0.806
ElCons	264	25.250	34.389	0.16	185.84
ElConsGrowth	252	0.055	0.088	−0.358	0.529
GDP	253	50,381.66	67,633.98	450	528,430
AV	219	0.348	0.145	0.14	0.84
NRA dummy	264	0.140	0.348	0	1
PopDensity	264	0.625	0.164	0.209	0.873
DistrLoss	264	0.157	0.087	0.034	0.518
Panel B: Institutional indicators					
Checks	248	1.290	0.614	1	4
Polity	237	5.992	3.839	1	17
PolStab	192	1.736	0.791	−0.68	3.31
Voice& Accountability	192	1.454	0.470	0.46	2.33
Rule	192	1.929	0.573	0.58	2.96
ControlCorr	192	1.926	0.465	0.92	3.05
Gov Tenure	262	14.901	11.780	1	46

4.3 Empirical Results: The Impact of NRAs on Investment in Electricity Capacity

The results of our test of Hypothesis 1 on the effect of the presence of an NRA on the growth of installed capacity are reported in Table 14.4.

Column 1 shows, comfortingly, that country-level controls are statistically significant, with investment in network capacity positively affected by growth in electricity demand (we find a plausible 2-year lag in network capital adjustment) and by the size of the economy. More importantly, we find that the dummy that accounts for the presence of the regulatory authority enters with a positive and significant coefficient, hence that investment in network capacity increases when an NRA is in place. The result holds when we control for urban population density (in Column 2 or 3) and for the percentage of distribution losses, even though these variables are statistically insignificant.

TABLE 14.4 The Impact of NRAs on Investment in Capacity

Dependent variable: *Capacity growth$_t$*	(1)	(2)	(3)	(4)
Electric consumption growth$_{t-1}$	−0.102 (0.123)	−0.51 (0.066)	−0.107 (0.126)	−0.107 (0.133)
Electric consumption growth$_{t-2}$	0.146** (0.049)	0.093 (0.065)	0.139** (0.052)	0.139** (0.054)
Log GDP$_t$	0.109* (0.054)	0.169* (0.079)	0.114** (0.048)	0.114 (0.065)
NRA$_t$	0.034* (0.019)	0.047* (0.022)	0.040* (0.018)	0.039* (0.020)
Value added industry$_{t-1}$		0.0003 (0.0004)	–	–
Urban population density$_{t-1}$			−0.464 (0.446)	−0.460 (0.539)
Distribution loss in %$_{t-1}$				0.005 (0.187)
Country dummies	Yes	Yes	Yes	Yes
Time dummies	Yes	Yes	Yes	Yes
R-squared	0.161	0.242	0.167	0.167
Observations	222	199	222	222
Number of countries	12	12	12	12

All regressors are defined in Table 14.2. Robust standard errors are given in parentheses.
***$p < 0.01$, **$p < 0.05$, *$p < 0.1$.

4.4 Institutional Determinants of NRAs in MENA Countries

This section conjectures that the establishment of NRAs could be the result of a maturation process within the institutions of the country. We test Hypothesis 2 by investigating whether the setup of the NRA can be associated with such country-specific characteristics. However, preliminary inspection of the correlation matrix in Table 14.5 reveals that most variables are highly correlated to the *NRA* dummy, particularly so as to the Rule of Law, the Control of Corruption, and the institutional checks and balances. However, the matrix also shows that they are all strongly crosscorrelated. Therefore, in Table 14.6 we estimate simple univariate logit regressions where the probability of having an NRA in place is related to each of these variables separately.

The results from the logit regressions show that only Rule of Law and Control of Corruption are positively related to the probability of establishing an NRA. Other variables, such as checks and balances, polity, and political stability report coefficients that are not far from significance, hence suggesting that

TABLE 14.5 Correlation Matrix

	NRA	Checks and balances	Polity	Rule of Law	Political stability	Voice and accountability	Control of Corruption.	Government tenure
NRA	1							
Checks and balances	0.290	1						
Polity	0.195	0.255	1					
Rule of Law	0.486	−0.254	−0.038	1				
Political stability	0.188	−0.355	−0.247	0.687	1			
Voice and accountability	0.139	0.065	0.508	0.535	0.146	1		
Control of Corruption	0.406	−0.143	0.095	0.839	0.610	0.630	1	
Government tenure	0.012	−0.030	−0.302	0.062	0.232	−0.236	−0.046	1

TABLE 14.6 The Institutional Determinants of NRA in MENA Countries

Dependent variable: NRA_t	(1)	(2)	(3)	(4)	(5)	(6)	(7)
Checks and balances$_{t-1}$	1.059 (0.730)	–	–	–	–	–	–
Polity$_{t-1}$		0.127 (0.092)	–	–	–	–	–
Political stability$_{t-1}$			0.687 (0.456)	–	–	–	–
Voice and accountability$_{t-1}$				0.757 (0.848)	–	–	–
Rule of Law$_{t-1}$					3.187** (1.356)	–	–
Control of Corruption$_{t-1}$						2.567** (1.099)	–
Government tenure$_{t-1}$							0.003 (0.041)
Constant	−3.139*** (1.179)	−2.455*** (0.776)	−2.646** (1.031)	−2.493* (1.391)	−8.320** (3.398)	−6.680*** (2.513)	−1.793** (0.832)
Observations	236	227	180	180	180	180	250
Number of countries	12	12	12	12	12	12	12

All regressors are defined in Table 14.2. Robust standard errors in parentheses.
***$p < 0.01$.
**$p < 0.05$.
*$p < 0.1$.

they are possibly relevant for the institutional design of regulatory agencies, but do not reach conventional levels.

4.5 NRA, Political Institutions, and Investment Capacity

In this section we expand our analysis of growth in generation capacity by examining the separate and joint effects of institutional endowment and the NRA. Table 14.6 shows that at least two variables, Rule of Law and Control of Corruption, influence the probability of setting up an NRA. So we now explore whether institutional variables might also affect *per se* capacity growth, absent the NRA. As previously discussed, recent literature has argued the positive role played by the quality of political and social institutions on economic outcomes and, particularly, infrastructure investment.

In particular, we include checks and balances, because the equilibrium between executive, legislative, and juridical power may favor investment by decreasing the likelihood of expropriation. In a similar vein, we add Rule of Law (i.e., the legal principle according to which law governs the nation to the exclusion of arbitrary systems), political stability since a predictable political situation may also favor investment, and Control of Corruption to measure the perception of both the extent to which public power is exercised to obtain private gains and the capture of public power by private interests. Finally, we include: voice and accountability since the transparency and responsibility of institutions are also viewed as positively influencing investment decisions; polity, an index related to the degree of democracy of the regime, which has been found to positively affect investment; and government tenure, which measures how long the country has been autocratic or democratic.

Table 14.7 reports the results of regressions in which the effect of individual institutional variables on growth in capacity are tested. The results show that none of these variables enters the investment equation, thus providing more ground to the role of a unique institution entirely devoted to regulation of the sector.

As a last piece of evidence, we test whether the positive impact of the NRA on capacity growth survives when we enter institutional variables. Table 14.8 gives estimated coefficients of a specification in which the usual investment model is extended to include not only the *NRA* dummy but also institutional variables separately.

In Table 14.8 we find that the main result (i.e., the presence of the NRA has a positive and significant effect on capacity growth of the electricity network) survives when we enter all institutional variables except for checks and balances and Government tenure (Columns 1 and 7). In the case of checks and balances, however, since the correlation with NRA is quite strong (Table 14.5), multicollinearity may explain why the coefficients of the two variables are imprecisely estimated. Conversely, NRA is not highly correlated with Government tenure, which enters significantly in Column 7 at the expense of the *NRA* dummy,

TABLE 14.7 Political Institutions and Investment in Capacity

Dependent variable: ***Capacity growth$_t$***	(1)	(2)	(3)	(4)	(5)	(6)	(7)
Electric consumption growth$_{t-1}$	−0.121 (0.151)	0.053 (0.047)	−0.149 (0.156)	−0.156 (0.177)	−0.151 (0.162)	−0.163 (0.177)	−0.178 (0.185)
Electric consumption growth$_{t-2}$	0.129** (0.057)	0.113 (0.077)	0.151** (0.057)	0.137* (0.073)	0.150** (0.060)	0.135* (0.071)	0.153** (0.055)
Log GDP$_t$	0.102 (0.060)	0.044* (0.020)	0.130 (0.076)	0.105* (0.058)	0.128 (0.075)	0.132 (0.079)	0.097* (0.051)
Urban population density$_{t-1}$	−0.752 (0.490)	0.008 (0.232)	−0.362 (0.854)	−0.449 (0.708)	−0.237 (0.789)	−0.444 (0.893)	−0.357 (0.522)
Distribution loss in %$_{t-1}$	−0.041 (0.265)	0.259** (0.106)	−0.020 (0.297)	−0.138 (0.414)	−0.008 (0.288)	−0.021 (0.300)	0.142 (0.106)
Checks and balances$_{t-1}$	0.040 (0.026)	–	–	–	–	–	–
Polity$_{t-1}$		0.003 (0.002)	–	–	–	–	–
Political stability$_{t-1}$			−0.011 (0.020)	–	–	–	–
Voice and accountability$_{t-1}$				0.095 (0.083)	–	–	–
Rule of Law$_{t-1}$					−0.032 (0.048)	–	–

(Continued)

TABLE 14.7 Political Institutions and Investment in Capacity *(cont.)*

Dependent variable: *Capacity growth$_t$*	(1)	(2)	(3)	(4)	(5)	(6)	(7)
Control of Corruption$_{t-1}$						0.014 (0.034)	–
Government tenure$_{t-1}$							0.001 (0.001)
Country dummies	Yes	Yes	Yes	Yes	Yes	Yes	Yes
Time dummies	Yes	Yes	Yes	Yes	Yes	Yes	Yes
R-squared	0.183	0.170	0.181	0.200	0.182	0.181	0.188
Observations	206	199	178	178	178	178	220
Number of countries	12	12	12	12	12	12	12

All regressors are defined in Table 14.2. Robust standard errors are given in parentheses. ***$p < 0.01$; **$p < 0.05$; *$p < 0.1$.

TABLE 14.8 NRA, Political Institutions, and Investment in Capacity

Dependent variable: ***Capacity growth$_t$***	(1)	(2)	(3)	(4)	(5)	(6)	(7)
Electric consumption growth$_{t-1}$	−0.119 (0.151)	0.062 (0.042)	−0.147 (0.158)	−0.158 (0.178)	−0.145 (0.162)	−0.164 (0.178)	−0.172 (0.183)
Electric consumption growth$_{t-2}$	0.133** (0.055)	0.129 (0.080)	0.166** (0.057)	0.148* (0.070)	0.172** (0.055)	0.147* (0.068)	0.159** (0.056)
Log GDP$_t$	0.114 (0.067)	0.051 (0.024)	0.156* (0.082)	0.131* (0.063)	0.154* (0.082)	0.158* (0.085)	0.108* (0.055)
Urban population density$_{t-1}$	−0.812 (0.529)	−0.052 (0.193)	−0.738 (0.819)	−0.855 (0.787)	−0.535 (0.696)	−0.778 (0.873)	−0.479 (0.514)
Distribution loss in %$_{t-1}$	−0.065 (0.260)	0.221* (0.103)	−0.114 (0.292)	−0.242 (0.418)	−0.105 (0.286)	−0.105 (0.293)	0.107 (0.099)
NRA$_{t-1}$	0.034 (0.025)	0.027* (0.014)	0.066* (0.035)	0.069* (0.036)	0.074* (0.039)	0.064* (0.031)	0.036† (0.021)
Checks and balances$_{t-1}$	0.037 (0.025)	–	–	–	–	–	–
Polity$_{t-1}$		0.003 (0.002)	–	–	–	–	–
Political stability$_{t-1}$			−0.014 (0.021)	–	–	–	–
Voice and accountability$_{t-1}$				0.101 (0.085)	–	–	–

(Continued)

TABLE 14.8 NRA, Political Institutions, and Investment in Capacity *(cont.)*

Dependent variable: *Capacity growth$_t$*	(1)	(2)	(3)	(4)	(5)	(6)	(7)
Rule of Law$_{t-1}$					(0.059 (0.057)	–	–
Control of Corruption$_{t-1}$						0.001 (0.029)	–
Government tenure$_{t-1}$							0.001* (0.0006)
Country dummies	Yes	Yes	Yes	Yes	Yes	Yes	Yes
Time dummies	Yes	Yes	Yes	Yes	Yes	Yes	Yes
R-squared	0.189	0.178	0.200	0.220	0.204	0.198	0.195
Observations	206	199	178	178	178	178	220
Number of countries	12	12	12	12	12	12	12

All regressors are defined in Table 14.2. Robust standard errors are given in parentheses. ***$p < 0.01$; **$p < 0.05$; *$p < 0.1$; †p-value = 11%.

suggesting that government tenure has a deeper influence than that of the NRA on the decision to augment generation capacity. In all other cases, however, the presence of the NRA seems to work like a catalyst for political and institutional forces by positively influencing investment decisions.

5 CONCLUSIONS

The presence of a regulatory authority in the energy sector that is *de jure*, if not *de facto*, detached from the executive power is widely recognized by both the theoretical literature and empirical evidence as pivotal to spurring investment. This is the reason the EU's authorities – concerned with achieving a unified regulatory framework across the Mediterranean Basin – have recently set up NRAs in the MENA region. This area is presently the object of particular interest worldwide on account of its fast-growing demand for energy, its ample endowment of natural resources, and its closeness to the European markets.

This chapter examined a sample of MENA countries and studied the rate of growth in their generation capacity over a time span of 22 years. Our aim was twofold. First, we wanted to verify the impact of the presence of an NRA on investments. Second, we investigated whether, and to what extent, the institutional landscape played a role in determining investment levels in electricity generation. As for the second objective, political institutions have an impact on economic decisions by affecting the degree of risk and instability of the environment. Factors such as checks and balances, type of regime, political stability, citizen participation and freedom of expression, Rule of Law, Control of Corruption, and government tenure may have an impact on investor decisions and interfere with the effectiveness of the regulatory system.

In our empirical model, we found that inception of a free-standing regulatory agency external to the ministry of concern seemed to have a positive effect on investments. Even when NRAs do not enjoy a high degree of independence from the executive, the establishment of a regulatory agency distinct from the ministry office may lend credibility to good regulatory practices. To support this conjecture, we found that the presence of an NRA is positively correlated with indicators measuring enforcement of the Rule of Law and of the Control of Corruption. This confirmed that establishment of a regulatory agency is often associated with the institutional endowment of the country. In this respect, it may be worth noticing that, not surprisingly, all institutional variables employed in the model exhibit a strong crosscorrelation.

The significant role of NRAs in promoting investment survived even when political variables were included in the investment model, although the quality of the institution itself did not appear to have a significant effect on capacity growth. This suggests that setting up a regulatory agency that is independent (at least formally if not substantially) of government control acts as a catalyst for a number of positive effects generated by the quality of political institutions.

REFERENCES

Andres, L., Foster, V., Guasch, J.L., 2006. The Impact of Privatization on the Performance of the Infrastructure Sector: the Case of Electricity Distribution in Latin American Countries. World Bank Policy Research Working Paper 3936.

Andres, L., Guasch, J.L., Diop, M., Lopez, S., 2007. 'Assessing the Governance of Electricity Regulatory Agencies in the Latin American and Caribbean Region: a Benchmarking Analysis.', World Bank Policy Research Working Paper Series, 4380.

Andres, L., Guasch, J.L., Azumendi, S.L., 2008. 'Regulatory Governance and Sector Performance: Methodology and Evaluation for Electricity Distribution in Latin America.' World Bank Policy Research Working Paper 4494.

Armstrong, M., Sappington, D.E.M., 2006. Regulation, competition and liberalization. J. Econ. Lit. XLIV, 325–366.

Bortolotti, B., Cambini, C., Rondi, L., 2013. Reluctant regulation. J. Comp. Econ. 41 (no.3), 804–828.

Cambini, C., Franzi, D., 2013. Independent regulatory agencies and rules harmonization for the electricity sector and renewables in the Mediterranean region. Energy Policy 60, 179–191.

Cambini, C., Rondi, L., 2011. Regulatory Independence, Investment and Political Interference: Evidence From the European Union. EUI Working Paper RSCAS 2011/42, Florence.

Cambini, C., Rondi, L., Spiegel, Y., 2012. Investment and the strategic role of capital structure in regulated industries: theory and evidence. In: Harrington, J., Katsoulacos, Y., Regibeau, P. (Eds.), Recent Advances in the Analysis of Competition Policy and Regulation. E. Elgar Publishing, Cheltenham, UK.

Correa, P., Pereira, C., Mueller, B., Melo, M., 2006. Regulatory Governance in Infrastructure Industries: Assessment and Measurement of Brazilian 193 Regulators: Public–Private Infrastructure Advisory Facility. World Bank, Washington, DC.

Cubbin, J., Stern, J., 2005. Regulatory effectiveness and the empirical impact of variations in regulatory governance: electricity industry capacity and efficiency in developing countries. World Bank Policy Research Working Paper No. 3535, Washington, DC.

Cubbin, J., Stern, J., 2006. The impact of regulatory governance and privatization on electricity industry generation capacity in developing economies. World Bank Econ. Rev. 20 (no. 1), 115–141.

EU Commission, 2010. Medring Study: Mediterranean Electricity Interconnections (2010 update).

Franzi, D., 2013. Make Energy From the Sun a Real Growth Perspective. LAP Lambert Academic Publishing, Saarbrucken, Germany.

Freedom House, 2011. Countries at the cross-roads.

Geradin, D., 2004. The development of European regulatory agencies: what the EU should learn from American experience. Columbia J. Eur. Law 11, 11–52.

Gilardi, F., 2005. The institutional foundations of regulatory capitalism: the diffusion of independent regulatory agencies in Western Europe. Ann. Am. Acad. Polit. Social Sci. 598 (no. 1), 84–101.

Gutiérrez, L.H., 2003. The effect of endogenous regulation on telecommunications expansion and efficiency in Latin America. J. Regul. Econ. 23 (no. 3), 257–286.

Joskow, P.L., 2005. Incentive Regulation in Theory and Practice: Electricity Distribution and Transmission Network. CEEPR, Cambridge, MA, pp. 1–95.

Joskow, P.L., 2007. Regulation of natural monopolies. Mitchell Polinsky, A., Shavell, S. (Eds.), Handbook of Law and Economics, vols. I & II, Elsevier Science Publishing, Amsterdam.

Keefer, P., Stasavage, D., 2003. The limits of delegation: veto players, central bank independence, and the credibility of monetary policy. Am. Polit. Sci. Rev. 97 (no. 3), 407–423.

Larsen, A., Holm Pedersen, L., Sørensen, E.M., Olsen, O.J., 2006. Independent regulatory authorities in European electricity markets. Energy Policy 34 (no. 17), 2858–2870.

Levi, B., Spiller, T., 1994. The institutional foundations of regulatory commitment: a comparative analysis of telecommunications regulation. J. Law Econ. Organ. 10 (no. 2), 201–246.

Levi, B., Spiller, T., 1996. Regulations, Institutions, and Commitment: Comparative Studies of Telecommunication. Cambridge University Press, Cambridge, UK.

Mehlum, H., Moene, K., Torvik, R., 2006. Institutions and the resource curse. Econ. J. 116 (no. 508), 1–20.

Newbery, D.M., 2001. Issues and Options for Restructuring Electricity Supply Industries. DAE Working Paper WP 0210.

North, D.C., 1990. Institutions, Institutional Change and Economic Performance. Cambridge University Press, Cambridge, UK.

North, D.C., Weingast, B.R., 1989. Constitutions and commitment: the evolution of institutions governing public choice in seventeenth-century England. J. Econ. Hist. 49 (no. 4), 803–832.

Pargal, S., 2003. Regulation and Private Sector Investment in Infrastructure: Evidence From Latin America. World Bank Policy Research Working Paper 3037.

Robinson, J.A., Torvik, R., Verdier, T., 2006. Political foundations of the resource curse. J. Dev. Econ. 79, 447–468.

Saal, D.S., 2002. Restructuring, Regulation, and the Liberalization of Privatized Utilities in the UK. Aston University, Birmingham.

Stigler, G., 1971. The theory of economic regulation. Bell J. Econ. Manag. Sci. 2 (no.1), 3–21.

Trillas, F., Montoya, M.A., 2011. Commitment and regulatory independence in practice in Latin American and Caribbean countries. Compet. Regul. Netw. Ind. 12 (no. 1), 27–57.

World Bank, 2013. Worldwide Global Indicators. World Bank, Washington, DC.

Zhang, Y.F., Parker, D., Kirkpatrick, C., 2005. Competition, regulation and privatisation of electricity generation in developing countries: does the sequencing of the reforms matter? Quart. Rev. Econ. Finance 45 (no. 2–3), 358–379.

Chapter 15

Financing Mediterranean Electricity Infrastructure: Challenges and Opportunities for an Interconnected Mediterranean Grid

Houda Ben Jannet Allal*, Matteo Urbani**
*General Directorate, OME; **Electricity Division, OME

1 INTRODUCTION: REGIONAL ENERGY CONTEXT AND OME VISION

Before addressing the question of financing electricity infrastructure in the Mediterranean, it seems appropriate to present an overview of the regional energy context in terms of energy policy and perspectives. Energy is not in itself a final good, but energy services are a key component of economic development, contributing to improving overall social welfare. In this respect, energy should be secure and sustainable, safe and affordable, efficient and competitive; it should take into account resource availability and environmental challenges. Thus, the question is how to cope with increasing energy needs, in particular electricity demand, both in the short and the long run. To that end, a sound analysis of the investment needed in the future seems to be crucial. This exercise requires a shared vision of energy system development in the region, within a cooperative approach.

According to the OME scenarios (OME, 2015), the demand in the Mediterranean region is growing rapidly as a result of population and economic growth. Most of the increase in Mediterranean energy demand is expected to be concentrated in the southern and eastern Mediterranean countries (SEMCs). Indeed, the OME Conservative Scenario[1] shows almost a doubling of energy demand

1. The Conservative Scenario prolongs current and past trends by taking them into account as part of undergoing projects and being cautious regarding announced measures and projects. Despite the current crisis, robust economic growth is expected in the south and east Mediterranean with an annual growth rate of 3.5% on average by 2040 (only 1.5% is forecast for the north). The Conservative Scenario does not include large-scale deployment of renewables and does not ▶

Regulation and Investments in Energy Markets. http://dx.doi.org/10.1016/B978-0-12-804436-0.00015-1

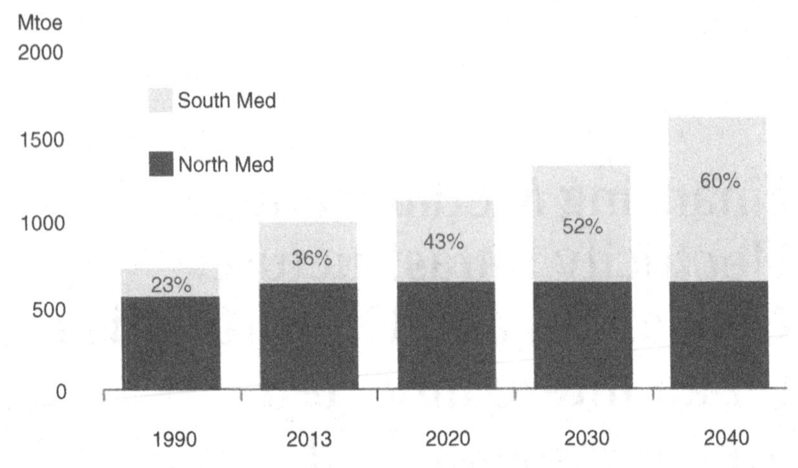

FIGURE 15.1 Energy demand by region.

from 2013 to 2030 (362–687 Mtoe), and something close to a tripling a decade later (up to 958 Mtoe in 2040). Turkey and Egypt alone are expected to reach 566 Mtoe in 2040, which is more than half of the total energy demand in the south. These two countries represent more than half of the total increase of energy demand in the entire region with some 367 Mtoe of additional energy consumption (Fig. 15.1).

As for electricity perspectives, demand is expected to remain almost stable in the north Mediterranean over the next two decades, while it would more than triple in the south and in the east during the same period (from 47 Mtoe in 2013 to about 146 Mtoe in 2040). In terms of electricity demand per capita, steady growth is expected in the south, even though this increase will remain far below levels in the north over the next two decades (Fig. 15.2).

From the supply side, energy dependency in the Mediterranean region is presently about 44%, and reaches 66% for fossil fuels, although regional production is still considerable. By 2040, dependency may increase despite recent discoveries and promising available resources, mainly because of sustained growth in demand in the region.

In all scenarios, both Conservative and Proactive, hydrocarbons will play a major role in the electricity generation mix at the regional level. According to the Conservative Scenario, 446 GW of natural gas-fired power plants are expected in 2040 (from 202 GW in 2013), of which 280 GW are in the south and east Mediterranean (SEM) (from 86 GW in 2013). Renewables will also play an important role and are expected to reach around 449 GW of installed capacity

◀ foresee specific and strong measures to enforce large-scale energy savings in the south and east Mediterranean. It assumes that all electricity needs will be met by currently used fuels and, to a lesser degree, by alternative fuels. As for nuclear, it assumes a later date of operation than that announced based on observed delays on these kinds of projects.

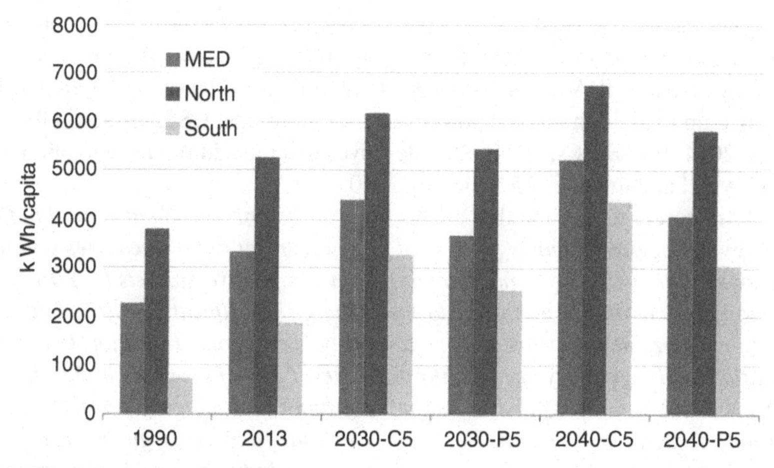

FIGURE 15.2 **Electricity demand per capita.**

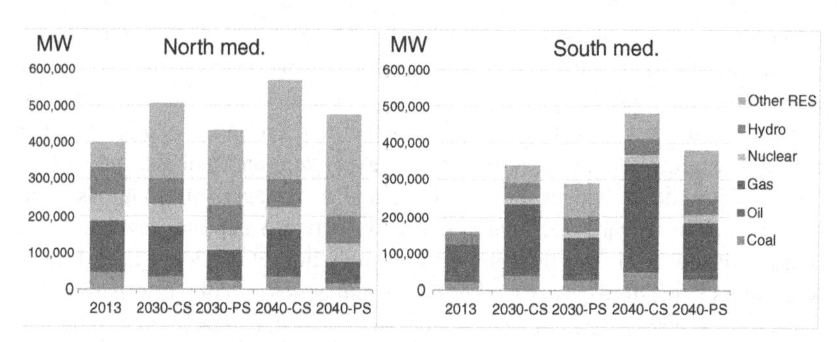

FIGURE 15.3 **Installed capacity by region.**

in 2040 (105 GW in the SEM), of which 331 GW are from nonhydro (61 GW in the SEM). Most global additions in terms of installed capacity are expected to take place in the south and east (313 GW, including nuclear), whereas 218 GW are expected in the north, where most additions will be from renewable energy sources (RES) (Fig. 15.3).

To achieve resilience at the regional level in this respect, taking into account current and future energy challenges, social concerns, and environmental risks, we consider the following five axes essential to making energy development more sustainable:

- energy efficiency and demand-side management;
- optimal use of all resources and technologies at the regional level;
- infrastructure investment;
- well-designed public policies;
- capacity building.

There is a clear need for long-term investment, both for development and for transformation of the Euro-Mediterranean energy sector. The International Energy Agency (IEA, 2014) estimates total amount of investment required in Europe up to 2035 in just the power sector at more than US$2 trillion (OECD/IEA, 2014). In SEMCs, OME estimates investment in additional generation capacity will amount to €715 billion by 2030.

In its recent report, the IEA argues that: "*decisions to commit capital to the energy sector are increasingly shaped by government policy measures and incentives, rather than by signals coming from competitive markets (...). Private sector participation is essential to meet energy investment needs in full, but mobilizing private investors and capital will require a concerted effort to reduce political and regulatory uncertainties (...). New types of investors in the energy sector (smaller market players and new entrants) are emerging, but the supply of long-term finance on suitable terms is still far from being guaranteed (...). The increase of small and distributed renewables reduces the share of utilities, and will rely more on debt financing.*"(OECD/IEA, 2014, p. 12).

All this fully applies to the Euro-Mediterranean region, where major efforts are needed to attract investment. As recognized by every international institution and actor alike, it requires: a stable political and economic system; transparent, consistent, and credible policies; an adapted institutional and legal framework; satisfactory market regulation based on competitive neutrality; and independent national regulatory authorities for the energy system and market competition. The latter includes budget independence and a management commissioning process. In this regard, regional cooperation between national regulatory authorities is essential, and the role of the Association of Mediterranean Energy Regulators (MEDREG) is crucial to promoting a clear, stable, transparent, and harmonized regulatory framework in the Mediterranean area.

In addition to this, reforming energy subsidy policies are also fundamental to creating a favorable and pertinent framework for investments. This has already started in several countries in the region and should proceed in a progressive and consistent manner. Such reform should be further complemented by significant and targeted social policies for access to modern energy.

As already highlighted, a shared view of the transition toward energy demand in light of technology innovation and deployment as well as understanding the new role of economic actors in a changing environment are of crucial importance. In such a situation, investment in infrastructure (particularly in electricity transmission and generation) should leverage all sources available – public and private, local and international – taking into account the specificity of every actor to tackle the main financing challenges.

The first of the remaining sections in this chapter (Section 15.2) presents and analyzes the challenges involved in financing infrastructure in the SEM, focusing on the role of public–private partnerships and multilateral financing. Section 15.3 deals with the important issue of setting a common regulatory and institutional framework. Some recommendations are proposed in order to establish a concrete pathway for progressive integration of electricity systems in the region. Section 15.4 concludes by identifying policy implications at both the institutional and regulatory level, which suggest proceeding by taking a gradual and more pragmatic bottom-up approach.

2 THE CHALLENGE OF FINANCING INFRASTRUCTURE IN SEMCS

The financing infrastructure in most SEMCs is strictly linked to ongoing transformations in the global electricity industry, the weakness of local capital markets, and the fragmentary reform process.

There is a need for a more balanced public–private approach and for innovative risk-sharing mechanisms to facilitate mobilization of the private sector. Financing infrastructure development in the electricity sector is not only a question of efficient allocation of resources, but also concerns the economic characteristics of infrastructure – high capital intensity, elements of natural monopoly, and location-specific investments – all of which affect private sector incentives to commit long-term capital.

Economic and financial expansion of the power sector in the SEM requires a larger share of private capital and governments to increase their focus on creating the necessary conditions to attract domestic and foreign capital. Creating appropriate economic signals is necessary for providing incentives for investments that not only can be financed, but also reward the ownership of assets.

As the IEA underlines, *"Even where states and state-owned companies take direct responsibility for energy investment, pressures on public funds and the need for new technology and expertise create room for greater private involvement."*(OECD/IEA, 2014, p. 12). There is a clear role here for institutional actors – multilateral and bilateral financing institutions, such as the European Investment Bank (EIB) and the European Bank for Reconstruction and Development (EBRD), and selected institutional investors such as sovereign wealth funds and pension funds – to act as intermediaries in the market and leverage additional funding notably for renewable energy projects and network investments, while reducing transaction costs and capital risks.

To foster such a process, regional cooperation would help to pool resources and mobilize private funding. With this end in mind governments should ensure there is no discrimination between actors regarding access to finance. In addition, it would be appropriate to strengthen domestic financial markets, increase the range of financial products available, and derisking policy schemes that reduce the interest rate on capital lending.

The bottlenecks to ensure a healthy flow of capital from international markets to SEM infrastructure projects subsist at the political, institutional, and regulatory level. There is a real need to unlock new sources of finance, via growth of bonds, securitization, and equity markets, and, potentially, by tapping into the large funds held by institutional investors. Multilateral financing can play a crucial role in providing risk mitigation instruments (including guarantees and political risk insurance) and promoting the development of local capital markets (OECD/IEA, 2014).

Foreign direct investment is important to finance new infrastructure, especially in markets where privatization reforms are implemented. Multilateral development banks, regional development banks, and import/export agencies have a variety of mechanisms and instruments available to facilitate foreign investor activity. However, governments will continue to have a key role to play in mitigating risks by creating a legal system capable of giving assured legal protection, planning for the establishment of financial sectors, and providing multilateral agreements on investments. The long lifetime of power infrastructure and long amortization periods inflate the risk of political change over the lifetime of a project (Cambini and Rubino, 2014).

At present, no single solution fits all countries. Public entities will remain major players in the financing, development, and delivery of infrastructure services in many SEMCs. Fundamental improvements in their creditworthiness will be essential to facilitate access to global and domestic capital markets, as well as to bring in private equity investments to a range of public–private partnerships. Corporate-level and sector-specific reforms will have to be pursued. At the corporate level, investment planning, financial reporting, and corporate governance should have to meet commercial standards, and public listing of companies may reduce the transaction costs of capital flows. This would allow them to tap capital markets more easily to raise money. With this end in mind, reforms in the regulatory environment will be essential to minimize regulatory risk. Furthermore, transparency and corporate standards may stimulate more efficient management.

Substantial investments in SEM infrastructure are unlikely to materialize unless there is a strong institutional framework that protects creditor rights and has effective covenants and reliable avenues of legal enforcement and remedy. Over the longer term, enhancing SEMC infrastructure access to international capital markets will also require developing an international mechanism to deal with crossborder investment regulation, competition rules, and consistency between national regulatory regimes. As technology increasingly interacts with economic pressures to regionalize infrastructure industries and open them up to international competition, the consistency and compatibility of national competition laws and policies will become more important for achieving gains. Where elements of competition and natural monopoly coexist and are complementary, the regulation of third-party access to essential facilities is essential.

The challenge involved in financing infrastructure requires a cooperative regional approach to foster investment and exploit energy complementarities between countries across the region. This can be done by progressive reform of market design and the regulatory framework in SEMCs, and by developing new interconnectors and supporting power exchanges, particularly from renewables, whose potential is huge.

The OME is trying to adapt such a regional integration process to the electricity sector by helping to identify the main issues to be tackled and proposing some recommendations expected to offer improvements in terms of policies and measures to implement. In OME and Medgrid (2013), a number of actions and recommendations have been identified that should deliver a more interconnected power system across the region.

3 TOWARD AN INTERCONNECTED MEDITERRANEAN GRID: SOME REGULATORY PERSPECTIVES

The idea of integrating electricity systems in the region is based on the assumption that regional integrated markets are more effective than fragmented national ones at delivering benefits in terms of a more secure, more stable, and more affordable power supply. In addition, the integration of power systems in the Mediterranean region will contribute to orient cooperation between Mediterranean systems toward open electricity markets characterized by transparency, reciprocity, and nondiscrimination in all countries.

The potential for regional electricity trade in the Mediterranean (and its "corridors") is particularly significant, given the great complementarity between supply and demand profiles across the region. However, despite this crossborder trade potential, the volume of electricity exchanged among Euro-Mediterranean countries is presently very low. This is evident in the south, where physical interconnections between North African countries exist. However, the utilization rate of these infrastructures is often very low. As for north–south exchanges, they are even lower and are limited to just the trans-Mediterranean interconnection between Spain and Morocco, through the Strait of Gibraltar.

Suboptimal exploitation of regional complementarities in terms of power exchanges is mainly due to lack of cooperation between countries, lack of transmission networks (both domestic and crossborder), lack of technical coordination in terms of system operations and measures (when the infrastructure exists but is not adequately exploited), and unsatisfactory regulatory harmonization. All these deficiencies represent a set of drawbacks that jeopardize the development of a more integrated regional electricity system by discouraging investment to the detriment of wider social welfare.

As a general rule, clear separation between politics, regulation, and system exploitation at the institutional level seems necessary to create a more attractive environment to foster investment in infrastructure. In countries where monopolies exist, there is often confusion between these roles. Introducing clear separation between *political authorities* (that remain responsible for defining the energy policy), *regulation* (preferably entrusted to a specialized authority independent from politics and participants of the electricity sector), and *exploitation* (entrusted to companies, either public or private, which take part in the national electricity market) is a fundamental precondition of the reform process.

As for market organization, in countries where national or regional monopolies exist, vertical splitting between different sectors (unbundling) may be necessary to allow the development of competition. In particular, effective dissociation of accounts and operations between "contestable" sectors and those that are regulated (as natural monopolies) is fundamental to guaranteeing effective regulation and adequate control by an independent authority, to eliminating general, indirect, and crossed subsidies, and to establishing tariffs that reflect costs and allow transparent financing of the sector.

The most important objectives of vertical unbundling refer to equitable access to the network for new entrants to ensure market transparency and promote in a nondiscriminatory way infrastructure investment. Moreover, unbundling is essential to avoid cross-subsidies between "contestable" and regulated activities, which affect fair competition negatively and prevent integrated companies from granting, to other branches of the same firm, preferential treatment compared to third competitors. These conditions are also necessary to implement other objectives of the energy policy (in particular, the development of renewable energies and energy efficiency measures).

Today, the process used to integrate Mediterranean electricity systems is based on three pillars: (1) *technical coordination* allows crossborder networks to be integrated; (2) *regulatory harmonization* of national regulations; and (3) *political cooperation* between Mediterranean countries. These three approaches are rooted in the national will of all Mediterranean countries wanting to autonomously implement reforms. This can be done by getting support from the Association of Mediterranean Transmission System Operators (Med-TSO) to promote closer and stronger coordination between Transmission System Operators (TSOs); MEDREG to promote better and institutionalized cooperation among national regulators; and the Union for the Mediterranean (UfM) to enhance political consensus.

Convergence between different institutional and regulatory frameworks across the Mediterranean is essential to move forward, but does not mean reproducing the European model. This is because grid maturity has yet to be achieved in SEMCs, the electricity industrial background is not the same, and energy policy targets are different, as are economic needs and energy priorities.

There is an urgent need for new infrastructure investment such that power systems can be regionally integrated. This requires the establishment of common rules for network operations and electricity trade, a *conditio sine qua non* for the achievement of this objective. In the aforementioned study (OME and Medgrid, 2013) some recommendations are presented for better identification and implementation of Mediterranean projects of common interest, as well as for establishment of a common regulatory and legal framework enabling mutually beneficial electricity exchanges. However, these recommendations do not aim to apply existing European regulations to all Mediterranean countries. In particular, a fully open and competitive electricity market in the Mediterranean region is both unrealistic in the short to medium term and not adapted to the strong load growth of SEMCs. A more pragmatic approach would allow regulations existing in each country/subregion to evolve progressively and allow north–south and south–south electricity exchanges to the benefit of all Mediterranean countries.

The following recommendations aim at establishing a concrete pathway for such an evolution and should be considered as interim steps toward market integration characterized by transparent, nondiscriminatory conditions and total reciprocity. The main actions to be undertaken are summarized in the remainder of this section.

1. *An unbundled TSO and an independent NRA should be established in each Mediterranean country.*
   In most SEMCs, electricity supply is still the responsibility of an integrated utility, covering generation, transmission, and distribution. To achieve the benefits of Mediterranean integration, new entrants are necessary (in particular, independent power producers) to diversify generation technologies and funding sources. A prerequisite to ensure that independent producers are treated in a nondiscriminatory way is to "unbundle" the TSO from the rest of the incumbent utility and to check the independence of the TSO at least in terms of management and accounting.
   An independent NRA should also be created in each Mediterranean country to ensure that all players in the electricity market are treated by the TSO in a nondiscriminatory way. In particular, the regulator will be responsible for checking that independent producers are subject to the same conditions as the generation department of the incumbent utility when it comes to access to the grid.
   The unbundled TSO and the independent regulator will have all the necessary capacities to work out a fair and transparent network tariff, ensuring nondiscriminatory grid access for all market players.
   The creation of an NRA in each SEMC should follow the principles set out by the MEDREG Institutional Ad Hoc Group, including clear legal status and real independence.
2. *Incumbent utilities in SEMCs should take advantage of participating in the European market (being interconnected and within a framework gives rise to fair sharing of costs and benefits for all interconnected countries). They should optimize their participation by establishing a mechanism that coordinates countries in the south.*
   A fully competitive market, such as the one that exists in the European Union, with open access for all generators and all customers to the grid can only be a long-term target for other Mediterranean countries. In the short term, such a competitive market is not considered a practical approach, especially when taking into account the need to cope with the rapid growth in electricity demand in SEMCs. Regulations in these countries should evolve step by step toward open access to the grid (with an unbundled TSO) for an increasing number of new players.

When an SEMC is first interconnected to an EU member state, the incumbent utility in the SEMC should become an active player in the European market. In this way the incumbent utility could submit demand and offer bids, based on the costs of its own generation units, thus allowing short-term north–south exchanges. As a result of proper sharing of congestion rents, benefits may appear on both sides. This step has already been completed in the Spain–Morocco interconnection with the Moroccan utility ONEE participating in the Iberian market, although balanced sharing of congestion rents on this interconnector remains pending.

Furthermore, incumbent utilities in other SEMCs that are not directly connected to Europe but only via a first SEMC should also be players in the European market (by transiting across the first SEMC). The possibility of setting up a coordination mechanism in the south by which incumbent utilities in several interconnected SEMCs participating in the same EU market could merge their bids prior to submission to the market should be explored. A coordination mechanism covering several SEMC utilities participating together in the same European market could allow them short-term exchanges, not only bilaterally between each of them and the EU market, but also directly between them. Pricing of south–south exchanges could follow the same rules as that of north–south exchanges (i.e., based on the same bid prices covering the extent of north–south interconnector capacity).

Further steps toward market integration will progressively lead to balanced sharing of congestion rents in north–south interconnectors, north–south reciprocity of conditions for access to the grid and to the market, and transparency of electricity prices in southern countries.

3. *Regulation in SEMCs should allow independent producers access to the grid and to north–south interconnectors under fair and transparent conditions.*

   In the short term, independent producers in SEMCs should not be limited to power purchase agreements (PPA) with the incumbent national monopoly. Access conditions should be published by the TSO for the internal grid of the SEMC and for interconnection with the EU. Access conditions – which must be fair, transparent, and nondiscriminatory – should cover:

   a. First connection to the national grid in the SEMC (cost, delay).
   b. Congestion management (guarantee of access).
   c. Pricing of access (e.g., through regulated tariffs, capacity auctioning, or other nondiscriminatory mechanisms).
   d. Settlement of imbalances (preferably with reference to actual system conditions, instead of a fixed tariff).

   More precisely, the access tariff for a north–south interconnector should separately show the price for accessing the interconnector itself, the price for accessing the adjacent main transmission grid, and the price for the inter-TSO compensation mechanism (if any). Once the costs and congestion rents have been shared in the case of a regulated interconnector, it is recommended that the national regulator in each country consider various possibilities for cost recovery by the TSO, namely:

a. Fee for actual use of the infrastructure.
b. Socialization of costs through a national grid tariff (this solution would remove barriers to crossborder trade).

More generally, the TSO in each SEMC should develop a grid code stipulating technical access rules in detail subject to approval by the NRA. Eventually, national grid codes should be harmonized, under supervision of MEDREG and Med-TSO, and coordinated by the European Network of Transmission System Operators for Electricity (ENTSO-E).

4. *A compensation mechanism should be developed for Mediterranean TSOs to cover the costs in their existing grids as a result of hosting crossborder transits (originating/ending in other countries).*
   Crossborder trade in energy between Mediterranean countries can induce transit flows into national grids that are neither the origin nor destination of the energy. The TSOs in transited grids need to be compensated for the costs incurred, including additional losses. For EU member states, an inter-TSO compensation (ITC) mechanism has been set up to resolve this issue, based on European Regulation (EU) No. 838/2010.
   It is recommended that a similar mechanism (separate from the European one) be set up to cover compensation for all Mediterranean countries once they are interconnected (with a possible contribution from Med-TSO and supervision by MEDREG). In the longer term, consideration could be given to merging the Mediterranean compensation mechanism with the European one.
5. *National legislations and regulations should specifically address north–south interconnector issues in all Mediterranean countries.*
   It must be emphasized that north–south interconnectors are not subject to EU legislation, but to national legislations in northern and southern countries. Depending on the country interconnected, the transmission infrastructure can be "regulated" (i.e., subject to national regulations concerning ownership and/or access rules) or (less often) "merchant" (i.e., partially or totally exempted from observing regulatory obligations in terms of third-party access, congestion rent, assets remuneration, etc.). However, in almost all cases national legislations concerning north–south interconnectors are rather vague. It is therefore recommended that it be made more precise.
   a. National legislation that only allows regulated north–south interconnectors could authorize TSOs to sell very long–term transmission rights to such interconnectors to satisfy the needs of producers (in coherence with the prescriptions of European grid codes).
   b. National legislation could also allow merchant north–south interconnectors that are exempted from regulations. The conditions for exemption could be defined in each country based on those already existing in the EU regulations.

Above all, it is recommended that SEMCs consider signing the Energy Charter Treaty, which provides more security to foreign investors.

6. *North–south interconnector projects should be considered relevant candidates in the selection process for "Projects of Common Interest" without encountering any form of discrimination.*
The Regulation on Guidelines for Trans-European Infrastructure (EU) No. 347/2013 introduces the possibility of selecting some transmission infrastructure as "Projects of Common Interest" (PCI), which benefit from several advantages: faster authorization procedures, access to special sources of funding, and possibly cost sharing between several beneficiary countries. These advantages would be highly beneficial to north–south interconnector projects; their promoters should apply to the PCI selection process and participate actively, together with the TSO of the third country concerned, in the corresponding regional group led by the EC. Two regional groups are relevant for north–south interconnector projects, corresponding to two priority corridors: "north–south electricity interconnections in western Europe" and "north–south electricity interconnections in central eastern and southeastern Europe."
Particular attention should be paid to showing that a north–south interconnector project involves at least two member states (a condition required for a project to be considered a PCI): it allows for delivery of energy from renewable sources in a third country not only to one member state, but to several EU countries. This can be checked, as required by the European regulation, by showing the capacity increase brought by the interconnector at an essential cross-section of a north–south priority corridor.
As required by the European Regulation on Energy Infrastructure, ENTSO-E has developed a methodology for cost–benefit analysis (CBA) of PCI candidates. It is recommended that some improvements to this methodology be investigated, to better show the contribution of north–south interconnector projects to electricity exchanges between countries in the priority corridor. Once an interconnector project is regulated, the results of CBA could be used to share interconnector project costs between the third country and the member states concerned according to their respective benefits.
7. *A first-ever Mediterranean-wide Ten-Year Network Development Plan should be drawn up by Mediterranean TSOs.*
To progress toward a more integrated grid between all Mediterranean countries, it is absolutely necessary to identify the need for new infrastructure, not only for new north–south interconnectors, but also for new south–south and north–north interconnectors. A common approach is required between all Mediterranean TSOs and their regulators.
It is recommended that Mediterranean TSOs, in close cooperation with ENTSO-E draw up a Mediterranean-wide Ten-Year Network Development Plan (TYNDP), with the participation of all interested stakeholders, possibly following the same principles as its EU equivalent. The Mediterranean

TYNDP, possibly updated every 3 years, should reflect the long-term needs of transmission grids around the Mediterranean Sea and identify bottlenecks, mainly in interconnectors (north–north, north–south, and south–south), but also possibly in internal grids.
Furthermore, CBA should be made possible for new infrastructure in the Mediterranean grid. Therefore, it is recommended to set up a model of the Mediterranean power system (north and south), including data over the years $n + 5$ to $n + 20$, for all Mediterranean countries concerned. This model could be based on that under construction for the European grid and then extended to all other Mediterranean countries. This task should be given to Med-TSO in close cooperation with ENTSO-E.

8. *A cost-sharing mechanism should be developed to cover the cost of grid reinforcement needed by Mediterranean countries to accommodate cross-border transits (based on a CBA approach).*
Grid reinforcement may be necessary in Mediterranean countries purely to accommodate larger numbers of crossborder transits. In such a case, corresponding costs should not be covered by national consumers, but by the final beneficiaries of crossborder transits in the countries of origin and/or destination. If grid reinforcement is located in an EU member state, it can be considered a PCI, and the cost issue can be resolved by Regulation on Energy Infrastructure (EU) No. 347/2013, since this introduces a regulatory procedure for cost sharing, based on Europe-wide CBA.
If grid reinforcement is located in an SEMC, or in an EU member state but has not been selected as a PCI,' the cost-sharing mechanism described in the regulation is not applicable and another similar mechanism has to be introduced. This mechanism must be agreed *ex ante* and remain stable in the long term, so as to minimize uncertainties for the investor in the new grid infrastructure as well as for those making use of it. Med-TSO and MEDREG could investigate the issue.
9. *For a new north–south interconnector project, governments, TSOs, and investors should commit themselves by getting together within an adequate contractual framework.*
The contractual framework for a new north–south interconnector project must be adapted to the diverse situations found in the north and in the south as well as to the diverse responsibilities of governments and TSOs on both sides. Therefore, governments, TSOs, and investors involved in a new north–south interconnector project (TSOs can be investors themselves in the case of a regulated interconnector) should adopt a contractual framework including the following agreements:
   a. An intergovernmental agreement to ensure international cooperation in the project and, in particular, the provision of land rights and permission for electricity transmission in the interconnector.
   b. For each part (north and south) of the project, an agreement between the local government (north and south, respectively) and investors. The

purpose of the agreement is to expand on issues dealt with by the intergovernmental agreement.

c. An interoperability agreement between TSOs and investors (submitted for regulatory approval), covering all issues related to system operation and market operation (including access conditions by third parties).

In the absence of an interoperability agreement for a regulated north–south interconnector, the two TSOs would have to compete against each other during the operational life of the infrastructure, each TSO trying to keep a larger part of the revenue from access by users of the interconnector, ultimately to the detriment of the latter. Furthermore, the interconnector project could be articulated in several "vehicles" (companies) should this be deemed necessary for a particular regulatory situation (in the north and in the south) or for a particular task (construction, asset ownership, and management operation).

10. *Long-term transmission rights should be available to mitigate the risks of congestion incurred by producers of electricity from renewable sources in SEMCs willing to export.*

The standard duration of long-term transmission rights given to interconnectors in the EU is presently 1 year (expected to be pluriannual with the new grid code "forward capacity allocation"). It is therefore recommended to offer at least the same duration to the whole Mediterranean region.

Furthermore, electricity producers from renewable sources in an SEMC may need transmission rights with longer durations, since their installations are characterized by high upfront investment costs. This makes them much more risky investments than conventional installations, which have lower investment costs. Therefore, investors in renewable installations need long-term visibility for their market arrangements to secure their investment. They should have the possibility to buy (very) long-term transmission rights (on timescales longer than 1 year, possibly 10 years or more).

Should very long-term transmission rights for all market players not be considered compatible with competition policy, the availability of such rights could be studied at least for those producers of renewable energy (in EU member states or in SEMCs) who do not receive support for their energy in their own country. Transmission rights over very long periods (10 years or more) would also be beneficial for investors in the interconnector. This is a way to secure long-term revenues, thus facilitating the financing of investment. Such an approach is already used for the gas infrastructure (open season) and should be considered for a new Mediterranean interconnector.

4 POLICY IMPLICATIONS AND CONCLUSIONS

Linking Mediterranean countries together through electricity highways is a long-standing idea introduced as far back as the 1930s as a promise of peace in the region.

Today, technological advances put this possibility within our reach: continuous current links can transport impressive amounts of energy under the sea and on land; converter stations for switching from AC to DC are able to interconnect different-sized networks; and sophisticated control means and new "smart" devices allow for responsive demand and supply. Technology is no longer an obstacle and should even continue to improve. However, many obstacles still exist; they are not technological but regulatory entailing consequences in terms of finance availability.

North Mediterranean EU countries needed over 20 years to define common rules of power exchange as part of an integrated electricity market. This work is still in progress, notably to develop a European "network code" and a long-term network development plan. In addition to this, there still exist insufficiently interconnected European countries such as Italy and Spain. However, large amounts of electricity are traded every day, every hour, between European countries in a transparent and nondiscriminatory way. These exchanges allow the provision of mutual aid between countries and contribute to the most efficient use of generation facilities in Europe, enabling lower power supply costs.

On the other hand, there is no general framework today for electricity trading in the SEM. There exist agreements between countries allowing for "technical" exchanges, specifically dedicated to mutual aid, and a few "commercial" exchanges entailing real benefits in terms of cost reduction and efficiency improvements.

Despite these two different systems and related regulatory frameworks, it is remarkable that north–south trade is regularly practiced in two precursor systems set up for it by Spain and Morocco through the submarine interconnection crossing the Strait of Gibraltar.

Both countries have succeeded in building a pragmatic and operational system that allows them both to share real trade benefits. In the present context, Spain could sell its excess power generation and Morocco can sometimes benefit from cheaper electricity than that produced by its own production facilities. According to our scenarios, energy transfer in the future could occur in both directions using the capacity of existing lines and even requiring creation of new interconnection links.

The recommendations we have made stem from propositions derived from our analysis. They are also based on a case study of the electricity trade between Morocco and Spain. These two countries, placed as they are in very different regulatory systems (an open and competitive market in Spain; a single-buyer, integrated business in Morocco), decided to proceed following a successful pragmatic approach which represents a good example to take into consideration.

In this respect, our recommendations aim to offer a pragmatic perspective without transferring the European model to the SEM. However, they highlight some fundamental issues that need to be dealt with in order to proceed toward a more interconnected regional grid in a more comprehensive and coordinated way.

A bottom-up approach has been considered a better way to proceed than the more traditional top-down method, which has proved inefficient and scarcely attractive to SEMCs. Regional institutions such as Med-TSO, MEDREG, and the UfM Secretariat are the expression of a common objective that domestic TSOs, national regulators, and governments share toward integration of electricity systems across the Mediterranean. This approach helps to identify common needs and draw regional perspectives in areas of action such as sector regulation and governance, regional TSO commitment, and legal frameworks. Such action might be framed in the newly established "Euro-Mediterranean Platform on Regional Electricity Market" (Box 15.1), with the aim of becoming a new

Box 15.1 UfM Energy Platforms

The final statement of the Italian Presidency of the EU and of the European Commission concluding the high-level conference on "Building a Euro-Mediterranean Energy Bridge: The Strategic Importance of Euromed Gas and Electricity Networks in the Context of Energy Security," held in Rome on November 19, 2014 proposed invigorating regional cooperation by establishing, in the context of the Union for the Mediterranean process, three thematic platforms dedicated to pursuing high-level dialog and to providing a permanent high-level forum for discussing energy policy objectives and measures, with a view to identify specific and concrete partnership actions and following up on their implementation:

- The "Euro-Mediterranean Platform on Gas" – to assist governments and industry operators to develop shared perspectives and propositions on natural gas issues in order to reinforce the security of gas supply and regional gas exchanges; to promote and exchange regulatory and commercial practices; and to allow all countries in the region to be part of an increasingly integrated gas market, with a view to creating a Mediterranean Gas Hub. In the conference it was proposed that such a platform be supported by the OME, building on its experience in promoting regional dialog and cooperation in the Mediterranean gas sector.
- The "Euro-Mediterranean Platform on Regional Electricity Market" – to assist governments and their partners in the gradual establishment of regional and subregional interconnected electricity markets; namely, by exploring the feasibility of appropriate options and arrangements between the EU and interested Mediterranean countries, as a means to provide a better framework for attracting investment. In the conference it was proposed that this platform be supported by the UfM Secretariat, with the assistance of regional associations of energy regulatory authorities (MEDREG), transmission systems operators (MEDTSO), and other key stakeholders.
- The "Euro-Mediterranean Platform on Renewable Energy and Energy Efficiency" – to assist governments and industry operators in the deployment of renewable energy and energy efficiency technologies and projects, the deployment of national energy efficiency action plans, and the creation of favorable conditions for private sector investments. In the conference it was proposed that this platform be supported by the UfM Secretariat, building on its experiences *inter alia* in relation to the Mediterranean Solar Plan.

catalyst for development of a more integrated regional electricity system on the basis of an inclusive, pragmatic, and bottom-up approach (MEDREG, 2014).

It is here that sector-based organizations, industrial associations, financing institutions, and scientific cooperation projects, at national and supranational level, will play a central role in undertaking actions and implementing measures. This will also contribute to enhancing capacity building and transfer of know-how to SEMCs, which are fundamental aspects for a coherent and harmonized development process. Associations and organizations with a regional remit are particularly functional for catalyzing the fundamental interests of private stakeholders, and represent an important link between the industry, the financing milieu, and the scientific community operating in the Mediterranean area. Furthermore, they work in cooperation with institutions such as the UfM Secretariat, MEDREG, and Med-TSO, which is essential to addressing issues and providing specific analysis on different topics.

Financing the energy infrastructure in the Mediterranean is a complex and articulated undertaking, where financial aspects are interrelated not only with economic ones but also with political, institutional, technical, and regulatory features, in a context that is geopolitically unstable and uncertain, implying more risks for investors. To build a more interconnected Mediterranean grid and proceed toward gradual market integration, it is imperative that nondiscriminatory conditions be applied, with transparency and total reciprocity between countries as the basis of a common regional will that needs to be translated into reality.

REFERENCES

Cambini, C., Rubino, A., 2014. Regional Energy Initiatives: MedReg and the Energy Community. Routledge, London.

MEDREG, 2014. Regulation and Investments: Solutions for the Mediterranean Region. MEDREG Paper No. 2.

OECD/IEA, 2014. World Energy Investment Outlook.

OME, 2015. Mediterranean Energy Perspectives. Preliminary results, Forthcoming.

OME and Medgrid, 2013. Towards an Interconnected Mediterranean Grid: Institutional Framework and Regulatory Perspectives.

LEGISLATION

Directive 2003/54/EC, June 26, 2003 repealing Directive 96/92/EC.

Directive 2009/28/EC, April 23, 2009 amending and subsequently repealing Directives 2001/77/EC and 2003/30/EC.

Directive 2009/72/EC, July 13, 2009 repealing Directive 2003/54/EC.

European Regulation (EU) No. 714/2009, of July 13, 2009 repealing Regulation (EC) No. 1228/2003.

European Regulation (EU) No. 838/2010, September 2010.

European Regulation (EU) No. 347/2013, April 17, 2013.

Chapter 16

New Regional and International Developments to Boost the Euro-Mediterranean Energy Sector

Ernesto Bonafé
Energy Charter Secretariat, Brussels, Belgium

1 INTRODUCTION

The European energy market is evolving toward a Euro-Mediterranean energy market. Middle East and North Africa (MENA) is important for the EU in terms of diversifying security of gas and oil supplies, being responsible for about 25% of today's EU imports of gas and 6% of crude oil. As for the future, the discoveries in the east Mediterranean region have considerably increased the potential to supply gas to the EU. Closer cooperation in the region is justified by complementary assets on both sides of the Mediterranean, the global geopolitics of oil and gas, the development of renewable and carbon-free energy sources, and the prospect of an energy sector that boosts the economy, develops a regional industry, and creates jobs.

The EU transition to a competitive, low-carbon economy is based on the ambitious objective to reduce its greenhouse gases emissions by 80% by 2050 (Communication, 2011). The MENA region is well placed to reap the benefits of a low-carbon transition thanks to its rich opportunities for exploiting renewable energy sources (RES) and energy efficiency. This also requires unlocking the potential for widening the EU-Mediterranean technological partnership in the area of RES.

An integrated electricity market all around the Mediterranean would be of benefit for all countries. It would allow complementarities between national systems experiencing different load profiles, fuel mix, and backup capacity for renewable electricity production. Moreover, a south Mediterranean market tied to the EU market would improve the overall stability and security of the electricity supply. A regional electricity market is also more attractive for private capital than segmented national markets as the huge investments needed in generation and networks cannot be conceived without private capital.

Regulation and Investments in Energy Markets. http://dx.doi.org/10.1016/B978-0-12-804436-0.00016-3

Electricity demand in the south Mediterranean is expected to double by 2030. Power capacity in 2009 amounted to 122 GW, while the planned new generation capacity by 2030 would represent between 160 GW and 200 GW. Installed renewable capacity would go from 2 GW in 2009 to some 40 GW in 2030 or 82 GW in a proactive green scenario (OME, 2012). Likewise, the growth in energy demand, which could triple in the south and east Mediterranean countries (SEMCs) by 2030, calls for a significant increase in installed electricity production capacity. Whatever the energy policies in place, increasing energy production capacities in the region would require investments of between US$310 and US$350 billion by 2030 (Moncef et al., 2013).

The mobilization of huge amounts of local and foreign investment requires a stable and predictable political, legal, and regulatory framework. International energy cooperation of this nature must be based on a sound regulatory framework in order to increase stability and transparency in the investment environment.

In 2011, the European Commission (EC) referred to a EU–Southern Mediterranean Energy Community starting with the Maghreb countries and possibly expanding progressively to the Mashreq (Joint European Commission, 2011). However, the situation in Mediterranean countries differs from the one that led to the Energy Community Treaty, which was conceived to reunify an energy system, which had disintegrated during the conflicts of the 1990s. In contrast to energy integration in the western Balkans, regulatory convergence with Mediterranean countries may have to follow a different policy and regulatory approach.

EU–Mediterranean integration is to be based on a differentiated and gradual approach to be implemented within the framework of a renewed European Neighbourhood Policy (ENP). In the energy sector, regional cooperation is needed to ensure a secure, affordable, and sustainable energy supply as a key factor for underpinning stability and shared prosperity in the Mediterranean area. This cooperation is to be invigorated by three thematic platforms for high-level dialog in the sectors of natural gas, the electricity market, and the promotion of RES and energy efficiency. At the same time, Mediterranean countries have the opportunity to embrace universal market-based principles and rules by adopting the new International Energy Charter and the Energy Charter Treaty.

This chapter starts with a general review of structural reforms in MENA countries, which is followed by a study on new Euro-Mediterranean platforms. The new ENP is then examined, and this section is followed by an overview of the Energy Charter. Finally, some conclusions and policy implications are provided.

2 ENERGY LEGAL REFORMS IN MENA COUNTRIES

All Mediterranean countries are engaged in defining and implementing national energy policies and legislation. Among the common objectives across the region, there are the promotion of RES and the development of electricity and gas regional markets allowing all countries to improve energy security, sustainability, and affordability. National market reforms in terms of industrial restructuring and incipient regulation of interconnections show the basic ground upon which to build, as a next step, a regional energy sector.

In Algeria, from 1947 until 2002, the electricity generation, transmission, and distribution was monopolized by Sonelgaz, the state-owned power company. In 2002, new legal reforms of the power market revoked the Sonelgaz monopoly and partially privatized the company despite the government remaining the main shareholder of Sonelgaz (Waqar, 2013). The reform also created the national regulatory authority, CREG, to supervise the power market and to ensure nondiscriminatory access to the networks. Moreover, Algeria has electrical interconnections with Morocco and Tunisia. According to Samborsky (2013), rules on interconnections are provided under Law 02-01 of 2002 on electricity and gas distribution. International electricity transactions have to be confirmed by CREG, which can refuse export activities if they have strongly negative impacts on the Algeria's national electricity supply (i.e., if demand cannot be satisfied). Power plants that have been constructed exclusively for export of electricity are exempt from this reservation.

In Egypt, the power market was controlled by the state-owned Egyptian Electricity Authority (EEA). In July 2000, seeking liberalization in the power sector, EEA was converted to a holding company called the Egyptian Electricity Holding Company (EEHC). Since July 2001, a series of restructuring steps took place for the affiliated companies, starting by unbundling generation, transmission, and distribution activities. Today EEHC has 16 affiliated companies: 6 generation, 9 distribution, and 1 transmission. Furthermore, Egypt has electrical interconnections with Jordan, Libya, and Syria (via Libya). However, access to interconnections by private operators needs further regulatory development. The main text governing the power sector is still a draft of 2008 that has not yet been approved by parliament. The objective of Egyptian market reform is to establish a fully competitive electricity market, where electricity generation, transmission, and distribution activities are fully unbundled (Bardolet, 2014a). The proposed market will adopt bilateral contracts with a balancing and settlement system. Under this reform, eligible customers will have the right to conclude bilateral contracts with present and future generation companies. Law 102/1986 established the New and Renewable Energy Authority (NREA), and the Presidential Decree No. 326/1997 established the Electric Utility and Consumer Protection Regulatory Agency (EgyptERA), which was reorganized by Presidential Decree No. 339/2000.

In Israel, the state-owned Israel Electric Corporation generates, transmits, distributes, and supplies most of the electricity used in the national market according to licenses granted under the Electricity Sector Law of 5756-1996 (Ministry of National Infrastructures, 2014). The law regulates production, system management, transmission, distribution, and supply or trade in electricity. The Public Services Authority and the Minister of Energy and Infrastructure are in charge of the national electricity policy, as well as connecting the electricity networks to those of neighboring countries. Licenses granted by the Authority come into force after approval by the minister. Details of the transmission license application process are in Electricity Market Regulations 5758-1997. Moreover, Israel is part of the EuroAsia interconnector, which is planning an interconnection between Greece, Cyprus, and Israel (EuroAsia Interconnector, 2015).

In Jordan, the Jordan Electricity Authority (JEA) was established in 1976. Twenty years later, in 1996, JEA was converted into the National Electric Power Company (NEPCO). In 1999, another restructuring transformed NEPCO into three companies: the National Electric Power Company (NEPCO) responsible for power transmission, which operates as a single buyer; the Central Electricity Generation Company (CEGCO), responsible for power generation, partly privatized, generating electricity along with Samra Electric Power distribution and independent power producers; and the privatized Electricity Distribution Company (EDCO), responsible for power distribution. In addition, Jordan has electrical interconnections with Egypt and Syria. The General Electricity Law 64-2003 gives the Ministry of Energy and Mineral Resources the power to cooperate with other countries for the purpose of electrical interconnection and trade in electric power. International transmission coordination is performed by NEPCO.

In Lebanon, the 2010 Policy Paper for the Electricity Sector (Bassil, 2010) pointed out that the legal framework for privatization, liberalization, and unbundling of the sector, Law 462-10, had not yet been applied. Instead, the law implemented by Decrees 16878/1964 and 4517/1972 giving *Electricité du Liban* (EDL) exclusive authority in the generation, transmission, and distribution areas was still being applied. However, Article 5 of Law 462-10 establishes that the transmission of electrical energy remains the property of the transmission company and it is possible by a decree of the council of ministers to ratify contracts for the management, operation, and development of transmission activities to the private sector, including any privatized or any company owned by the private sector.

In Morocco, the *Office National de l'Electricité et de l'Eau* (ONEE) has been in charge of the monopoly of electricity transport since 1963. ONEE is the grid operator and as such the grid expansion and reinforcement is under its sole responsibility. Morocco is synchronously connected with Algeria and Spain. Moreover, power transmission is organized and conducted by ONEE under a transmission contract. ONEE is the owner of 50% of the Spain–Morocco interconnection, and it holds the monopoly on the Moroccan side (Bardolet, 2014b). Law 13-09, on renewable energy, indicates that power produced from renewable sources is for national and international markets. In that sense, supply is guaranteed through access to high, medium, and lower tension national networks, under agreed terms and conditions between the operator and the manager of the national electricity grid transportation. Moreover, to achieve its ambitious renewable energy targets, the government created two dedicated agencies: the Moroccan Agency for Solar Energy (MASEN), in charge of implementing the Moroccan Solar Plan, and the National Agency for the Development of Renewable Energy and Energy Efficiency (ADEREE). Morocco is preparing a new national independent regulator authority, in charge of defining tariffs and conditions for transportation and interconnection access.

In Palestine, Law 13 of 2009 authorizes the creation of a national electric transmission company in charge of exporting and importing electricity from and to Palestine through connecting the grid, after signing the relevant agreements with the Palestinian Energy and Natural Resources Authority (PENRA); all relevant agreements are submitted for the approval of the Minister's Cabinet (Palestinian Electricity Regulatory Council, 2011).

In Syria, the Public Establishment for Electricity Generation and Transmission is responsible for electricity exchanges with neighboring countries. Before the conflict, the electrical systems were operated in parallel with Syria, Jordan, Egypt, and Libya synchronously as one electrical system, and electrical power exchange contracts were concluded with those countries. The interconnection between Syria and Turkey on a 400 kV level is ready; however, it is not yet operational.

The power market in Tunisia is controlled and operated by the state-owned company *Société Tunisienne de l'Electricité et du Gaz* (STEG). Law 96-27 withdrew STEG's monopoly for power generation to allow private investment in the generation sector. STEG is responsible for transmission and distribution, along with control of existing power generation plants. Moreover, Tunisia is connected to Europe through Morocco (Bardolet, 2014c). Tunisia has interconnections with Algeria and Libya, and is planning another with Italy. New network infrastructure is based on an authorization procedure handled by STEG, the different relevant ministries, and the Prime Minister.

The foregoing overview of regulatory milestones in MENA countries, from unbundling to private generation and interconnections to neighboring countries, reveals the general common ground of the power sector. This common vision is being reinforced by national strategies to promote RES. Technical solutions to cope with intermittency and ensure reliable system operation will put the region on track for further cooperation among neighboring countries.

3 THE NEW EURO-MEDITERRANEAN ENERGY PLATFORMS

The UfM was created in 2008 to revamp the Barcelona Process of 1995 and to give a new political impulse to economic and social cooperation between the EU and the Mediterranean. All 28 EU member states and 15 Mediterranean countries are part of the UfM.[1] Energy is one of the priority sectors. The UfM

---

1. From the north shore: Albania, Bosnia and Herzegovina, Monaco, Montenegro, and Turkey. From the south shore: Algeria, Egypt, Morocco, Tunisia, Israel, Jordan, Lebanon, Mauritania, Palestine, and Syria.

was launched at the same time as the Mediterranean Solar Plan (MSP) and announcement of the political target to develop 20 GW of new installed renewable capacity in the south Mediterranean by 2020. Since then a number of regional institutions have been active in contributing to develop RES and electricity exchanges between the EU and Mediterranean countries.[2] In fact, the UfM and the MSP have favored the establishment of political, economic, regulatory, and industrial regional institutions in the energy sector (Moncef et al., 2013).

Despite important developments since the establishment of the UfM, some difficulties and challenges persist in developing a Euro-Mediterranean energy sector, as was evident at the Energy Ministerial Meeting in December 2013 which failed to endorse the Master Plan for the MSP that the UfM Secretariat had developed in collaboration with all stakeholders. One of the problems undermining an effective regional strategy is the lack of common understanding and acceptance regarding the consumption patterns and generating capacities on the south and north shores of the Mediterranean. Stakeholders and countries alike might be pursuing specific goals that respond to different industrial interests and national solutions that undermine the benefits of a sensible regional approach.

Two specialized bodies gathering technical expertise have a direct impact on market conditions: namely, the Association of Mediterranean Energy Regulators (MEDREG) (OME and Medgrid, 2013) and the Association of the Mediterranean Transmission Operators for Electricity (Med-TSO). Aware of their specific and complementary mission, the two associations signed a cooperation protocol in Algiers in September 2013 (MEDREG, 2013). Their envisaged mandate is to play, in the Euro-Mediterranean region, a similar role that the Cooperation of European Energy Regulators (CEER)[3] and the European Network of Transmission System Operators (ENTSO) effectively play in the EU internal energy market.

MEDREG aims to improve the transparency, attractiveness, and investment climate by forging a regulatory framework in the fields of electricity and gas. It has demonstrated the need for a EU–MENA backbone grid using high-voltage direct current (HVDC) power transmission to accommodate increasing electricity generation from renewable sources, it designed a scenario to verify the effectiveness of cooperation mechanisms under Article 9 of Directive 2009/28/EC, and it prepared two studies on external dependence and security of supply and the improvement of national data transparency in the gas sector (Mediterranean

2. These institutions include the League of Arab States (LAS), the Regional Center for Renewable Energy and Energy Efficiency (RCREEE), the *Association Méditerranéenne des Agences Nationales de Maîtrise de l'Énergie* (MEDENER), the International Renewable Energy Agency (IRENA), the Parliamentary Assembly of the Mediterranean (PAM), the *Observatoire Méditerranéen de l'Énergie* (OME), the Arab Union of Electricity (AUE), the Euro-Mediterranean Electricity Cooperation (MEDELEC), the *Comité Maghrébin de l'Electricité* (COMELEC), Dii (DESERTEC), and Medgrid and RES4MED.

3. The EU Third Energy Legislative Package of 2009 created the Agency for the Cooperation of Energy Regulators (ACER), which repealed the European Regulatory Group of Electricity and Gas (ERGEG) that was composed of the CEER and the European Commission.

Energy Regulators, 2012). At the beginning of 2015, MEDREG held a consultation on the document " Interconnection Infrastructure in the Mediterranean: A Challenging Environment for Investments" (MEDREG, 2015). It focused on existing and projected interconnection infrastructure for electricity and gas in the Mediterranean Basin with a view to pointing out the main barriers impeding use of efficient existing infrastructures and impeding the financing of new projects.

Med-TSO was established in April 2012 as the Association of the Mediterranean Transmission Networks for Electricity. It should benefit from the work undertaken by the European Network of Transmission System Operators for Electricity (ENTSO-E) and, where necessary, adapt to the specific needs of the south Mediterranean. Med-TSO tasks include developing a Mediterranean transport and dispatching system, analyzing and proposing common technical rules for the interoperability of interconnected systems, providing technical assistance, promoting research and development, facilitating the transfer of know-how, exchanging relevant information between transmission system operators, enhancing transparency, promoting public acceptability by addressing environmental concerns regarding transmission infrastructure, consulting stakeholders, and sharing experiences and solutions regarding grid operation.

In November 2014, during the high-level conference "Building a Euro-Mediterranean Energy Bridge: The Strategic Importance of Euromed Gas and Electricity Networks in the Context of Energy Security," a Memorandum of Understanding was signed between the Directorate General for Energy of the European Commission, MEDREG, and Med-TSO with the aim of establishing a Euro-Mediterranean platform on regional electricity markets (European Commission, 2014). The high-level conference launched three Mediterranean energy platforms.

- *Platform on regional electricity market*. This will center on removing technical, regulatory, and infrastructure barriers to the free trade of electricity across international borders. This will boost energy security by diversifying supply and increasing competition between power generators. It is proposed that this platform be supported by the UfM Secretariat, with the assistance of MEDREG, MED-TSO, and other key stakeholders.
- *Platform on renewable energy sources and efficiency*. It will analyze how to promote regulatory frameworks and markets to enable investments in renewables and energy-efficient practices. The platform will also exchange best practices in areas including measuring energy efficiency and developing national energy efficiency plans: namely, via relevant policies and subsidies enforcing codes for buildings and defining minimum energy performance standards for appliances. It is proposed that this platform be supported by the UfM Secretariat and involve stakeholders such as the Regional Center for Renewable Energy and Energy Efficiency (RCREEE) and the *Association Méditerranéenne des Agences Nationales de Maîtrise de l'Énergie* (MEDENER).

- *Mediterranean gas hub*. It will discuss how to develop gas production in MENA countries for their domestic markets and for export to the EU. The platform will also debate regulatory, financing, and infrastructure issues, such as new pipelines and liquefied natural gas facilities in order to create a Mediterranean gas hub. The platform should promote the environmentally sustainable exploration and exploitation of hydrocarbon resources in line with common offshore safety and security rules. It is proposed that this platform be supported by the OME.

Other stakeholders have already joined the debate, as Eurelectric's proposal to secure energy investments in an "8-step Action Plan": (1) make regulatory agencies fit for purpose to boost investor confidence in generation and infrastructure projects; (2) enable cost-reflective energy prices and the phasing out of domestic electricity subsidies; (3) formulate sound and transparent energy policies; (4) facilitate technical and political coordination and cooperation on grid investments to urgently develop and enhance new and existing interconnections; (5) open market structures, in particular for renewable energy generation projects to allow free entry, new players, and competition among providers; (6) design and implement the right financing mechanisms to simulate exploring the renewable energy potential and to profit from technology transfers; (7) improve education and technological transfer; and (8) enhance EU–MENA energy cooperation and extend the energy community concept toward the south (Zvolikevich, 2015).

4 TOWARD A NEW EUROPEAN NEIGHBOURHOOD POLICY

The ENP was designed in 2003 (European Commission, 2003) to develop progressive integration between the EU and its neighboring countries by implementing political, economic, and institutional reforms, and committing to common values. The plan was to achieve closer economic integration and the prospect of increased access to the EU internal market.

At that time the objective was to avoid the emergence of new dividing borders between the enlarged EU and its neighbors on the basis of common values: democracy, the rule of law, respect for human rights, and social cohesion. The ENP is currently proposed to 10 southern Mediterranean neighbors: Algeria, Egypt, Israel, Jordan, Lebanon, Libya, Morocco, Palestine, Syria, and Tunisia.[4] The ENP was reviewed in 2011 to give a response to the events in the Arab world by strengthening reforms toward democracy and the rule of law.

The Treaty on European Union, Article 8(1), provides the legal basis for a stronger Europe when it comes to foreign policy. It states that "the Union shall develop a special relationship with neighbouring countries, aiming to establish an area of prosperity and good neighbourliness, founded on the values of the Union and characterised by close and peaceful relations based on cooperation."

4. Six Eastern neighboring countries: Armenia, Azerbaijan, Belarus, Georgia, Moldova, and Ukraine.

Over the past 10 years, the Mediterranean has become politically less stable: a civil war in Syria, a conflict in Libya, difficult changes in Egypt, and hostilities in the Middle East. The ENP has failed due to a lack of adequate responses to those major challenges and, on the other hand, to the changing aspirations of Mediterranean countries in different policy sectors. Moreover, the reform agenda has stalled because of competing interests and as not all partners are equally interested in the same model of partnership with the EU. The EU itself has experienced a major economic crisis that has had an impact on its neighbors.

The European Commission's new President Jean-Claude Juncker instructed that the ENP be reviewed within the first year of its mandate. It has appeared necessary to undertake a review of the principles on which the policy is based as well as its scope and how instruments should be used. To frame the debate, a consultation paper was adopted on March 4, 2015 setting out key questions for discussion with Mediterranean partners and stakeholders. A Commission Communication will follow in the autumn 2015 setting out proposals for the future direction of the ENP.

According to the ENP "more for more" principle, partners that are embarking on more ambitious democratic reforms are offered increased market access (e.g., Deep and Comprehensive Free Trade Agreements, DCFTA), people crossborder mobility partnerships, and further financial support. In its consultation document, the European Commission acknowledges that the "more for more" approach underlines the EU's commitment to its core values. However, it has not always been successful in providing incentives for further reforms and, more generally, it may not contribute to an atmosphere of equal partnership.

Therefore, the European Commission has decided to explore how a new policy can reflect better the interests and aspirations of the EU and its Mediterranean partners. The Commission's consultation is based on four priorities.

- A new ENP should face the challenges of differentiation, which means being aware and respecting increasing divergence in the aspirations of Mediterranean partner countries. While for some countries the Arab Spring in 2011 has led to positive political developments, others are undergoing complex transitions and instability arising from armed conflicts.
- Cooperation between EU and ENP partners is to be more focused. Action Plans set out very broad cooperation programs. As a result the agenda of the EU and its partners is not truly shared. Areas of focus must be ensuring the rule of law, protecting human rights, and deepening democratic principles.
- A differentiated and focused approach to a new ENP is to be supported with a more flexible toolbox; currently, the ENP is based on association agreements, action plans, reporting requirements, bilateral dialogs, and financial support.
- There is a need to improve communication of the objectives and results of the ENP, which should also lead to improving the ownership of the ENP by partner countries.

As for sectoral policies, there is a shared interest in increasing energy security and efficiency, as well as energy safety. As for governance challenges, the areas of focus are ensuring the rule of law, protecting human rights, and promoting democracy. By enhancing legal certainty, they also address issues that are important for domestic and foreign investors, such as fighting corruption and fraud and strengthening public finance management.

In 2014 the EU signed a Memorandum of Understanding with Jordan, Lebanon, and Morocco for the period 2014–2017, with an emphasis on promotion of the rule of law, employment and private sector development, and renewable energy and energy efficiency enhancement.[5]

In the energy sector, efforts have focused on energy security, market reform, and regional integration. The priorities have been to develop infrastructure, improve energy efficiency, and use more RES. In June 2013, the EU and Algeria launched a political dialog in the field of energy (EC-MEMO, 2013). In May 2015, the dialog was relaunched to boost investments in Algerian gas, thus improving Europe's energy security. Similarly, the EU has agreed to share its expertise in support of Algeria's renewable energy and efficiency goals.

The new European Neighbourhood Instrument (ENI) has set aside a budget of €15 billion for the period 2014–2020. It represents the bulk of funding to the 16 ENP partner countries. An incentive-based approach provides for flexibility in modulating financial assistance taking account of progress of individual countries toward democracy and respect of human rights. A midterm review is scheduled for 2017 in order to adjust the allocation and implementation of funding from the ENI to rapidly changing developments in the region.

In May 2015, the European Commission signed a €43 million financing agreement for one of the world's largest solar energy projects in Morocco. It is a project backed by the European Investment Bank and it is part of the Ouarzazate Solar Complex that will see the installation of a 100–150 MW concentrated solar power (CSP) plant. The overall objective is to reach 560 MW of solar power capacity by 2016. This aims to scale up the production of renewable energy, and thus will contribute to the country's energy security, diversify energy sources, cut carbon emissions, and create jobs.

On the other hand, there is the question of extending the ENP beyond the current framework of 16 neighboring countries. Many challenges, including those related to the energy sector, cannot be adequately addressed without taking into account and cooperating with the neighbors of EU neighbors. A new approach to a broader geographical area should allow more flexibility to work and interact with new neighboring countries that have been excluded from the ENP. Furthermore, a new ENP requires greater involvement of member states in addition to EU action.

5. Memorandum of Understanding was signed with Jordan on 13.10.2014, Lebanon on 14.10.2014, and Morocco on 5.11.2014.

5 THE ENERGY CHARTER TREATY AND THE NEW INTERNATIONAL ENERGY CHARTER

In 1991, after the fall of the Berlin Wall, the European Energy Charter was signed as a political foundation for east–west energy cooperation. It was based on common objectives and principles such as the development of open and efficient energy markets, the stimulus of private investments, nondiscrimination among participants, respect for state sovereignty over natural resources, and recognition of the importance of environmentally and energy-efficient policies.

The European Energy Charter was a political document that also emphasized the need for the establishment of an appropriate international legal framework for energy relations, leading to the signing, in 1994, of the Energy Charter Treaty (ECT), which entered into force in 1998 (Cambini and Rubino, 2014). Moreover, the new International Energy Charter of 2015 is a political declaration that updates the 1991 European Energy Charter and reflects today's global energy challenges.

5.1 The Energy Charter Treaty

The ECT's original purpose was to facilitate energy collaboration in transition economies. Nevertheless, the ECT offers a legal framework and policy forum that is open to all countries along the energy chain: producers, consumers, and transit states, as well as industrialized, transition, and developing economies. The ECT process expanded beyond its west–east traditional borders and gained a global dimension, including countries from Asia and Central Asia. To date, the ECT has 53 members, including the EU as a whole, whereas another 25 countries have observer status.

In an increasingly global and interconnected energy sector, the ECT sets an international level playing field in the energy sector. Its principles provide for a long-term global model of energy cooperation within the framework of a market economy based on mutual assistance and the principle of nondiscrimination. By joining the ECT, countries across the world will set the legal and policy standards for international energy relations on a global basis. In 2011, the European Commission stated, as part of the EU External Energy Policy, that "the Energy Charter Treaty should seek to extend membership towards North Africa and the Far East" (European Commission, 2003).

Pursuant to the Road Map for the Modernization of the Energy Charter Process, in 2012 the Energy Charter Conference adopted a policy of consolidation, expansion, and outreach aiming to promote the accession of new countries across the world and of course of MENA countries. The policy objective is to enhance the rule of law in the areas of investment, trade, and transit. Accession to the ECT would send out clear and strong signals indicating political willingness to attract foreign investments. With regard to the current ECT membership status in the south Mediterranean, Jordan in 2007 and Morocco in 2012 have signed the European Energy Charter, which is the first step toward accession to the ECT.

In MENA, energy consumption has risen by 5.2% each year since 2000, and energy demand in the region is expected to continue to rise above the world's average, by around 3% each year from 2010 to 2030 – electricity demand is forecast to rise by 6% each year over the same period. This is the result of rapid economic expansion, the energy-intensive nature of the region's extractive industries, and a rapidly growing population. As analyzed and recommended by the OECD, well-targeted government regulatory and financial mechanisms can support private investment in renewable energy (OECD, 2013).

However, simply providing development assistance will not be enough to bring about the necessary changes. Private investment flows remain insufficient compared with the energy needs of the region. More finance is needed to develop infrastructure, and the role of the private sector in mobilizing investments is crucial. The ECT is well equipped to attract private investments. For an open and private sector, confidence, predictability, and a stable energy framework are fundamental. This is the important added value that the ECT can bring to south Mediterranean countries.

The perceived degree of political risks in the host country considerably affects the decision of foreign companies regarding whether to make an investment or not and what level of return it would imply. The lower the perceived risk, the more capital is likely to be invested and the more potential revenue the host country will attract. By reducing political risks, the ECT seeks to improve investor confidence and contribute to an increase in international investment flows.

The Treaty provides a legal framework for energy relations as well as dispute settlement mechanisms. Beyond the resolution of particular disputes, ECT provisions have the effect of strengthening the rule of law and transparency, thereby contributing to an improvement in the general investment and business climate. At the same time, the ECT explicitly recognizes national sovereign rights over energy resources.

The ECT aims to promote long-term cooperation in the energy field, based on complementarities and mutual benefits and it imposes the obligation to promote access to international markets on commercial terms, and generally to develop an open and competitive market in the energy sector. The ECT contains a set of rights and obligations of "hard law" nature in the areas of investment protection, trade, and transit, which are enforceable in legally binding arbitration mechanisms. The ECT is supplemented with a Protocol on Energy Efficiency and Related Environmental Aspects that entered into force simultaneously with the ECT in 1998.

The ECT also contains "soft law" provisions related to competition, technology transfer, and access to capital. However, confirmation of the principle of national sovereignty over energy resources means that the ECT does not prescribe the structure of the domestic energy sector, the ownership of energy companies, or oblige member countries to open up their energy sector to foreign investors.

The investment chapter is a cornerstone of the ECT. Its provisions aim to promote and protect foreign investments. To this end, the treaty grants a number of rights to foreign investors with regard to their investment in the host country. Foreign investors are protected against the most important political risks, such as discrimination, expropriation and nationalization, breach of individual investment contracts, damages due to war and similar events, and unjustified restrictions on the transfer of funds. Investor rights are further protected by the dispute settlement provisions of the ECT, covering both interstate arbitration and investor–state dispute settlement.

A distinctive feature of the ECT is that it provides a set of rules that covers the entire energy chain, including not only investments in gas production and electricity generation but also the terms under which energy can be traded and transported across various national jurisdictions to international markets. ECT provisions on trade and transit are based on those of the World Trade Organization (WTO). WTO trade rules are thus extended to ECT contracting parties that are not yet members of the WTO.

The ECT addresses in more detail than the WTO the strategic issue of energy transit. In accordance with the principle of freedom of transit, ECT contracting parties undertake to facilitate the transit of energy through their territory. This includes the prohibition of discrimination in terms of origin, destination, and ownership of energy as well as equal treatment of transit and transport facilities in terms of governmental measures. A general obligation to secure established flows of energy implies that transit countries must not interrupt or reduce existing transit flows, even if they have disputes with any other country concerning this transit, nor must they place obstacles in the way of new capacity being established.

ECT market-based rules do not mean that a particular model of energy market structure is imposed on national governments at the international level. The ECT fully respects the sovereign right of each of its signatory states to determine the system of property ownership of its national energy resources. Moreover, each state continues to hold the right to decide *inter alia* the geographical areas to be made available for exploration and development of its energy resources and to determine the rate at which such energy resources may be depleted or exploited. The explicit recognition of national sovereign rights over energy resources by the ECT means that, at all times, the balance between protection of investors and sovereignty of host states needs to be maintained.

Furthermore, if countries are to integrate large amounts of electricity from RES into the network, they need to share not only interconnections but also common principles and rules. The creation of a Euro-Mediterranean energy market will result from initiatives such as the UfM (political), MEDREG (regulatory), Med-TSO (operational), and OME (industrial).

However, today those institutions provide policy guidance, coordinated action, guidelines, studies, and recommendations for best practices for an investment framework and crossborder energy exchanges. The regulatory outcomes are to be approved and implemented on a voluntary basis. A consensus-based approach to energy regulation is suitable to reach compromises among countries with different energy sectors. However, those incipient practices need to be endorsed by a series of legally binding provisions. The ECT contains a minimum set of rules that may contribute to enhance the rule of law in the Mediterranean energy market according to international legal standards. The importance of the ECT and the rule of law in the energy sector is emphasized by a new political declaration, the International Energy Charter (IEC).

5.2 The International Energy Charter

The IEC is a political declaration aimed at strengthening energy cooperation between its signatories and does not bear any legally binding obligation. The IEC is an updated version of the European Energy Charter. As a result of the increasingly global and interconnected energy sector, the IEC is intended to expand beyond traditional borders to reach out to new countries, regions, and international organizations with the aim of enhancing international cooperation to address today's global energy challenges. The IEC was adopted at a Ministerial Conference in The Hague, Netherlands, on May 20, 2015 by 75 countries including Israel, Jordan, Lebanon, Morocco, and Palestine from all continents.

The objectives of the IEC are development of a sustainable energy sector, improvement of energy security, and maximizing economic efficiency, while recognizing the sovereignty of countries over their energy resources and their rights to regulatory energy transmission and transportation. To achieve these objectives, the signatories are determined to create a climate favorable to the operation of enterprises and to the flow of investments and technologies.

IEC signatories underline the need to define energy policies and have regular exchanges of views on action taken, taking full advantage of the experience of existing international organizations and institutions. They also decide to foster private initiatives to make full use of the potential of enterprises, institutions, and all available financial sources as well as technical cooperation.

The IEC objectives are pursued by strengthening regional energy markets and enhancing the efficient functioning of the global energy market by joint or coordinated action in the following fields.

1. *Access to and development of energy sources.* Signatories will formulate relevant rules for the development of energy resources in economic and environmentally sound conditions, and will coordinate their actions in this area. They will ensure they are publicly available and transparent in consistence with domestic legislation and international obligations. Operators should not be discriminated in terms of ownership of resources, internal operation of companies, and taxation.
2. Access to national, regional, and international markets for energy products, taking into account the need to facilitate the operation of market forces and promote competition.

3. Liberalization of trade in energy by removing barriers to trade in energy products, equipment, and services in a manner consistent with the provision of WTO rules. IEC signatories should develop market-oriented energy prices. They recognize the importance of transit for the liberalization of trade in energy products, which should take place under economic and environmentally sound conditions. They should also cooperate in the development of international energy transmission networks and their interconnections, including crossborder oil, gas, and electricity.
4. *Promotion and protection of investments.* IEC signatories will remove all barriers to investment and provide a national level for a stable and transparent legal framework for foreign investments. They aim to stress the importance of bilateral and/or multilateral agreements on promotion and protection of investments as well as of full access to adequate dispute settlement mechanisms, including national mechanisms and international arbitration according to national laws and relevant international agreements. Moreover, signatories recognize the right to repatriate profits relating to an investment and recognize the importance of avoiding double taxation.
5. *Safety principles and guidelines and the protection of health and the environment.* Signatories will cooperate in the implementation, development, and mutual recognition of safety principles and guidelines.
6. Research, technological development, technology transfer, innovation, and dissemination in the fields of energy production, conversion, transport, distribution, and the efficient and clean use of energy.
7. *Energy efficiency, environmental protection, and sustainable and clean energy.* This includes consistency between relevant energy policies and environmental agreements, ensuring market-oriented price formation reflecting environmental costs and benefits, the exchange of know-how regarding RES, efficient use of energy, and framework conditions for profitable investment in energy efficiency.
8. *Access to sustainable energy.* The signatories underline the importance of access to sustainable, modern, affordable, and cleaner energy, in particular in developing countries, which may contribute to energy poverty alleviation. To this end, they will strengthen their cooperation, and support initiatives and partnerships at international levels.
9. *Education and training.* Signatories recognize industry's role in promoting vocational education and training in the energy field, and decide to cooperate in such activities.
10. Diversification of energy sources and supply routes in order to enhance energy security.

By signing the IEC, countries confirm and enhance established principles of energy cooperation, participate in a governmental platform to address contemporary energy challenges, and contribute to global energy governance by facilitating new accessions to the ECT, without containing any obligations in this respect.

6 CONCLUSIONS

The opportunities for economic, industrial, and social development in the Mediterranean countries will materialize in the realization of a Euro-Mediterranean regional energy market. The objective of secure, sustainable, and competitive energy in the region needs continuous progress and adaptations in terms of new policies, legislation, and institutions. Structural market reforms undertaken at the national level provide the common ground and starting point to build a regional vision shared by all countries. The recent establishment of new Euro-Mediterranean energy platforms that are to play an essential role in the framework of a renewed ENP, which will consider energy as a priority sectoral policy.

From the good governance perspective, the ENP aims to promote the rule of law in a way that accommodates the challenges of differentiation, a focused approach, a more flexible toolbox, and better communication. Moreover, the new policy does not only target its neighbors but it will also consider the neighbors of neighbors. Precisely, the first-ever attempt to build the rule of law at the international level has been the ECT, of which the EU and its eastern neighbors are members. Therefore, the ECT can enhance a level playing field in the Euro-Mediterranean region and make it a more friendly environment for investors.

Committing to legally binding international rules and standards of good governance may be a demanding exercise for some countries, including Mediterranean ones. Therefore, as a preparatory step, a new IEC was adopted on May 20, 2015 in The Hague as a political (nonbinding) declaration for intergovernmental cooperation. The IEC is an opportunity for all countries to express clear political willingness to share market principles and therefore contribute to a level playing field. It has been adopted by 75 countries, including the EU and its member states and, in the south Mediterranean, Israel, Jordan, Lebanon, Morocco, and Palestine. The IEC provides an umbrella for intergovernmental energy cooperation between the EU, MENA, Africa, and countries from all continents in an increasingly global energy sector.

DISCLAIMER

The views expressed by the author may not reflect those of the Energy Charter.

REFERENCES

Association of Mediterranean, Energy Regulators, (MEDREG), 2013. ACTION PLAN, 2014-2015-2016, pp. 4 http://www.medreg-regulators.org/Portals/45/forum/PRESS/pressmaterial/MEDREG_ActionPlan2014-2016.pdf (accessed 05.10.2015.).

Association of Mediterranean, Energy Regulators (MEDREG), 2015. Interconnection Infrastructures in the Mediterranean: A Challenging Environment for Investments, MEDREG Public Consultation, Evaluation of Responses. http://www.medreg-regulators.org/Portals/45/questionnaire/1_MEDREG_Public_Consultation_Interconnection_Infrastructures_Evaluation_of_Responses_public.pdf (accessed 05.10.2015.).

Bardolet, M., 2014a. Regulatory Overview – Egypt. http://www.desertenergy.org/fileadmin/Daten/RegulatoryOverview/Regulatory%20Overview%20%20Egypt.pdf (accessed 08.07.2015.).

Bardolet, M., 2014b. Regulatory Overview – Morocco. http://www.desertenergy.org/fileadmin/Daten/RegulatoryOverview/Regulatory%20Overview%20%20Morocco.pdf (accessed 09.07.2015.).

Bardolet, M., 2014c. Regulatory Overview – Tunisia. http://www.desertenergy.org/fileadmin/Daten/RegulatoryOverview/Regulatory%20Overview%20%20Tunisia2.pdf (accessed 09.07.2015.).

Bassil, G., 2010. Policy Paper for the Electricity Sector, Ministry of Energy and Water. http://climatechange.moe.gov.lb/viewfile.aspx?id=121 (accessed 05.10.2015.).

Cambini, C., Rubino, A. (Eds.), 2014. Regional Energy Initiatives: MedReg and the Energy Community. Routledge, New York.

Commission of European Communities, 2003. Wider Europe – Neighbourhood: A New Framework for Relations with our Eastern and Southern Neighbours. 104 final. Available from: http://eeas.europa.eu/enp/pdf/pdf/com03_104_en.pdf (accessed 08.07.2015.).

Electricity Sector Law, 5756-1996, updated on May 1, 2007, http://energy.gov.il/English/LegislationLibraryE1/ElectricitySectorLaw.doc (accessed 05.10.2015.).

EuroAsia Interconnector, 2015. Progress Meeting for Implementation of the EuroAsia Interconnector. http://www.euroasia-interconnector.com/News-Progress_Meeting_for_Implementation_of_the_EuroAsia_Interconnector,1042 (accessed 30.10.2015.).

European Commission, 2011. High Representative of the Union, Joint Communication To The European Council, The European Parliament, The Council, The European Economic And Social Committee And The Committee Of The Regions A Partnership For Democracy And Shared Prosperity With The Southern Mediterranean, Reference /* COM/2011/0200 final */, http://eur-lex.europa.eu/legal-content/EN/NOT/?uri=CELEX:52011DC0200 (accessed: 05.10.2015).

European Commission – MEMO/13/665, 2013. http://ec.europa.eu/energy/sites/ener/files/documents/dialogue.pdf (accessed 08.07.2015.).

European Commission, 2014. Joint Staff Working Document. Implementation of the European Neighbourhood Policy Partnership for Democracy and Shared Prosperity with the Southern Mediterranean Partners Report Accompanying the Document Joint Communication to The European Parliament. The Council, The European Economic and Social Committee and The Committee of the Regions, Implementation of the European Neighbourhood Policy in 2014, Reference /* SWD/2015/0075 final */ http://eur-lex.europa.eu/legal-content/ES/NOT/?uri=CELEX:52015SC0075 (accessed 05.10.2015.).

Mediterranean Energy Regulators, 2012. MEDREG Annual Report. (accessed 08.07.2015.).

Moncef, S., Jaques, M., Morgan, M., 2013. Towards a Euro-Mediterranean Energy Community: Moving from Import–Export to a New Regional Energy Model. IPEMED, Institut de Prospective Économique du Monde Méditerranéen, France, http://www.ipemed.coop/adminIpemed/media/fich_article/1373880614_Towards%20a%20Euro-Mediterranean%20Energy%20Community.pdf (accessed 09.07.2015.).

Organisation for Economic Corporation Development (OECD), 2013. Renewable Energies in the Middle East and North Africa: Policies to Support Private Investment. Paris. http://www.oecd.org/mena/investment/renewable-energies-mena-2013.htm (accessed 09.07.2015.).

Observatoire Méditerranéen de l'Energie (OME) and Medgrid, 2013. Towards an Interconnected Mediterranean Grid. Institutional Framework and Regulatory Perspectives.

Palestinian Electricity Regulatory Council, 2011. Electricity Sector Annual Report. http://www.perc.ps/ar/files/publications/annuareport2011en.pdf (accessed 09.07.2015.).

Samborsky, B., et al., 2013. Arab Future Energy Index™ (AFEX), Renewable Energy, Regional Center for Renewable Energy and Energy Efficiency (RCREEE), pp. 26. http://www.rcreee.org/sites/default/files/reportsstudies_afex_re_report_2012_en.pdf (accessed 05.10.2015.).

Walde, T., 1996. The Energy Charter Treaty. An East-West Gateway for Investment and Trade. Kluwer Law International, The Netherlands.

Waqar, A., 2013. Analysis of the Mediterranean states, power markets in connection to DESERTEC. Int. J. Res. Dev. Eng. 1 (3), Available from: http://ijrde.com/attachments/article/86/2012010304.pdf.

Zvolikevich, A., 2015. Securing energy investments in the MENA region. A EURELECTRIC 8-step Action Plan. http://www.eurelectric.org/media/172033/eurelectric_recommendations_mena__region_final-2015-2640-0002-01-e.pdf (accessed 08.07.2015.).

Chapter 17

Investing in Infrastructures: What Financial Markets Want

Alberto Ponti
Pan-European Utilities Team, Societé Générale – Cross Assets Research, London

> *...given the low interest rate environment and volatile stock markets of recent years, institutional investors are increasingly looking for new source of long-term, inflation-protected returns. Investments in real, productive assets, such as infrastructure could potentially provide the type of income which these investors require, supporting investments and driving growth.*
>
> *OECD (2013).*

In this chapter, we want to answer a simple question: What is it that makes investors want to put their money in new (existing) infrastructures? However simple this question might be, the answer needs to be more articulated. Hence, we have structured this chapter into four sections.

In Section 1, we give a historical perspective of the utilities sector in Europe over the past 15 years to see what lessons can be learnt, which mistakes have been made, and why investors think the way they do.

In Section 2, we frame the institutional investors' market in terms of size and internal organization so as to gauge their appetite for risk as well as the limits and benefits for regulators and countries wanting to attract them.

In Section 3, we set out the four key conditions (or indeed minimum requirements) that any regulatory regime should have in order to release the maximum benefits to consumers and the lowest burden to tax payers.

In Section 4, based on the European experience, we draw some conclusions for the Middle East and North African (MENA) countries. In essence, given the need for strong investments in the region, we suggest long-term contracts should be employed as opposed to regulated asset-based (RAB) tariffs that seem to better suit the more mature markets (for infrastructures). However, transparency and consistency of regulation should always be achieved as a prerequisite to minimizing investors' required rate of return and hence the cost for society as a whole.

Regulation and Investments in Energy Markets. http://dx.doi.org/10.1016/B978-0-12-804436-0.00017-5

1 THE UTILITIES SECTOR – A HISTORICAL PERSPECTIVE

The "real" liberalization of energy markets in Europe was finally achieved toward the beginning of the last decade thanks to implementation of a number of EU Directives.[1]

Before liberalization, utilities in Europe were regulated via an integrated tariff covering the costs of generation, transmission, distribution, and supply. Apart from the different services rendered unbundled, most utilities were also not listed. Utilities, as an investment class, did not really exist.

Because of the presence of the integrated tariff (which was typically adjusted by inflation on an yearly basis), fluctuation of commodity prices (chiefly oil and coal) was not a major concern. Moreover, as electricity and gas were managed in monopolistic regimes, competition was also not an issue for service providers.

But liberalization changed all this. Unbundling of networks was introduced to allow so-called third-party access (TPA). Furthermore, the gradual introduction of spot and then forward electronic exchanges for the main commodity utility use (i.e., gas and electricity) helped to introduce competition and more transparency regarding prices for consumers.

Meanwhile, utilities were being privatized such that investors finally had access to this category of stocks. However, since the end of 1999, when liberalization was introduced and privatizations launched, the definition of utilities itself changed.

Typically institutional and retail investors associate with the definition of "utilities" the concept of "safety," "stability," and "predictability." Some define utilities as a "safe harbor" type of investment.

1.1 Are Utilities Safe Investments?

No, despite some considering them to be. They are not for, at least, two reasons:

1. *Exposure to commodities*. As already pointed out, the liberalization of power and gas markets led to electronic exchanges for these two commodities and their prices were finally visible. During the period January 2005 to July 2008 all commodity markets entered into what was then called a "commodity bubble" of unprecedented magnitude. For example, at the climax of the bubble, German 1-year forward power prices were €90/MWh whereas they are a tad above €30/MWh as we write this chapter; coal prices were then at c. $220/t whereas now they are less than $60/t.

   The problem is that (Table 17.1) only c. 30% of the Utilities Index is represented by "Regulated" utilities (i.e., networks that have no exposure to commodities), whereas the vast majority of the sector (e.g., generators) do have partial or total exposure to commodities (Ponti, 2015a).

1. For example, Directives 96/92/EC and 98/30/EC.

TABLE 17.1 Weights of STXE Index Constituents (in %, as at end of March 2015)

	Regulated	Nonregulated
National Grid	14.6	
Iberdrola		10.0
GDF Suez		9.2
EON		8.5
ENEL		8.3
Scottish & Southern		6.4
Centrica		5.3
RWE		3.6
SNAM	2.9	
Fortum		2.8
EDP		2.8
United Utilities	2.7	
Red Electrica	2.5	
Veolia Environnement		2.4
Gas Natural		2.3
Severn Trent	2.1	
EDF		2.0
Endesa		1.8
Terna	1.8	
Enagas	1.8	
Suez Environnement		1.7
Pennon	1.4	
CEZ		1.1
ENEL Green Power		0.9
Drax		0.7
Rubis		0.6
	29.6	70.3

Source: SG Cross Asset Research/Equity, Bloomberg.

2. *Governmental intervention.* The second reason for utilities not being defensive, or *as* defensive as they typically are thought to be, relates to governmental intervention. The market adage goes that when governments are in trouble utilities are the first to be hit with new taxes (or freeze/cut in tariffs) because their assets *cannot be shipped elsewhere.* Typically taxing

TABLE 17.2 Levered "Raw" Beta (Daily Return March 2013 to March 2015)

	Regulated	Nonregulated
National Grid	0.59	
Iberdrola		0.84
GDF Suez		0.87
EON		0.90
ENEL		0.86
Scottish & Southern		0.46
Centrica		0.52
RWE		0.93
SNAM	0.49	
Fortum		0.51
EDP		0.94
United Utilities	0.55	
Red Electrica	0.89	
Veolia Environnement		1.14
Gas Natural		0.94
Severn Trent	0.81	
EDF		0.82
Endesa		n.m.
Terna	0.64	
Enagas	0.65	
Suez Environnement		1.08
Pennon	0.57	
CEZ		0.74
ENEL Green Power		0.83
Drax		0.67
Simple average	0.65	0.81

Source: SG Cross Asset Research/Equity, Bloomberg.

manufacturing activities will lead companies to move their factories to other countries; but utilities, for obvious reasons, cannot move their networks or power stations, hence their higher exposure to political intervention.

Table 17.2 shows levered raw betas for utilities in the sector index (Ponti, 2015b). Regulated and nonregulated utilities show a pretty different level of correlation with the (European) equity market. Moreover, for regulated utilities a

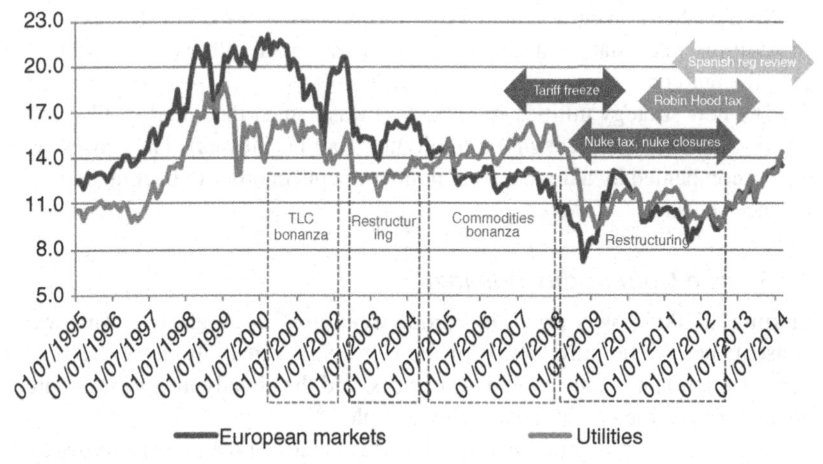

FIGURE 17.1 Price/earnings (P/E) 12-month trailing earnings per share (EPS). *(Source: SG Cross Asset Research/Equity.)*

beta of 0.65 would certainly signal that even the network companies are not as defensive as might have been thought (e.g., the Robin Hood tax[2] in Italy or the tariff reforms in Spain).

Utilities have themselves experienced a number of (short) cycles, as shown in Fig. 17.1, and these explain in our view the relative high level of the beta.

1.1.1 The Telecom Bonanza and the Multiutility Strategy

At the beginning of the last decade, most utilities benefited from what was then called the "Telecom bubble." As telecommunication services were expanding quickly, it was thought that utilities could have used electricity cables to provide its customer base with voice and data services. In the event the technology did not work. However, most utilities bought telecom companies at the time, participated in Universal Mobile Telecommunication System (UMTS) license auctions, and invested heavily in fiber optic cables to provide fast internet access for their customers.

Generally, it was the idea that utilities "owned the customers" via electricity, gas, and water bills, and therefore had direct access to them. This led them to diversify their businesses away from the provision of traditional utility services.

1.1.2 The Restructuring Phase and the Dual-Fuel Strategy

As the telecom bubble burst most utilities sought to restructure their balance sheets by selling assets, most of which had been acquired over the years and had little to do with the gas and electricity businesses.

2. During the summer of 2011, the Italian Government decided to introduce a tax on utlities hitting regulated utilities, in particular.

So the (then) recently bought/built telecom assets were being disposed of. In addition, real estate and media assets (among many others) were offloaded onto the market.

The new strategy utilities embarked on was named the "dual fuel" strategy (2003–2005) (i.e., refocusing on traditional core businesses). The process was also accompanied by a dramatic cut in investments in order to beef up company balance sheets.

1.1.3 The Commodity Bonanza

Starting in 2005 power prices in Europe went up in a straight line. This was a consequence of the introduction of the CO_2 market that, all else equal, had a direct (inflationary) impact on power prices, and the already mentioned unprecedented rise in the overall commodity complex.[3]

Utilities were piling up cash so quickly that a new spree of acquisitions took place and a new, intense, wave of investments was launched chiefly relating to new power stations (gas and coal mainly) while new investments in nuclear were also being considered.

1.1.4 Postcommodity Bonanza

However, it all came to an abrupt halt. As it is known, the commodity bubble burst in the middle of July 2008 and the descent that ensued was far more rapid than the ascent; in a few months commodity prices erased the gains recorded over the previous 3 years.

Utilities had to adjust once more but a number of new challenges were emerging. First of all, power stations that were being decided upon and approved just a few months earlier now had to be built in a much lower power price environment and their profitability therefore was being questioned by investors and analysts.

Due to the chronic delays in realizing investments in power stations, a number of them were delayed and as a consequence came online at times of even lower power prices. In fact, after strong correction in 2009, power prices have continued to fall until this year (2015).

To make things worse, the EU introduced in 2009 new targets for the development of a renewable capacity, which amidst an already ailing demand for electricity and gas, had the effect of generating massive overcapacity in most wholesale power markets in Europe, thus depressing power prices even more.

Utilities reacted (and still are reacting) by selling assets and cutting costs. During the commodity bubble utilities expanded their presence beyond the home's confines and those assets now had to be sold. Major efficiency programs were launched and vast redundancies plans were implemented.

---

3. Oil, coal, gas, and CO_2 are the commodities that are mainly relevant for power prices and therefore for utilities.

1.2 From Necessary Evils to Just "Evils"

Utilities are an integral part of any society in that the services they provide fulfill many of the primary needs like heating and cooking. So, utilities are "necessary" and they are "evils" because of course these services need to be paid for. On top of this, generation of electricity is one of the main contributors to pollution but Europe now has clearly stated objectives in terms of carbon emission reductions.

In the past all governments knew they could not do without utilities, but this has started to change. Distributed generation is emerging as a valid alternative to traditional large power stations and, if this continues, utilities may not (to any extent) be necessary for the system at all.

We think this and high commodity prices have been at the origin of the large wave of taxes, tariff cuts, and changes in regulation that we have seen in Europe since the beginning of this decade.

The current level of profitability of European utilities is such that it is unlikely new taxes will be introduced, yet the value of these companies was greatly affected and so were those investors who invested in them.

1.3 Takeaways for Regulators/Policy Makers

This brief recount of the past 15 years of history in the European utility sector shows that utilities are not as stable and safe as they appear at first glance.

Investors now know this full well and therefore the risk they attach to utility businesses is not much lower than for other sectors; hence, expected returns are higher than in the past. Risks associated with a single (physical) project are likely to be far higher than those associated with investing in a utility given the lack of diversification in the former case, thus calling for even higher returns. All this, of course, translates into a higher cost of capital, and regulators/policy makers need to be aware of this.

Utilities are subject to a large number of risks like many other sectors or industries. But some risks are different from others. Operational risks, like those on commodity prices or demand growth, can be predicted or indeed hedged/diversified. Typically, investors do not have a problem with these. Where they are at odds instead is with certain risks that typically cannot be diversified such as political, regulatory, and management risks.

1.3.1 Political Risks

Even the safest part of the sector, that of regulated utilities, has not been immune from sudden materialization of risks. This was the case in Italy of the Robin Hood tax that originally intended to hit generators and commodity traders, but in August 2011 was extended to networks.[4] On the day of its extension, SNAM

4. On February 12, 2015, the Italian Constitutional Court declared the tax illegal and therefore it was cancelled.

and Terna (two transmission companies in Italy – gas and electricity, respectively) saw their share prices fall 10% in just one day.

In 2013 and 2014, three rounds of regulatory changes in Spain led, among the other things, to cuts in the return on renewable assets with cuts applied retrospectively; although these cuts are being challenged in a number of courts, they are still applied. Changes in tariffs for distribution of electricity also led to cuts in revenues of the order of hundreds of millions of euros.

Little wonder that investors hold "political risk" high on their agenda. In most cases, this is the root cause behind good or bad performances of portfolios and eventually may influence their decision to invest in a certain asset or country in the future. It goes without saying that political intervention in a regulated market is to be avoided whenever possible.

1.3.2 Regulatory Risk

Regulatory risk is indeed similar to political risk but comes from a different source (i.e., the power to impose changes on tariffs or return on assets contrary to what was initially established). However, regulatory risk takes a slightly different form and typically relates to regulator independence.

We will describe this later in the chapter, suffice it here to say that if the regulator is perceived as being just "an agency" at the service of the government, the cost of capital investors will apply to a project or an asset is likely to be higher than if the regulator is independent.

1.3.3 Management Risk

History shows that when utilities are cash rich their desire to build upon the existing asset base by way of acquisitions (external growth) or the rollout of new infrastructures (organic growth) in a number of cases has led to shareholders' value being destroyed.

This is why investors place so much emphasis on "meeting with management." The same principle applies to assets, projects, or companies that are not listed but for which a country wants to attract investors.

2 THE ROLE OF INSTITUTIONAL INVESTORS

The "money management" industry is huge (World Economic Forum, 2014). It is calculated that institutional investors (IIs) manage around US$70 trillion worldwide. Of this US$25 trillion are in the Europe, Middle East, and Africa (EMEA) region (Fig. 17.2).

Focusing on EMEA, insurance funds (IFs) take the lion's share with 53% of assets under management (AUM) while pension funds (PFs) are the second largest with 29% (Figs 17.3–17.5). Sovereign Wealth Funds (SWFs) and Family Offices (FOs) have a lower but growing share of the market.

The investable universe available to IIs is equally huge though (Fig. 17.6). Typically, these invest in traditional asset classes such as money market instruments, fixed income (e.g., sovereign or corporate bonds), and equity (shares).

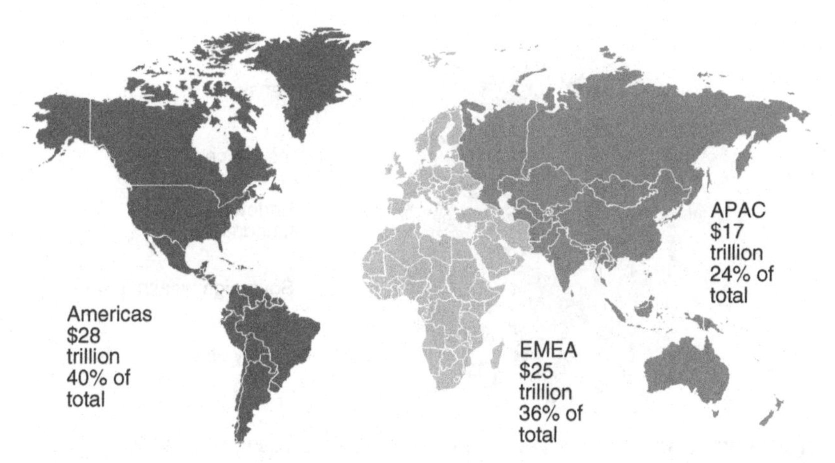

FIGURE 17.2 Funds under management (US$ trillion, 2013) – geographical split. *(Source: SG Cross Asset Research/Equity, Oliver Wyman.)*

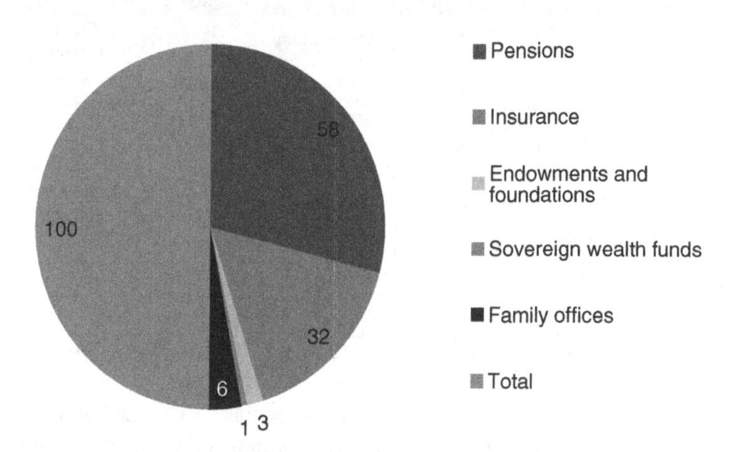

FIGURE 17.3 Americas. *(Source: SG Cross Asset Research/Equity, Oliver Wyman.)*

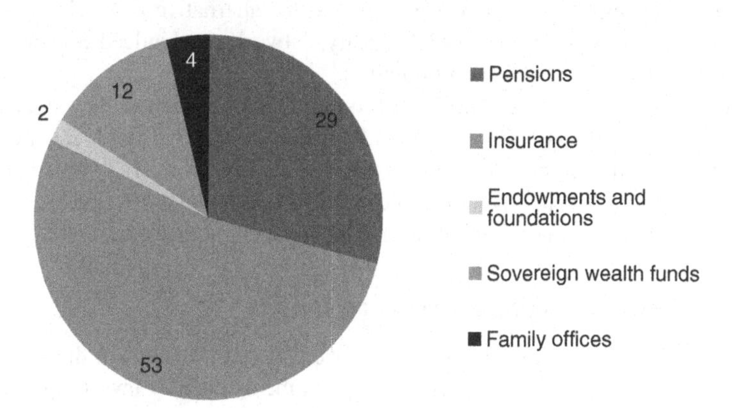

FIGURE 17.4 EMEA. *(Source: SG Cross Asset Research/Equity, Oliver Wyman.)*

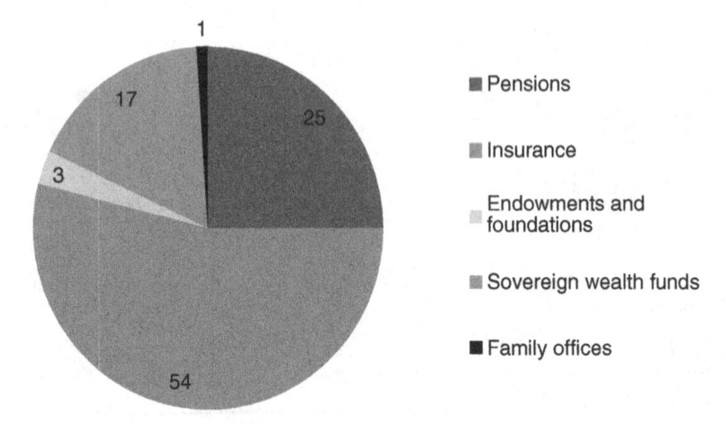

FIGURE 17.5 **APAC (%).** *(Source: SG Cross Asset Research/Equity, Oliver Wyman.)*

Traditional asset classes	Alternative assets				
	Traditional alternatives				Emerging alternatives
• Cash/money markets • Fixed income • Equities • Futures and options • Commodities	• Real estate equity – Public (REITs) – Private • Private real estate debt	• Infrastructure equity • Timber • Natural resources	• Leveraged buyout • Growth capital • Venture • Mezzanine	• Long/short • Global macro • Distressed debt • Etc.	• Loanfunds • Infrastructure debt • Risk transfer deals • Life insurance assets • Equipment leasing • Intellectual property and royalties

FIGURE 17.6 **Investable universe.** *(Source: SG Cross Asset Research/Equity, Oliver Wyman.)*

Futures and options can be seen as complementary to fixed income/equity investments or can be done in isolation. Some funds are instead specialized in commodity trading.

However, there is an increasing appetite for alternative asset classes, and certainly investments in renewable capacity, networks, and indeed conventional generation capacity are gaining ground.

IFs overall manage c. US$31 trillion, and it is estimated they could invest in alternative assets up to US$1.9 trillion by 2015 (Fig. 17.7), PFs up to US$4.7 trillion, and SWFs up to US$2.5 trillion while FOs invest up to US$1.2 trillion. In total, including endowments and foundations (E&Fs) investments in alternative assets could reach US$11.1 trillion this year (2015).

2.1 Decision-Making Process and Targets

The decision-making process has varied over the years and so has the structure of IIs. That IIs conduct their own research on the assets they invest in is a relatively new thing that started toward the end of the 1990s.

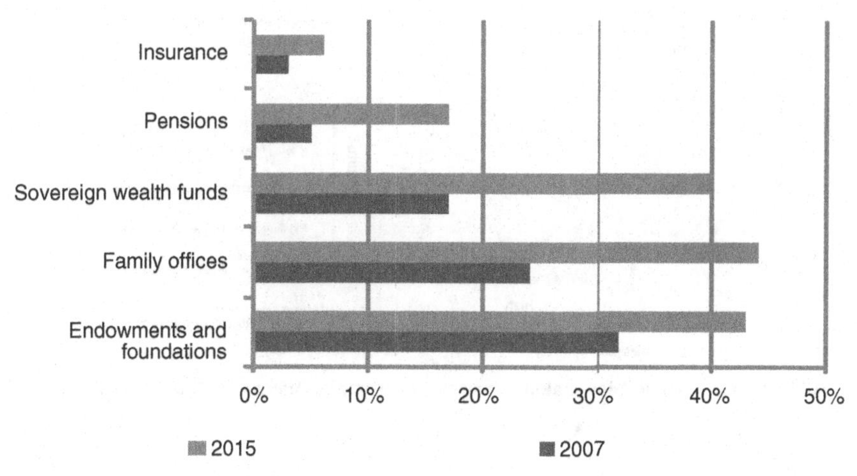

FIGURE 17.7 Estimated allocation to alternative assets. *(Source: SG Cross Asset Research/ Equity, Oliver Wyman.)*

The process certainly accelerated after the telecom bubble burst and the number of scandals that hit the sell-side industry[5] as a consequence. IIs (or the "buy-side") invested heavily in inhouse research departments. Today, these in-house research departments are still very much in place although most of them have been downsided, reflecting the prolonged crisis of financial markets (at least in Europe).

These research capabilities, of course, are not only used to invest in traditional assets classes, as it certainly was originally the intention, but are now also deployed for investing in alternative investments.

Within these, investments in infrastructure equity are certainly gathering pace. A number of relevant transactions have taken place in Europe over the past 2–3 years. To name just a few: Macquarie and the acquisition of Open Grid Europe (one of the leading gas transmission companies in Germany) or the recent acquisition by Credit Agricole of 10% of the French company Total Infrastructures Gaz France (TIGF) (gas transmission).

Direct investment and mergers and acquisitions (M&A) activities are normally directed at existing assets or "Brownfield projects." Analysis of Brownfield projects requires different levels of skill depending on the type of asset traded. In all cases, analysis comprises both a top-down and a bottom-up approach.

The presence of IIs in Greenfield projects is admittedly smaller as here the execution risk in realizing a new infrastructure can be a risk too high for certain IIs to bear. The technical knowledge required (e.g., for a hydro station or an offshore wind park) is a set of competencies that usually rests with a utility not

5. In other words, equity brokers and their analysts.

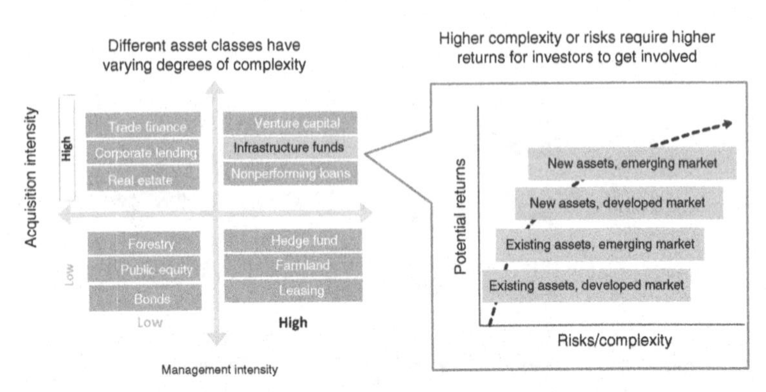

FIGURE 17.8 II's decision-making process – an example. *(Source: SG Cross Asset Research/ Equity; Oliver Wyman.)*

with a financial investor. The situation may change in the future although it is a struggle to foresee major developments in this sector of the market (Fig. 17.8).

2.1.1 Top-Down Analysis

Top-down, investors will look at the political setting of a country, economic indicators, and demographic trends to gauge the future need for infrastructures.

The political setting is particularly relevant because investing in infrastructures that stay in place for many years is like "writing a long-term contract" with the government, the problem being governments typically do not last as long as the infrastructure itself.

A stable political and economic environment is also a precondition to granting investors a future exit, which is often needed. Even pension funds, which have a long investment horizon, may end up needing to sell assets (e.g., during a period of financial crisis).

For example, the existence of a tariff deficit in the Spanish electricity market has been one of the main investors' concerns for many years, eventually resulting in a hefty cut of tariffs for all regulated activities and taxes for unregulated ones.

The fragile state finances of Italy, Spain, and Portugal have been at the center of investor discussions for a long time – until recently when a wave of liquidity hit the European bond markets with sovereign bond yields hitting all-time lows.

2.1.2 Bottom-Up Analysis

Bottom-up analysis typically consists in setting up a financial model for the assets to be purchased/built.

Analysis of revenues requires a detailed analysis of regulation or wholesale markets. As we explain later, the clearer and the more transparent the regulation, the easier it will be for investors to build a financial model that

TABLE 17.3 Target Sectors

Water	Attracting all types of institutional investors, either on listed or nonlisted target companies. In the United Kingdom, private equity funds have been very active mainly with the aim of releveraging the acquired business
Gas	There is an increasing number of assets that have been listed over the years with greater emphasis on transmission rather than distribution. Some transactions have occurred involving nonlisted entities
Electricity	The same as with gas. Less M&A, save for a number of deals with municipal utilities (e.g., in Germany and Italy)
Motorways	Listed companies have been active in M&A of smaller entities or concessions
Solar	Initially a residential-led business but gradually attracting financial investors in large projects, mainly Greenfields
Airports	Infrastructure funds have an active role although most of the M&A is dealt with by large corporates
Wind	A number of transactions have occurred in Europe of late, the buyers being chiefly pension funds. M&A mainly focused on onshore assets as the operational risks of offshore are still perceived as being too high

Source: SG Cross Asset Research/Equity.

eventually translates into using a lower discount rate for evaluating the project or the assets.

For networks, RAB regulation is usually the one investors understand best, although all aspects of regulation (calculation of the RAB and constituents of the allowed return) need to be published (AEEGSI, 2015a,b). For deregulated generation assets, long-term purchasing power agreements (PPAs) are usually the preferred form because they ensure long-term visibility of earnings and cash flows (Table 17.3).

2.2 A Very Lively industry

Utility assets are a very lively industry indeed where M&A is concerned. It has been calculated that the total value of M&A deals in 2015 reached US$177 billion, getting close to the peak reached in 2010, US$184 billion (Fig. 17.9).

Apart from integrated and other tariffs (Table 17.4) the one sector in the market where most transactions took place was in transmission and distribution (T&D) reaching US$42 billion, the Americas taking the lion's share with a total deal value of US$25 billion.

It is worth noting that financial investors contributed 24% of total deal value (i.e., c. US$42 billion), thus reaching their highest participation ever in the M&A of global utilities.

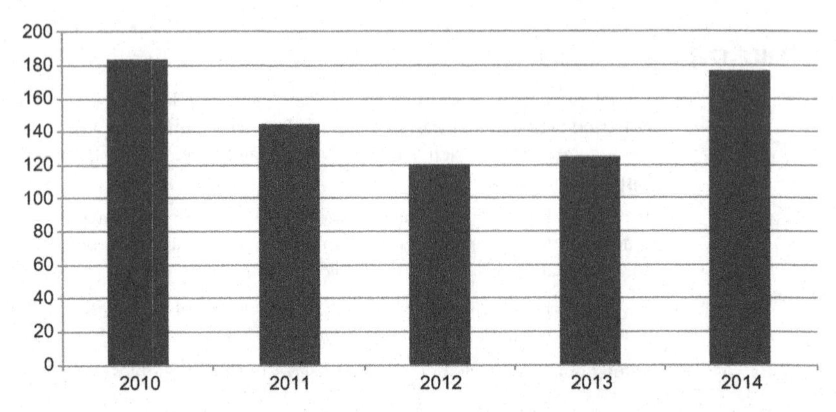

FIGURE 17.9 Global power and utility deal value (US$ billion) 2010–2014. *(Source: SG Cross Asset Research/Equity, EY – Power Transactions and Trends (EY, 2014).)*

2.3 Regulatory Regimes

There are many models regulators can refer to when deciding how to regulate assets under their jurisdiction. Table 17.5 summarizes some of them.

The question often asked is whether one model is better than another; a question that is difficult to answer as in most cases the answer depends on the use of a given model. However, as far as financial investors are concerned, certain models appear superior to others.

For example, a PPA model often suits corporates wishing to invest in power generation. It is not well suited to financial investors, particularly in a Greenfield project, because it would force them to meddle with the intricacies of often complex tolling agreements, which may go beyond their areas of expertise.

ROE and IRR models are often used for motorways or water concessions; financial investors are more akin to this type of regulation. However, regulated

TABLE 17.4 Global Power and Utility Deal Value (US$ billion) by Geography and Market Sector in 2014

	Generation	Transmission and distribution	Renewables	Integrated, water, and other tariffs
Americas	14.2	25.4	21.2	35.0
Europe	8.6	10.9	8.5	20.4
Asia-Pacific	11.1	5.6	8.2	7.4
Total	33.9	41.9	37.9	62.8

Source: SG Cross Asset Research/Equity. EY – Power Transactions and Trend (EY, 2014).

TABLE 17.5 Examples of Regulatory Regimes

IRR (un)levered	Return on equity	CAPs	RAB	PPAs
Typically with inflation link and no specific regulatory review	Typically with inflation link, financial costs are a passthrough	Price or revenues cap	Value of the RAB fixed but not always disclosed	Value of revenues or tolling agreement decided
IRR not always specified or officially announced	ROE often published	Typically long-term contracts	Allowed return published but not always its constituent parts	Very few details typically released
For example, wind, motorways	For example, certain electricity distribution networks		For example, water and gas/electricity networks	For example, power generation

Source: SG Cross Asset Research/Equity.

asset based (RAB) regulation is probably the one that best conjugates the need for attractive returns and long-term visibility of earnings.

In Europe, not all networks are regulated with tariffs based on RAB but in those countries where they are utilities usually enjoy higher market valuations. Moreover, where tariffs based on RAB have been in place for longer, and regulation proved to be stable, premia to RAB are higher.

Figure 17.10 shows our estimates for premia to RAB of a sample of utilities in Europe. The United Kingdome, where RAB regulation has been in place the

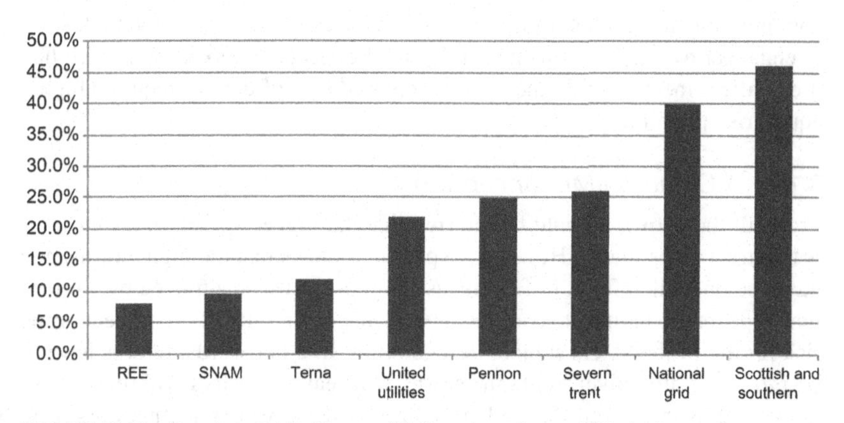

FIGURE 17.10 Premia to RAB. *(Source: SG Cross Asset Research/Equity.)*

longest, enjoys a higher premium than in Italy. In Spain, where RAB regulation has just been introduced, the premium is lower.

Of course, premia to RAB are a function of many factors and not just time consistency of regulation or its age; yet investors over the years have attributed higher valuations to UK assets versus Italian assets, for example.

3 THE FOUR KEY CONDITIONS TO STIMULATE INVESTMENTS IN INFRASTRUCTURES

As mentioned earlier, there are a number of regulatory regimes in Europe and each has its own attractions and features. Each regime meets different investors' expectations; it is beyond the scope of this chapter to say which regime is better or which delivers the ideal setting.

However, experience shows that any regulatory regime should at least have four characteristics. Each of these, if met, proves to be effective in attracting investments to a country's infrastructures but ultimately proves to be beneficial to consumers too, which is the primary goal of any regulatory body. Investments in infrastructures, at the same time, allow regulators to achieve the other goal of paramount importance (i.e., security of supply).

3.1 Attractiveness

For a regulatory regime to be attractive (from the perspective of investors) returns on investments need to be aligned or above the returns of other investments (opportunity cost, risk adjusted). Typically, returns are defined using the weighted average cost of capital (WACC) formula, and it is advisable to do so as this is the main reference for financial markets.

Returns so defined need to take into account the long duration that investments in infrastructures usually require and should allow for a level of leverage that can be high (e.g., debt to total investments) of up to 60–70% (this is common practice in most European regulatory regimes), but not too high so as to endanger overall investment in the case of exogenous shocks or so high that it can affect the cost of financing via increased cost of debt or required higher equity risk premium.

3.1.1 A Common Misconception

From all the above, it could be inferred that the higher the return the easier it is to attract investments. However, experience shows that too high returns may have the opposite effect. In fact, investors will hardly be attracted by a return that appears too high if it is not justified by the economics of the project or the risk profile of the assets subject to regulation. In other words, returns have to be *fair* to be attractive: too high returns will leave investors with the fear of possible cuts in the future; too low returns instead discourage investors from the start.

Of course, return is not the only element of a regulatory setting, but it is the key driver of profitability investors and financial markets usually focus on. Other elements are equally important. For example, certain regulations stimulate new investments allowing a base return plus a premium and this can extend for a number of years; others define standard investment costs and if the asset operator beats the regulator assumption it keeps all (or part) of the outperformance.

3.2 Transparency

Regulatory regimes need to be transparent. This means, for example, that the constituents of the WACC formula all need to be published. Furthermore, it is advisable that the criteria used to set the parameters in the WACC formula (e.g., the beta, gearing, etc.) are also laid out in full.

For example, in some cases it is not clear how regulators estimate the beta in the WACC formula (e.g., returns regressed against what index and over what period) or what sample of comparators they use. In some other cases it is not specified whether rational and/or peer comparison is used to determine gearing levels and how the regulator sees this parameter evolving in the future.

As regulation matures in Europe, incentives to investments are also changing. In the early stages of regulation, when Europe had a strong need for investments, an incentive based on the simple rule of a premium return on top of the base return was often used. This is a rule that is well understood by investors. As mentioned earlier, provided the return is not too high, it will be perceived as sustainable, so investors will "buy into it."

However, as markets and regulation mature, incentives are now becoming output based (as is the case of RIIO in the United Kingdom). In Italy it is being proposed that future incentives to investments will also be output based, the simple principle being that investments are remunerated at a rate that is commensurate to the actual (or estimated *ex ante*) benefits brought to consumers and/or the system. In this case though, it is necessary that key performance indicators (KPIs) are introduced, published, and consistently applied. It is also advisable that these KPIs are simple in their formulation and possibly allow quick and simple *ex post* control.

There are cases in Europe where the value of the RAB is published and its evolution over time is estimated by the regulator (e.g., in the United Kingdom and France), where the RAB is published but not its future evolution (e.g., Italy), where the RAB is not entirely recognized in existing tariffs (e.g., Portugal and Belgium), where the RAB is communicated by utilities to the market but not by the regulator (Spain), or indeed instances where the RAB is not part of the regulatory formula (e.g., Germany).

Consultations are also a central part in ensuring transparency. Through consultations, stakeholders have the opportunity to participate in the formation of a particular regulatory decision. While any regulatory decision may result in an appeal against it, a wide consultation process is likely to minimize litigation risks. A regulatory regime with a high level of litigation often results in a higher cost of capital required by investors because litigation means uncertainty in the application of tariffs and, eventually, in the returns investors expect.

3.3 Consistency

Regulatory regimes need to be applied consistently over time. If there is something that markets and investors really despise it is unexpected change. Changes can manifest themselves in many forms but typically these boil down to new taxes or changes in returns on assets.

One important caveat is that returns *can* indeed change provided that these changes are within the regulatory framework that is defined *ex ante*. For example, most regulatory regimes allow for a regulatory period that lasts for a number of years (usually 4 or 6).

At the end of the regulatory period the return and the value of the asset base are typically reestimated. This is not a problem for investors, of course, since they will be able to anticipate the changes and in some cases hedge the risks if regulation is transparent.

3.4 Independence

The structure and corporate governance of regulatory bodies in Europe varies quite substantially. In some cases the government sets the tariffs directly and the regulatory body as such does not exist. In other cases the regulatory body exists, but it only has an advisory role to the government (e.g., Spain) or it has some autonomy (e.g., France where the regulator is pretty independent in setting network tariffs but the government retains full power on generation and end customers tariffs). Other countries boast a fully independent regulator (e.g., the United Kingdom or Italy).

Ultimately, all regulators are administrative bodies, so a government will always be able to overrule a regulator's decision with a new law. However, if the regulator is independent and is set up with a dedicated law (which also defines its powers and jurisdiction) the government may find it difficult to do so.

Independence can be defined in many ways, but policy makers should take particular care in the definition of at least a few elements in our view. First of all the prerequisites to candidates being appointed members of the board of the regulatory body should be the right mix of different skills where possible. For example, with the notable exception of the United Kingdom, representatives of financial markets are hardly ever considered possible members of any board in continental Europe.

The duration of the board is also important, and it is advisable that its renewal does not overlap with political cycles so as to keep possible interference to a minimum.

Independence also lays in a clear set of powers that are attributed to the regulator. While it remains clear that it is within the mandate of the government to set the energy policy of a country, it is equally important that implementation of that policy is left to the regulator. The regulator also needs sanctioning powers, but this is not always the case.

4 CONCLUSIONS

Investments in European utilities or European infrastructure assets are facilitated by the existence of liquid markets for their equity and debt and by a rather lively M&A market. On top of this, regulatory frameworks have been converging over the past, say, 10 years.

Initially based on the "cost-plus" notion, European regulatory frameworks have gradually developed toward RAB regulations. The latter has the benefit of defining a precise (although not always published) value for assets to be regulated, which constitutes the basis for calculating the tariffs consumers have to pay.

The return "allowed" by the regulator[6] is also published and the most recent experience shows gradual alignment on this front too. For example, regulators' betas are today more aligned than in the past or the equity risk premia (ERP) are between 4.5% and 5.5%.

Most regulators in continental Europe also define a level of operating expenses (OPEX), and utilities are allowed to recover depreciation charges via the application of tariffs charged to consumers. However, in the UK regulators are moving away from this approach toward what is called TOTal EXpenses (TOTEX), whereby utilities are given the flexibility of choosing how to allocate their expenses (i.e., charging them on their profit and loss (P&L) statements or capitalizing them in their balance sheet) in an attempt to push utilities to find the most efficient combination between the two while at the same time making them more accountable for investments in existing and new infrastructures. Recently, the Italian regulator has proposed moving to TOTEX regulation starting in 2019.

Therefore, as experience shows, RAB and then TOTEX-based regulations typically appear in mature markets after a long investment cycle has been executed, often coinciding with either a major refurbishment/upgrade of the existing infrastructures or when a substantial increase in existing capacity is planned.

4.1 The Use of Long-Term Contracts

With this in mind, RAB or TOTEX-based regulations may not be ideal in the MENA countries, and empirical evidence also suggests they are not necessary to attract investments. Instead, what is important is long-term visibility and stability of the rules set by national governments.

---

6. This in essence is the same formula as weighted average cost of capital (WACC), where the national regulator defines certain parameters (e.g., the beta or gearing levels) or sets a methodology for calculating them (e.g., risk-free rate or equity risk premium).

For example, long-term contracts are valid substitutes as the experience of many Latin American countries shows. Such countries as Brazil, Chile, or Mexico were able in the past, and still are today, to attract substantial investments from all over the world thanks to the use of long-term contracts.

Certain long-term contracts for networks define an overall amount of revenue that is earned by the developer/operator over a predefined life of the asset – updated yearly by inflation. In the case of generation, developers typically want to hedge their fuel exposure with offtake or tolling agreements, but the long-term nature of the supply contract remains the basis for building the new asset.

4.2 A Clear "Energy Policy"

Governments in MENA countries seeking financing for existing/new infrastructures would need a clear and official energy policy,[7] better if substantiated in a national law. Investors, in fact, typically start their assessment of an infrastructural investment by looking at a country's need for a specific asset (e.g., a gas pipeline or a power station), since the actual need that country has for that asset is perhaps the strongest guarantee investors have on future earnings.

4.3 Country Risk

So-called "country risk" is an unavoidable complication for investors in countries that are perceived either less politically or financially stable, or both. A typical example was that of Italy and Spain in 2011 whose 10-year bond yield spreads versus the German Bund increased more than 500 bps during the sovereign crisis of 2011–2012, and this despite the two countries being in the Eurozone.

Investors can hedge their currency risk although such hedging is usually limited to a much shorter horizon than the life of the assets, and in any case it cannot eliminate the operational and political risks an infrastructure is structurally subject to. So it is fair to assume that returns granted to investors in MENA countries will on average be higher so as to offset the higher country risk.

4.4 The Four Key Conditions

As a partial offset to country risk and other risks, the four key conditions previously mentioned should be seen as mitigating factors and this is regardless of the regulatory setting (i.e., whether they are long-term contracts, RAB tariffs, or any other regulatory regime).

As a simple rule of thumb, the more transparent the tariffs and the more independent the regulator, and the less politicians have a track record for meddling with regulation/tariffs, the lower the return required by investors for investing in a country/infrastructure.

---

7. That is, publicly announced.

4.5 A Typical Trade-Off and the "Usual Temptation"

Transparent regulation usually presents regulators or national administrative agencies with a *trade-off*.

As mentioned, the upside from transparent regulation is lower overall cost of the infrastructure to be financed. The downside is that politicians/regulators have both hands tied to the regulatory regime chosen, with little flexibility to change tariffs in due course. This, for example, may not be optimal (for them) at times when economic conditions change rapidly (for the better or worse).

For example, when tariffs are based on a published value for the allowed return (WACC) and its constituents, a change in tariffs becomes possible only if regulators/politicians explain the change with sound financial empirical evidence. Of course, they can decide to change it anyway, but the loss in reputation that would ensue could have substantial repercussions on projects yet to be financed.

The *usual temptation* therefore is that of withholding some of the information as to what is the basis for allowing a certain level of return so that this can be changed at any point in time.

However, reputational losses often lead to much bigger costs than the short-term benefits of deviating from an announced tariff structure. So, in conclusion, transparency and consistency in regulation are ultimately what maximizes benefits to consumers and minimizes cost to tax payers from financing new/existing infrastructures in any given country.

REFERENCES

AEEGSI, 2015a. Criteri per la determinazione e l'aggiornamento del tassi di remunerazione del capital investito per le regolazioni infrastrutturali dei settori elettrico e gas, orientamenti iniziali. 275/2015/R/com. Available from: *www.autorita.energia.it*

AEEGSI, 2015b. Criteri per la fissazione del costo riconosciuto, per i servizi di trasmissione, distribuzione e misura dell'energia elettrica, nel quinto periodo regolatorio. 335/2015/R/eel. Available from: *www.autorita.energia.it*

EY, 2014. Power transaction and trends. 2014 Review and 2015 Outlook. Available from: *www.ey.com/Publication*

European Commission, 1996. Directive 96/92/EC concerning common rules for the internal market in electricity.

European Commission, 1998. Directive 98/30/EC concerning common rules for the internal market in natural gas.

OECD, 2013. The role of banks, equity markets and institutional investors in long-term financing for growth and development.

Ponti, A., 2015a. The PathFinder – Bonds, Utilities and... a few surprises. Societe Generale, Cross Assets Research/Equity.

Ponti, A., 2015b. Two battles worth fighting. Societe Generale, Cross Assets Research/Equity.

World Economic Forum, 2014. Direct Investing by Institutional Investors: Implications for investors and policy-makers.

Subject Index

D

H

I

J

K

L

M

N

O

P

Q

R

S

T